Lecture Notes in Mathematics 2219

More information about this series at http://www.springer.com/series/304

Fondazione C.I.M.E., Firenze

C.I.M.E. stands for *Centro Internazionale Matematico Estivo*, that is, International Mathematical Summer Centre. Conceived in the early fifties, it was born in 1954 in Florence, Italy, and welcomed by the world mathematical community: it continues successfully, year for year, to this day.

Many mathematicians from all over the world have been involved in a way or another in C.I.M.E.'s activities over the years. The main purpose and mode of functioning of the Centre may be summarised as follows: every year, during the summer, sessions on different themes from pure and applied mathematics are offered by application to mathematicians from all countries. A Session is generally based on three or four main courses given by specialists of international renown, plus a certain number of seminars, and is held in an attractive rural location in Italy.

The aim of a C.I.M.E. session is to bring to the attention of younger researchers the origins, development, and perspectives of some very active branch of mathematical research. The topics of the courses are generally of international resonance. The full immersion atmosphere of the courses and the daily exchange among participants are thus an initiation to international collaboration in mathematical research.

C.I.M.E. Director (2002 – 2014)
Pietro Zecca
Dipartimento di Energetica "S. Stecco"
Università di Firenze
Via S. Marta, 3
50139 Florence
Italy
e-mail: zecca@unifi.it

C.I.M.E. Director (2015 –)
Elvira Mascolo
Dipartimento di Matematica "U. Dini"
Università di Firenze
viale G.B. Morgagni 67/A
50134 Florence
Italy
e-mail: mascolo@math.unifi.it

C.I.M.E. Secretary
Paolo Salani
Dipartimento di Matematica "U. Dini"
Università di Firenze
viale G.B. Morgagni 67/A
50134 Florence
Italy
e-mail: salani@math.unifi.it

CIME activity is carried out with the collaboration and financial support of INdAM (Istituto Nazionale di Alta Matematica)

For more information see CIME's homepage: **http://www.cime.unifi.it**

Angela Kunoth • Tom Lyche • Giancarlo Sangalli •
Stefano Serra-Capizzano

Splines and PDEs: From Approximation Theory to Numerical Linear Algebra

Cetraro, Italy 2017

Tom Lyche, Carla Manni, Hendrik Speleers

Editors

 Springer

Authors
Angela Kunoth
Mathematical Institute
University of Cologne
Cologne, Germany

Giancarlo Sangalli
Department of Mathematics
University of Pavia
Pavia, Italy

Tom Lyche
Department of Mathematics
University of Oslo
Oslo, Norway

Stefano Serra-Capizzano
Department of Science and High
Technology
University of Insubria
Como, Italy

Editors
Tom Lyche
Department of Mathematics
University of Oslo
Oslo, Norway

Carla Manni
Department of Mathematics
University of Rome Tor Vergata
Rome, Italy

Hendrik Speleers
Department of Mathematics
University of Rome Tor Vergata
Rome, Italy

ISSN 0075-8434   ISSN 1617-9692   (electronic)
Lecture Notes in Mathematics
C.I.M.E. Foundation Subseries
ISBN 978-3-319-94910-9   ISBN 978-3-319-94911-6   (eBook)
https://doi.org/10.1007/978-3-319-94911-6

Library of Congress Control Number: 2018954553

Mathematics Subject Classification (2010): Primary: 65-XX; Secondary: 65Dxx, 65Fxx, 65Nxx

This Springer imprint is published by the registered company Springer Nature Switzerland AG
The registered company address is: Gewerbestrasse 11, 6330 Cham, Switzerland

Preface

The four chapters of this book collect the main topics that have been lectured at the C.I.M.E. summer school "Splines and PDEs: Recent Advances from Approximation Theory to Structured Numerical Linear Algebra" held in Cetraro, July 2–7, 2017. The aim of the summer school has been to give an introduction to the most advanced mathematical developments and numerical methods originating in the numerical treatment of PDEs based on spline functions.

A renewed interest in spline methods has been stimulated in the last decade by the success of isogeometric analysis. The large research activity around isogeometric methods shows that splines yield a powerful tool to PDE discretizations. In this general perspective, the progress of isogeometric analysis went hand in hand with the formulation of new problems requiring new techniques to address them properly. This gave rise to novel (spline) results in different areas of classical numerical analysis, ranging from approximation theory to structured numerical linear algebra. These developments motivated the topics of the summer school.

The first chapter "Foundations of Spline Theory: B-Splines, Spline Approximation, and Hierarchical Refinement" is written by us. It provides a comprehensive and self-contained introduction to B-splines and their properties, discusses the approximation power of spline spaces, and gives a review on hierarchical spline bases.

The second chapter "Adaptive Multiscale Methods for the Numerical Treatment of Systems of PDEs" by Angela Kunoth is devoted to numerical schemes based on B-splines and B-spline-type wavelets as a particular multiresolution discretization methodology, in the context of control problems.

The third chapter "Generalized Locally Toeplitz Sequences: A Spectral Analysis Tool for Discretized Differential Equations" by Carlo Garoni and Stefano Serra-Capizzano presents the theory of generalized locally Toeplitz sequences, a framework for computing and analyzing the spectral distribution of matrices arising from the numerical discretization of differential equations.

The last chapter "Isogeometric Analysis: Mathematical and Implementational Aspects, with Applications" by Thomas J.R. Hughes, Giancarlo Sangalli, and Mattia Tani provides an overview of the mathematical properties of isogeometric analysis, discusses computationally efficient isogeometric algorithms, and presents some isogeometric benchmark applications.

We express our deepest gratitude to all the people who have contributed to the success of this C.I.M.E. summer school: the invited lecturers, the seminar and contributed talk speakers, and the authors who have contributed to this C.I.M.E. Foundation Subseries book. In addition, we thank all the participants, from 11 countries, that enthusiastically contributed to the success of the school. Last but not least, we also thank C.I.M.E., in particular Elvira Mascolo (the C.I.M.E. director) and Paolo Salani (the C.I.M.E. scientific secretary) for their continuous support in the organization of the school.

Oslo, Norway   Tom Lyche
Rome, Italy   Carla Manni
Rome, Italy   Hendrik Speleers

Acknowledgments

This C.I.M.E. activity was carried out with the collaboration and financial support of INdAM (Istituto Nazionale di Alta Matematica) and EMS (European Mathematical Society). It was also partially supported by the MIUR "Futuro in Ricerca 2013" Programme through the project "DREAMS."

Contents

Chapter 1
Foundations of Spline Theory: B-Splines, Spline Approximation, and Hierarchical Refinement

Tom Lyche, Carla Manni, and Hendrik Speleers

Abstract This chapter presents an overview of polynomial spline theory, with special emphasis on the B-spline representation, spline approximation properties, and hierarchical spline refinement. We start with the definition of B-splines by means of a recurrence relation, and derive several of their most important properties. In particular, we analyze the piecewise polynomial space they span. Then, we present the construction of a suitable spline quasi-interpolant based on local integrals, in order to show how well any function and its derivatives can be approximated in a given spline space. Finally, we provide a unified treatment of recent results on hierarchical splines. We especially focus on the so-called truncated hierarchical B-splines and their main properties. Our presentation is mainly confined to the univariate spline setting, but we also briefly address the multivariate setting via the tensor-product construction and the multivariate extension of the hierarchical approach.

1.1 Introduction

Splines, in the broad sense of the term, are functions consisting of pieces of smooth functions glued together in a certain smooth way. Besides their theoretical interest, they have application in several branches of the sciences including geometric modeling, signal processing, data analysis, visualization, numerical simulation, and probability, just to mention a few. There is a large variety of spline species, often referred to as the *zoo of splines*. The most popular species is the one where the pieces are algebraic polynomials and inter-smoothness is imposed by means of equality of

T. Lyche
Department of Mathematics, University of Oslo, Oslo, Norway
e-mail: tom@math.uio.no

C. Manni · H. Speleers (✉)
Department of Mathematics, University of Rome Tor Vergata, Rome, Italy
e-mail: manni@mat.uniroma2.it; speleers@mat.uniroma2.it

© Springer Nature Switzerland AG 2018   1
T. Lyche et al. (eds.), *Splines and PDEs: From Approximation Theory to Numerical Linear Algebra*, Lecture Notes in Mathematics 2219,
https://doi.org/10.1007/978-3-319-94911-6_1

derivatives up to a given order. This species will be the topic of the chapter. Several other species can be found in [35, 45] and references therein.

To efficiently deal with splines, one needs a suitable basis for their representation. B-splines turn out to be the most useful spline basis functions because they possess several properties that are important from both theoretical and computational point of view. The construction of B-splines is not confined to the algebraic polynomial case but can be done for many species in the zoo of splines. As it is often the case for important tools or concepts, B-splines have a long history in the sciences. They were already used by Laplace in the early nineteenth century [33], and many of their relevant properties were derived by Chakalov and Popoviciu in the 1930s; see [10] and [37]. However, the modern B-spline theory roots in the seminal works by Schoenberg; see [41, 42] and [15, 16]. There are several ways to define B-splines, based on recurrence, differentiation, divided differences, etc. Each of those definitions has certain advantages according to the problem one has to face. It is impossible to trace all modern works on B-splines, but we refer the reader to Schumaker's book [45] for an extended bibliography on the topic also beyond the polynomial setting.

This chapter provides an introduction to (polynomial) B-splines, starting from their definition via a recurrence relation. Furthermore, we establish some spline results of interest within the isogeometric analysis (IgA) paradigm. More precisely, the chapter contains

- a self-contained overview of splines and B-splines;
- a constructive exploration of approximation properties of spline spaces;
- a discussion on adaptive spline representations based on hierarchical refinement.

There exists a huge amount of literature about the first two items including some well-established books; see, e.g., [6, 26, 45] and references therein. The hierarchical spline setting received only recently a lot of attention; see, e.g., [22, 51, 53]. The novelties of the chapter can be essentially summarized as follows.

- Our introduction to B-splines differs somewhat from the standard presentations of the topic. It is mainly based on properties of the dual polynomial functions in the local Marsden identity.
- Our proof of the approximation properties of a given spline space relies on the explicit construction of a spline quasi-interpolant based on local integrals. For this quasi-interpolant we show error estimates of optimal order to any smooth function and its derivatives.
- Our presentation of the hierarchical spline setting provides a rather complete and unified treatment of the main properties of both the hierarchical and the truncated hierarchical B-spline basis.

The chapter does not address the geometric modeling aspects of B-splines, explaining why they form the mathematical core of current computer aided design (CAD) systems. For this we refer the reader to the books [13, 27, 38].

Our presentation is mainly confined to the univariate spline setting. Nevertheless, this is the building block of the multivariate setting via the tensor-product construction. Tensor-product B-splines are currently the most common tool in CAD systems and IgA. It is worth mentioning that there are also many other important extensions of the univariate B-spline concepts to the multivariate setting, not restricted to a tensor-product grid; see, for example, [31, 35] and references therein.

The remaining part of the chapter is divided into six sections. The next section is devoted to the definition of B-splines and their main properties, including differentiation and integration formulas, local representation of polynomials, and local linear independence. In Sect. 1.3 we analyze the space spanned by a set of B-splines, and we consider the representation of its elements, knot insertion, and the stability of the B-spline basis. Cardinal B-splines, i.e., B-splines with uniform knots, are of prominent interest in practical applications. They are addressed in Sect. 1.4 where, in particular, the evaluation of their inner products and uniform knot insertion are discussed. In Sect. 1.5, after a general discussion about quasi-interpolants, we present the construction of a new spline quasi-interpolant based on local integrals and we use it to show the approximation properties of the considered spline space. The hierarchical spline approach is the topic of Sect. 1.6, which is mainly devoted to the construction of the truncated hierarchical B-spline basis and the derivation of its main properties, including the so-called preservation of coefficients and the construction of hierarchical quasi-interpolants. Finally, tensor-product B-splines and their hierarchical extension are briefly discussed in Sect. 1.7.

1.2 B-Splines

In this section we introduce one of the most powerful tools in computer-aided geometric design and approximation theory: B-spline functions (in short, B-splines).[1] They are piecewise polynomials with a certain global smoothness. The positions where the pieces meet are known as knots.

1.2.1 Definition and Basic Properties

In order to define B-splines we need the concept of knot sequences.

Definition 1 A **knot sequence** ξ is a nondecreasing sequence of real numbers,

$$\xi := \{\xi_i\}_{i=1}^m = \{\xi_1 \le \xi_2 \le \cdots \le \xi_m\}, \quad m \in \mathbb{N}.$$

The elements ξ_i are called **knots**.

---

[1]The original meaning of the word "spline" is a flexible ruler used to draw curves, mainly in the aircraft and shipbuilding industries. The "B" in B-splines stands for basis or basic.

Provided that $m \geq p + 2$ we can define B-splines of degree p over the knot-sequence $\boldsymbol{\xi}$.

Definition 2 Suppose for a nonnegative integer p and some integer j that $\xi_j \leq \xi_{j+1} \leq \cdots \leq \xi_{j+p+1}$ are $p + 2$ real numbers taken from a knot sequence $\boldsymbol{\xi}$. The j-th **B-spline** $B_{j,p,\boldsymbol{\xi}} : \mathbb{R} \to \mathbb{R}$ of degree p is identically zero if $\xi_{j+p+1} = \xi_j$ and otherwise defined recursively by[2]

$$B_{j,p,\boldsymbol{\xi}}(x) := \frac{x - \xi_j}{\xi_{j+p} - \xi_j} B_{j,p-1,\boldsymbol{\xi}}(x) + \frac{\xi_{j+p+1} - x}{\xi_{j+p+1} - \xi_{j+1}} B_{j+1,p-1,\boldsymbol{\xi}}(x), \quad (1.1)$$

starting with

$$B_{i,0,\boldsymbol{\xi}}(x) := \begin{cases} 1, & \text{if } x \in [\xi_i, \xi_{i+1}), \\ 0, & \text{otherwise.} \end{cases}$$

Here we used the convention that fractions with zero denominator have value zero.

We start with some preliminary remarks.

• For degree 0, the B-spline $B_{j,0,\boldsymbol{\xi}}$ is simply the characteristic function of the half open interval $[\xi_j, \xi_{j+1})$. This implies that a B-spline is continuous except possibly at a knot ξ. We have $B_{j,p,\boldsymbol{\xi}}(\xi) = B_{j,p,\boldsymbol{\xi}}(\xi_+)$, where

$$x_+ := \lim_{\substack{t \to x \\ t > x}} t, \quad x_- := \lim_{\substack{t \to x \\ t < x}} t, \quad x \in \mathbb{R}.$$

Thus a B-spline is **right continuous**, i.e., the value at a point x is obtained by taking the limit from the right.

• We also use the notation

$$B[\xi_j, \ldots, \xi_{j+p+1}] := B_{j,p,\boldsymbol{\xi}},$$

showing explicitly on which knots the B-spline depends.

• We say that a knot has **multiplicity** μ if it occurs exactly μ times in the knot sequence. A knot is called **simple**, **double**, **triple**, ... if its multiplicity is equal to $1, 2, 3, \ldots$, and a **multiple knot** in general.

---

[2]The recurrence relation is due to de Boor, Cox and Mansfield [4, 14]. However, it appears already in works by Popoviciu and Chakalov in the 1930s; see [8] for an account of the early history of splines. For the modern theory of splines we refer the reader to the seminal papers by Schoenberg [41–43] and Curry/Schoenberg [15, 16]. In their works, B-splines were defined by divided differences of truncated power functions.

Example 1 A B-spline of degree 1 is also called a **linear B-spline** or a **hat function**. The recurrence relation (1.1) takes the form

$$B_{j,1,\xi}(x) = \frac{x - \xi_j}{\xi_{j+1} - \xi_j} B_{j,0,\xi}(x) + \frac{\xi_{j+2} - x}{\xi_{j+2} - \xi_{j+1}} B_{j+1,0,\xi}(x),$$

resulting in

$$B_{j,1,\xi}(x) = \begin{cases} \dfrac{x - \xi_j}{\xi_{j+1} - \xi_j}, & \text{if } x \in [\xi_j, \xi_{j+1}), \\[2mm] \dfrac{\xi_{j+2} - x}{\xi_{j+2} - \xi_{j+1}}, & \text{if } x \in [\xi_{j+1}, \xi_{j+2}), \\[2mm] 0, & \text{otherwise.} \end{cases} \tag{1.2}$$

A linear B-spline is discontinuous at a double knot and continuous at a simple knot.

Example 2 A B-spline of degree 2 is also called a **quadratic B-spline**. Using the recurrence relation (1.1), the three pieces of the quadratic B-spline $B_{j,2,\xi}$ are given by

$$B_{j,2,\xi}(x) = \begin{cases} \dfrac{(x - \xi_j)^2}{(\xi_{j+2} - \xi_j)(\xi_{j+1} - \xi_j)}, & \text{if } x \in [\xi_j, \xi_{j+1}), \\[2mm] \dfrac{(x - \xi_j)(\xi_{j+2} - x)}{(\xi_{j+2} - \xi_j)(\xi_{j+2} - \xi_{j+1})} \\[2mm] + \dfrac{(x - \xi_{j+1})(\xi_{j+3} - x)}{(\xi_{j+2} - \xi_{j+1})(\xi_{j+3} - \xi_{j+1})}, & \text{if } x \in [\xi_{j+1}, \xi_{j+2}), \\[2mm] \dfrac{(\xi_{j+3} - x)^2}{(\xi_{j+3} - \xi_{j+1})(\xi_{j+3} - \xi_{j+2})}, & \text{if } x \in [\xi_{j+2}, \xi_{j+3}), \\[2mm] 0, & \text{otherwise.} \end{cases} \tag{1.3}$$

Example 3 Figure 1.1 illustrates several sets of B-splines of degree $p = 1, 2, 3$. The same knot sequence is chosen for the different degrees, with only simple knots.

(a)   (b)   (c)

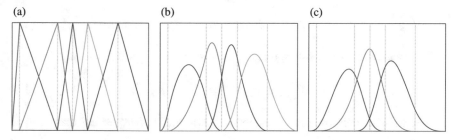

Fig. 1.1 Several sets of B-splines of degree $p = 1, 2, 3$. The knot positions are visualized by vertical dotted lines. (a) $p = 1$. (b) $p = 2$. (c) $p = 3$

The general explicit expression for a B-spline quickly becomes complicated. Applying the recurrence relation repeatedly we find

$$B_{j,p,\xi}(x) = \sum_{i=j}^{j+p} B_{j,p,\xi}^{\{i\}}(x) B_{i,0,\xi}(x), \quad p \geq 0, \tag{1.4}$$

where each $B_{j,p,\xi}^{\{i\}}$ is a polynomial of degree p, assumed to be zero if $\xi_i = \xi_{i+1}$. Note that if $\xi_i = \xi_{i+1}$ then $B_{i,0,\xi} = 0$ and the corresponding polynomial piece is not used. In particular, for the nontrivial cases we have

$$B_{j,0,\xi}^{\{j\}}(x) = 1, \quad B_{j,1,\xi}^{\{j\}}(x) = \frac{x - \xi_j}{\xi_{j+1} - \xi_j}, \quad B_{j,1,\xi}^{\{j+1\}}(x) = \frac{\xi_{j+2} - x}{\xi_{j+2} - \xi_{j+1}}.$$

Furthermore, for the nontrivial cases it follows from Definition 2 that the first and last polynomial pieces in (1.4) are given by

$$B_{j,p,\xi}^{\{j\}}(x) = (x - \xi_j)^p \Big/ \prod_{i=1}^{p} (\xi_{j+i} - \xi_j),$$

$$B_{j,p,\xi}^{\{j+p\}}(x) = (\xi_{j+p+1} - x)^p \Big/ \prod_{i=1}^{p} (\xi_{j+p+1} - \xi_{j+i}). \tag{1.5}$$

Using induction on the recurrence relation (1.1), we deduce immediately the following basic properties of a B-spline.

- **Local Support.** A B-spline is locally supported on the interval given by the extreme knots used in its definition. More precisely,

$$B_{j,p,\xi}(x) = 0, \quad x \notin [\xi_j, \xi_{j+p+1}). \tag{1.6}$$

- **Nonnegativity.** A B-spline is nonnegative everywhere, and positive inside its support, i.e.,

$$B_{j,p,\xi}(x) \geq 0, \quad x \in \mathbb{R}, \quad \text{and} \quad B_{j,p,\xi}(x) > 0, \quad x \in (\xi_j, \xi_{j+p+1}). \tag{1.7}$$

- **Piecewise Structure.** A B-spline has a piecewise polynomial structure, i.e.,

$$B_{j,p,\xi}^{\{i\}} \in \mathbb{P}_p, \quad i = j, \dots, j + p, \tag{1.8}$$

where $\mathbb{P}_p$ denotes the space of algebraic polynomials of degree less than or equal to p.

- **Translation and Scaling Invariance.** A B-spline is invariant under a translation and/or scaling transformation of its knot sequence, i.e.,

$$B_{j,p,\alpha\boldsymbol{\xi}+\beta}(\alpha x + \beta) = B_{j,p,\boldsymbol{\xi}}(x), \quad \alpha, \beta \in \mathbb{R}, \quad \alpha \neq 0, \tag{1.9}$$

where $\alpha\boldsymbol{\xi} + \beta := \{\alpha\xi_i + \beta\}_i$.

Further properties will be considered in the next sections.

1.2.2   Dual Polynomials

To each B-spline $B_{j,p,\boldsymbol{\xi}}$ of degree p, there corresponds a polynomial $\psi_{j,p,\boldsymbol{\xi}}$ of degree p with roots at the interior knots of the B-spline. We define $\psi_{j,0,\boldsymbol{\xi}} := 1$ and

$$\psi_{j,p,\boldsymbol{\xi}}(y) := (y - \xi_{j+1}) \cdots (y - \xi_{j+p}), \quad y \in \mathbb{R}, \quad p \in \mathbb{N}. \tag{1.10}$$

This polynomial is called **dual polynomial**. Many of the B-spline properties can be proved in an elegant way by exploiting a recurrence relation for these dual polynomials.

Theorem 1 *For $p \in \mathbb{N}$, $x, y \in \mathbb{R}$ and $\xi_{j+p} > \xi_j$, we have the **dual recurrence relation***

$$(y - x)\psi_{j,p-1,\boldsymbol{\xi}}(y) = \frac{x - \xi_j}{\xi_{j+p} - \xi_j}\psi_{j,p,\boldsymbol{\xi}}(y) + \frac{\xi_{j+p} - x}{\xi_{j+p} - \xi_j}\psi_{j-1,p,\boldsymbol{\xi}}(y), \tag{1.11}$$

*and the **dual difference formula***

$$\psi_{j,p-1,\boldsymbol{\xi}}(y) = \frac{\psi_{j-1,p,\boldsymbol{\xi}}(y)}{\xi_{j+p} - \xi_j} - \frac{\psi_{j,p,\boldsymbol{\xi}}(y)}{\xi_{j+p} - \xi_j}. \tag{1.12}$$

Proof For fixed $y \in \mathbb{R}$ let us define the function $\ell_y : \mathbb{R} \to \mathbb{R}$ given by $\ell_y(x) := y - x$. By linear interpolation, we have

$$\ell_y(x) = \frac{x - \xi_j}{\xi_{j+p} - \xi_j}\ell_y(\xi_{j+p}) + \frac{\xi_{j+p} - x}{\xi_{j+p} - \xi_j}\ell_y(\xi_j).$$

By multiplying both sides with $\psi_{j,p-1,\boldsymbol{\xi}}(y)$ we obtain (1.11). Moreover, (1.12) follows from (1.11) by differentiating with respect to x. □

Proposition 1 *The r-th derivative of the dual polynomial $\psi_{j,p,\boldsymbol{\xi}}$ for $0 \leq r \leq p$ can be bounded as follows:*

$$|D^r \psi_{j,p,\boldsymbol{\xi}}(y)| \leq \frac{p!}{(p-r)!}(\xi_{j+p+1} - \xi_j)^{p-r}, \quad \xi_j \leq y \leq \xi_{j+p+1}. \tag{1.13}$$

Moreover,

$$|D^r \psi_{j,p,\xi}(y)| \le \frac{p!}{(p-r)!}(\xi_{j+p} - \xi_{j+1})^{p-r}, \quad \xi_{j+1} \le y \le \xi_{j+p}. \tag{1.14}$$

Here we define $0^0 := 1$ *if* $r = p$ *and* $\xi_{j+p} = \xi_{j+1}$.

Proof Clearly (1.13) holds for all $p \in \mathbb{N}_0$ if $r = 0$. Using induction on r, p and the product rule for differentiation, we get

$$|D^r \psi_{j,p,\xi}(y)| = |D^r(\psi_{j,p-1,\xi}(y)(y - \xi_{j+p}))|$$

$$= |(D^r \psi_{j,p-1,\xi}(y))(y - \xi_{j+p}) + r D^{r-1} \psi_{j,p-1,\xi}(y)|$$

$$\le \left(\frac{(p-1)!}{(p-1-r)!} + r \frac{(p-1)!}{(p-r)!} \right)(\xi_{j+p+1} - \xi_j)^{p-r},$$

and (1.13) follows. The proof of (1.14) is similar.   □

1.2.3 Local Marsden Identity and Linear Independence

In this and the following sections (unless specified otherwise) we will extend the knots $\xi_j \le \cdots \le \xi_{j+p+1}$ of $B_{j,p,\xi}$ by defining p extra knots at each end, and we will assume

$$\xi := \{\xi_{j-p} \le \cdots \le \xi_{j-1} < \xi_j \le \cdots \le \xi_{j+p+1} < \xi_{j+p+2} \le \cdots \le \xi_{j+2p+1}\}. \tag{1.15}$$

These extra knots can be defined in any way we like. One possibility is

$$\xi_{j-p} = \cdots = \xi_{j-1} := \xi_j - 1, \quad \xi_{j+p+1} + 1 =: \xi_{j+p+2} = \cdots = \xi_{j+2p+1}. \tag{1.16}$$

On such a knot sequence $2p + 1$ B-splines $B_{i,p,\xi} = B[\xi_i, \dots, \xi_{i+p+1}]$, $i = j - p, \dots, j + p$ are well defined.

The following identity was first proved by Marsden [36] and simplifies many dealings with B-splines.

Theorem 2 (Local Marsden Identity) *For* $j \le m \le j + p$ *and* $\xi_m < \xi_{m+1}$, *we have*

$$(y - x)^p = \sum_{i=m-p}^{m} \psi_{i,p,\xi}(y) B_{i,p,\xi}(x), \quad x \in [\xi_m, \xi_{m+1}), \quad y \in \mathbb{R}. \tag{1.17}$$

If $B_{i,p,\xi}^{\{m\}}$ *is the polynomial which is equal to* $B_{i,p,\xi}(x)$ *for* $x \in [\xi_m, \xi_{m+1})$ *then*

$$(y - x)^p = \sum_{i=m-p}^{m} \psi_{i,p,\xi}(y) B_{i,p,\xi}^{\{m\}}(x), \quad x, y \in \mathbb{R}. \tag{1.18}$$

Proof Suppose $x \in [\xi_m, \xi_{m+1})$. The equality (1.17) can be proved by induction. It is clearly true for $p = 0$. Let us now assume it holds for degree $p - 1$. Then, by means of the dual recurrence (1.11) and the B-spline recurrence relation we obtain

$$(y - x)^p = (y - x)(y - x)^{p-1} = (y - x) \sum_{i=m-p+1}^{m} \psi_{i,p-1,\xi}(y) B_{i,p-1,\xi}(x)$$

$$= \sum_{i=m-p+1}^{m} \left(\frac{x - \xi_i}{\xi_{i+p} - \xi_i} \psi_{i,p,\xi}(y) + \frac{\xi_{i+p} - x}{\xi_{i+p} - \xi_i} \psi_{i-1,p,\xi}(y) \right) B_{i,p-1,\xi}(x)$$

$$= \sum_{i=m-p}^{m} \left(\frac{x - \xi_i}{\xi_{i+p} - \xi_i} B_{i,p-1,\xi}(x) + \frac{\xi_{i+p+1} - x}{\xi_{i+p+1} - \xi_{i+1}} B_{i+1,p-1,\xi}(x) \right)$$

$$\times \psi_{i,p,\xi}(y)$$

$$= \sum_{i=m-p}^{m} \psi_{i,p,\xi}(y) B_{i,p,\xi}(x).$$

Here we used that $\frac{x-\xi_i}{\xi_{i+p}-\xi_i} B_{i,p-1,\xi}(x) = 0$ for $i = m - p, m + 1$. □

The local Marsden identity immediately leads to the following properties, where we suppose $\xi_m < \xi_{m+1}$ for some $j \leq m \leq j + p$.

- **Local Representation of Monomials.** We have for $p \geq k$,

$$x^k = \sum_{i=m-p}^{m} \left((-1)^k \frac{k!}{p!} D^{p-k} \psi_{i,p,\xi}(0) \right) B_{i,p,\xi}(x), \quad x \in [\xi_m, \xi_{m+1}). \quad (1.19)$$

Proof Fix $x \in [\xi_m, \xi_{m+1})$. Differentiating $p - k$ times with respect to y in (1.18) results in

$$\frac{(y - x)^k}{k!} = \sum_{i=m-p}^{m} \left(\frac{1}{p!} D^{p-k} \psi_{i,p,\xi}(y) \right) B_{i,p,\xi}(x), \quad y \in \mathbb{R}, \quad (1.20)$$

for $k = 0, 1, \ldots, p$. Setting $y = 0$ in (1.20) results in (1.19). □

- **Local Partition of Unity.** Taking $k = 0$ in (1.19) gives

$$\sum_{i=m-p}^{m} B_{i,p,\xi}(x) = 1, \quad x \in [\xi_m, \xi_{m+1}). \quad (1.21)$$

- **Local Linear Independence.** The two sets $\{B_{i,p,\xi}\}_{i=m-p}^{m}$ and $\{\psi_{i,p,\xi}\}_{i=m-p}^{m}$ form both a basis for the polynomial space $\mathbb{P}_p$ on any subset of $[\xi_m, \xi_{m+1})$ containing at least $p+1$ distinct points.

Proof Let A be a subset of $[\xi_m, \xi_{m+1})$ containing at least $p+1$ distinct points. From (1.20) we see that on A every polynomial of degree at most p can be written as a linear combination of the $p+1$ polynomials $B_{i,p,\xi}^{\{m\}}$, $i = m-p, \ldots, m$. Since the dimension of the space $\mathbb{P}_p$ on A is $p+1$, these polynomials must be linearly independent and a basis. The result for $\{\psi_{i,p,\xi}\}_{i=m-p}^{m}$ follows by symmetry. □

1.2.4 Smoothness, Differentiation and Integration

The derivative of a B-spline can be expressed by means of a simple difference formula. In the following, we denote the right derivative by D_+ and the left derivative by D_-.

Theorem 3 (Differentiation) *We have*

$$D_+ B_{j,p,\xi}(x) = p\left(\frac{B_{j,p-1,\xi}(x)}{\xi_{j+p} - \xi_j} - \frac{B_{j+1,p-1,\xi}(x)}{\xi_{j+p+1} - \xi_{j+1}}\right), \quad p \geq 1, \tag{1.22}$$

where fractions with zero denominator have value zero.

Proof If $\xi_{j+p+1} = \xi_j$ then both sides of (1.22) are zero, so we can assume $\xi_{j+p+1} > \xi_j$. We continue to use the extra knots (1.15). If $x < \xi_j$ or $x \geq \xi_{j+p+1}$ then both sides of (1.22) are zero. Otherwise $x \in [\xi_m, \xi_{m+1})$ for some m with $j \leq m \leq j+p$ and it is enough to prove (1.22) for such an interval. Differentiating both sides of (1.17) with respect to x gives

$$-p(y-x)^{p-1} = \sum_{i=m-p}^{m} DB_{i,p,\xi}(x)\psi_{i,p}(y), \quad x \in [\xi_m, \xi_{m+1}). \tag{1.23}$$

On the other hand, using the local Marsden identity (1.17) for degree $p-1$ and the difference formula for dual polynomials (1.12) results in

$$-p(y-x)^{p-1} = -p \sum_{i=m-p+1}^{m} \psi_{i,p-1}(y)B_{i,p-1,\xi}(x)$$

$$= p \sum_{i=m-p+1}^{m} \left(\frac{\psi_{i,p}(y)}{\xi_{i+p} - \xi_i} - \frac{\psi_{i-1,p}(y)}{\xi_{i+p} - \xi_i}\right)B_{i,p-1,\xi}(x)$$

$$= \sum_{i=m-p}^{m} p\left(\frac{B_{i,p-1,\xi}(x)}{\xi_{i+p} - \xi_i} - \frac{B_{i+1,p-1,\xi}(x)}{\xi_{i+p+1} - \xi_{i+1}}\right)\psi_{i,p}(y).$$

When comparing this with (1.23) and using the linear independence of the dual polynomials, it follows that (1.22) holds for $i = m - p, \ldots, m$. In particular, since $m - p \leq j \leq m$, (1.22) holds for $i = j$. $\qquad\qquad\qquad\qquad\qquad\square$

Example 4 The differentiation formula (1.22) for $p = 2$ together with the expression (1.2) immediately gives the piecewise form of the derivative of the quadratic B-spline $B_{j,2,\xi}$:

$$D_+ B_{j,2,\xi}(x) = \begin{cases} \dfrac{2(x - \xi_j)}{(\xi_{j+2} - \xi_j)(\xi_{j+1} - \xi_j)}, & \text{if } x \in [\xi_j, \xi_{j+1}), \\[2ex] \dfrac{2(\xi_{j+2} - x)}{(\xi_{j+2} - \xi_j)(\xi_{j+2} - \xi_{j+1})} \\ \quad - \dfrac{2(x - \xi_{j+1})}{(\xi_{j+3} - \xi_{j+1})(\xi_{j+2} - \xi_{j+1})}, & \text{if } x \in [\xi_{j+1}, \xi_{j+2}), \\[2ex] -\dfrac{2(\xi_{j+3} - x)}{(\xi_{j+3} - \xi_{j+1})(\xi_{j+3} - \xi_{j+2})}, & \text{if } x \in [\xi_{j+2}, \xi_{j+3}), \\[2ex] 0, & \text{otherwise.} \end{cases}$$

This is in agreement with taking the derivative of the piecewise expression (1.3) of $B_{j,2,\xi}$ given in Example 2.

Proposition 2 *The r-th derivative of the B-spline $B_{j,p,\xi}$ for $0 \leq r \leq p$ can be bounded as follows. For any $x \in [\xi_m, \xi_{m+1})$ with $j \leq m \leq j + p$ we have*

$$|D^r B_{j,p,\xi}(x)| \leq 2^r \frac{p!}{(p-r)!} \prod_{k=p-r+1}^{p} \frac{1}{\Delta_{m,k}}, \tag{1.24}$$

where

$$\Delta_{m,k} := \min_{m-k+1 \leq i \leq m} h_{i,k}, \quad h_{i,k} := \xi_{i+k} - \xi_i, \quad k = 1, \ldots, p. \tag{1.25}$$

Proof This holds for $r = 0$ because of the nonnegativity of $B_{j,p,\xi}$ and the partition of unity property (1.21). By the differentiation formula (1.22) and the local support property (1.6) we have

$D^r B_{j,p,\xi}(x)$

$$= p \begin{cases} -D^{r-1} B_{j+1,p-1,\xi}(x)/h_{j+1,p}, & \text{if } m = j + p, \\ D^{r-1} B_{j,p-1,\xi}(x)/h_{j,p} - D^{r-1} B_{j+1,p-1,\xi}(x)/h_{j+1,p}, & \text{if } j < m < j + p, \\ D^{r-1} B_{j,p-1,\xi}(x)/h_{j,p}, & \text{if } m = j. \end{cases}$$

It follows that

$$|D^r B_{j,p,\xi}(x)| \leq 2p \max_{m-p+1 \leq i \leq m} |D^{r-1} B_{i,p-1,\xi}(x)|/\Delta_{m,p},$$

and by induction on r we obtain (1.24).   □

Note that the upper bound in (1.24) is well defined since $\Delta_{m,k} \geq \xi_{m+1} - \xi_m > 0$.

Theorem 4 (Smoothness) *If ξ is a knot of $B_{j,p,\xi}$ of multiplicity $\mu \leq p+1$, then*

$$B_{j,p,\xi} \in C^{p-\mu}(\xi), \tag{1.26}$$

i.e., its derivatives of order $0, 1, \ldots, p - \mu$ are continuous at ξ.

Proof Suppose ξ is a knot of $B_{j,p,\xi}$ of multiplicity μ. We first consider the smoothness property when $\mu = p+1$. For $x \in [\xi_j, \xi_{j+p+1})$ it follows immediately from (1.4) and (1.5) that

$$B_{j,p,\xi}(x) = \frac{(x - \xi_j)^p}{(\xi_{j+p+1} - \xi_j)^p}, \quad \xi_j < \xi_{j+1} = \cdots = \xi_{j+p+1}, \tag{1.27}$$

$$B_{j,p,\xi}(x) = \frac{(\xi_{j+p+1} - x)^p}{(\xi_{j+p+1} - \xi_j)^p}, \quad \xi_j = \cdots = \xi_{j+p} < \xi_{j+p+1}. \tag{1.28}$$

These two B-splines are discontinuous with a jump of absolute size one at the multiple knot showing the smoothness property for $\mu = p+1$.

Let us now consider the case where $B_{j,p,\xi}$ has an interior knot of multiplicity equal to $\mu = p$, i.e., $\xi_j < \xi_{j+1} = \cdots = \xi_{j+p} < \xi_{j+p+1}$. For $x \in [\xi_j, \xi_{j+p+1})$ it follows from (1.4) and (1.5) that

$$B_{j,p,\xi}(x) = \frac{(x - \xi_j)^p}{(\xi_{j+p} - \xi_j)^p} B_{j,0,\xi}(x) + \frac{(\xi_{j+p+1} - x)^p}{(\xi_{j+p+1} - \xi_{j+1})^p} B_{j+p,0,\xi}(x). \tag{1.29}$$

The two nontrivial pieces have both value one at the center knot $\xi_{j+1} = \xi_{j+p}$, and $B_{j,p,\xi}$ is continuous on $\mathbb{R}$. Moreover, the first derivative has a nonzero jump at the center knot.

For the remaining cases we use induction on p to show that $B_{j,p,\xi} \in C^{p-\mu}(\xi)$. The case $p = 1$ follows from Example 1. Suppose for some $p \geq 2$ that $B_{j,p-1,\xi} \in C^{p-1-\mu}(\xi)$ at a knot ξ of multiplicity μ. For the multiplicity p case $\xi = \xi_j = \cdots = \xi_{j+p-1} < \xi_{j+p} \leq \xi_{j+p+1}$ we use the recurrence relation

$$B_{j,p,\xi}(x) = \frac{x - \xi_j}{\xi_{j+p} - \xi_j} B_{j,p-1,\xi}(x) + \frac{\xi_{j+p+1} - x}{\xi_{j+p+1} - \xi_{j+1}} B_{j+1,p-1,\xi}(x).$$

The first term vanishes at $x = \xi = \xi_j$. Since $B_{j+1,p-1,\xi}$ has a knot of multiplicity $p - 1$ at ξ, it follows from the induction hypothesis that it is continuous there.

We conclude that $B_{j,p,\xi}$ is continuous at ξ. The case where the right end knot of $B_{j,p,\xi}$ has multiplicity p is handled similarly. Finally, if $\mu \le p-1$ then both terms in the differentiation formula (1.22) have a knot of multiplicity at most μ at ξ and by the induction hypothesis we obtain $D_+ B_{j,p,\xi} \in C^{p-1-\mu}(\xi)$. Moreover, by the recurrence relation and the induction hypothesis it follows that $B_{j,p,\xi}$ is continuous at ξ, and so we also conclude that $B_{j,p,\xi} \in C^{p-\mu}(\xi)$ if $\mu \le p-1$. This completes the proof. □

The B-spline $B_{j,p,\xi}$ is supported on the interval $[\xi_j, \xi_{j+p+1}]$. Hence, Theorem 4 implies that $B_{j,p,\xi}$ is continuous on $\mathbb{R}$ whenever $\xi_{j+p} > \xi_j$ and $\xi_{j+p+1} > \xi_{j+1}$. Similarly, $B_{j,p,\xi}$ is C^r-continuous on $\mathbb{R}$ whenever $\xi_{j+p-r+i} > \xi_{j+i}$ for each $i = 0, \ldots, r+1$ and $-1 \le r < p$.

Theorem 5 (Integration) *We have*

$$\gamma_{j,p,\xi} := \int_{\xi_j}^{\xi_{j+p+1}} B_{j,p,\xi}(x)\,dx = \frac{\xi_{j+p+1} - \xi_j}{p+1}. \tag{1.30}$$

Proof This time we define $p+1$ extra knots at each end, and we assume

$$\xi := \{\xi_{j-p-1} = \cdots = \xi_{j-1} < \xi_j \le \cdots \le \xi_{j+p+1} < \xi_{j+p+2} = \cdots = \xi_{j+2p+2}\}.$$

On this knot sequence we consider $p+1$ B-splines $B_{i,p+1,\xi}, i = j-p-1, \ldots, j-1$ of degree $p+1$. From Theorem 4 we know that these B-splines are continuous on $\mathbb{R}$. Therefore, we get for $i = j-p-1, \ldots, j-1$,

$$0 = B_{i,p+1,\xi}(\xi_{i+p+2}) - B_{i,p+1,\xi}(\xi_i) = \int_{\xi_i}^{\xi_{i+p+2}} D_+ B_{i,p+1,\xi}(x)\,dx = E_i - E_{i+1},$$

where by the local support and the differentiation formula (1.22),

$$E_i := \frac{p+1}{\xi_{i+p+1} - \xi_i} \int_{\xi_i}^{\xi_{i+p+1}} B_{i,p,\xi}(x)\,dx, \quad i = j-p-1, \ldots, j.$$

This means that $E_j = E_{j-1} = \cdots = E_{j-p-1}$. Moreover, since $\xi_{j-p-1} = \cdots = \xi_{j-1}$, we obtain from (1.28) that

$$E_{j-p-1} = \frac{p+1}{\xi_j - \xi_{j-p-1}} \int_{\xi_{j-p-1}}^{\xi_j} \frac{(\xi_j - x)^p}{(\xi_j - \xi_{j-p-1})^p}\,dx = 1,$$

and the integration formula (1.30) follows. □

1.3 Splines

A spline function (in short, spline) is a linear combination of B-splines defined on a given knot sequence with a fixed degree. In this section we analyze the space of splines and discuss several of their properties.

1.3.1 The Spline Space $\mathbb{S}_{p,\xi}$ and Some Spline Properties

Suppose for integers $n > p \geq 0$ that a knot sequence

$$\xi := \{\xi_i\}_{i=1}^{n+p+1} = \{\xi_1 \leq \xi_2 \leq \cdots \leq \xi_{n+p+1}\}, \quad n \in \mathbb{N}, \quad p \in \mathbb{N}_0,$$

is given. This knot sequence allows us to define a set of n B-splines of degree p, namely

$$\{B_{1,p,\xi}, \ldots, B_{n,p,\xi}\}. \tag{1.31}$$

We consider the space

$$\mathbb{S}_{p,\xi} := \left\{ s : [\xi_{p+1}, \xi_{n+1}] \to \mathbb{R} : s = \sum_{j=1}^{n} c_j B_{j,p,\xi}, \; c_j \in \mathbb{R} \right\}. \tag{1.32}$$

This is the space of **splines** spanned by the B-splines in (1.31) over the interval $[\xi_{p+1}, \xi_{n+1}]$, which is called the **basic interval**.

We now introduce some terminology to identify certain properties of knot sequences which are crucial in the study of the space (1.32).

- A knot sequence ξ is called $(p+1)$-**regular** if $\xi_j < \xi_{j+p+1}$ for $j = 1, \ldots, n$. By the local support (1.6) such a knot sequence ensures that all the B-splines in (1.31) are not identically zero.

- A knot sequence ξ is called $(p+1)$-**basic** if it is $(p+1)$-regular with $\xi_{p+1} < \xi_{p+2}$ and $\xi_n < \xi_{n+1}$. As we will show later, the B-splines in (1.31) defined on a $(p+1)$-basic knot sequence are linearly independent on the basic interval $[\xi_{p+1}, \xi_{n+1}]$.

- A knot sequence ξ is called $(p+1)$-**open** on an interval $[a, b]$ if it is $(p+1)$-regular and it has end knots of multiplicity $p + 1$, i.e.,

$$a := \xi_1 = \cdots = \xi_{p+1} < \xi_{p+2} \leq \cdots \leq \xi_n < \xi_{n+1} = \cdots = \xi_{n+p+1} =: b. \tag{1.33}$$

This sequence is often used in practice. In particular, it turns out to be natural to construct open curves, clamped at two given points. Note that $(p + 1)$-open implies $(p + 1)$-basic.

Some further preliminary remarks are in order here.

- We consider B-splines on a closed basic interval $[\xi_{p+1}, \xi_{n+1}]$. In order to avoid the asymmetry at the right endpoint we define the B-splines to be **left continuous** at the right endpoint, i.e., their value at ξ_{n+1} is obtained by taking the limit from the left:

$$B_{j,p,\xi}(\xi_{n+1}) := \lim_{\substack{x \to \xi_{n+1} \\ x < \xi_{n+1}}} B_{j,p,\xi}(x), \quad j = 1, \ldots, n. \tag{1.34}$$

  Note that for a $(p+1)$-open knot sequence the end condition (1.34) means that $B_{n,p,\xi}(\xi_{n+p+1}) = 1$ and (1.6) has to be modified for this B-spline.

- We define a **multiplicity function** $\mu_\xi : \mathbb{R} \to \mathbb{N}_0$ given by $\mu_\xi(\xi_i) = \mu_i$ if $\xi_i \in \xi$ occurs exactly $\mu_i \geq 1$ times in ξ, and $\mu_\xi(x) = 0$ if $x \notin \xi$. If ξ and $\tilde{\xi}$ are two knot sequences we say that $\xi \subseteq \tilde{\xi}$ if $\mu_\xi(x) \leq \mu_{\tilde{\xi}}(x)$ for all $x \in \mathbb{R}$.

- Without loss of generality, we can always assume that the end knots have multiplicity $p + 1$. If this is not the case, then we can add extra knots at the ends and assume the extra B-splines to have coefficients zero. This observation simplifies many proofs.

Example 5 Figure 1.2 illustrates all the B-splines of degree $p = 3$ on a $(p+1)$-open knot sequence, where the interior knots are simple.

From the properties of B-splines, we immediately conclude the following properties of the spline representation in (1.32).

- **Smoothness.** If ξ is a knot of multiplicity μ then $s \in C^r(\xi)$ for any $s \in \mathbb{S}_{p,\xi}$, where $r + \mu = p$. This follows from the smoothness property of the B-splines (Theorem 4). The relation between smoothness, multiplicity and degree is as follows:

$$\text{``smoothness} + \text{multiplicity} = \text{degree''.} \tag{1.35}$$

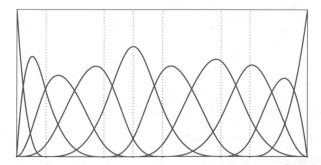

Fig. 1.2 The B-spline basis of degree $p = 3$ on a $(p + 1)$-open knot sequence. The knot positions are visualized by vertical dotted lines

- **Local Support.** The local support (1.6) of the B-splines implies

$$\sum_{j=1}^{n} c_j B_{j,p,\xi}(x) = \sum_{j=m-p}^{m} c_j B_{j,p,\xi}(x), \ x \in [\xi_m, \xi_{m+1}), \ p+1 \le m \le n,$$

(1.36)

and if $\xi_m < \xi_{m+p}$ then

$$\sum_{j=1}^{n} c_j B_{j,p,\xi}(\xi_m) = \sum_{j=m-p}^{m-1} c_j B_{j,p,\xi}(\xi_m), \quad p+1 \le m \le n+1.$$

(1.37)

- **Minimal Support.** From the smoothness properties it can be proved that if the support of $s \in \mathbb{S}_{p,\xi}$ is a proper subset of $[\xi_j, \xi_{j+p+1}]$ for some j then $s = 0$. Therefore, the B-splines have minimal support.

- **Coefficient Recurrence.** For $x \in [\xi_{p+1}, \xi_{n+1}]$, by the recurrence relation (1.1) we have

$$\sum_{j=1}^{n} c_j B_{j,p,\xi}(x) = \sum_{j=2}^{n} \check{c}_j(x) B_{j,p-1,\xi}(x),$$

(1.38)

where

$$\check{c}_j(x) := \frac{x - \xi_j}{\xi_{j+p} - \xi_j} c_j + \frac{\xi_{j+p} - x}{\xi_{j+p} - \xi_j} c_{j-1},$$

(1.39)

and $\check{c}_j(x) B_{j,p-1,\xi}(x) = 0$ if $\xi_{j+p} = \xi_j$.

- **Differentiation.** For $x \in [\xi_{p+1}, \xi_{n+1}]$, by the differentiation formula (1.22) we have

$$D_+ \left(\sum_{j=1}^{n} c_j B_{j,p,\xi}(x) \right) = \sum_{j=2}^{n} c_j^{(1)} B_{j,p-1,\xi}(x), \quad p \ge 1,$$

(1.40)

where

$$c_j^{(1)} := p \left(\frac{c_j - c_{j-1}}{\xi_{j+p} - \xi_j} \right),$$

(1.41)

and fractions with zero denominator have value zero.

- **Linear Independence.** If ξ is $(p+1)$-basic, then the B-splines in (1.31) are linearly independent on the basic interval. Thus, the spline space $\mathbb{S}_{p,\xi}$ is a vector space of dimension n.

Proof We must show that if $s(x) = \sum_{j=1}^{n} c_j B_{j,p,\xi}(x) = 0$ for $x \in [\xi_{p+1}, \xi_{n+1}]$ then $c_j = 0$ for all j. Let us fix $1 \leq j \leq n$. Since ξ is $(p+1)$-regular, there is an integer m_j with $j \leq m_j \leq j + p$ such that $\xi_{m_j} < \xi_{m_j+1}$. Moreover, the assumptions $\xi_{p+1} < \xi_{p+2}$ and $\xi_n < \xi_{n+1}$ guarantee that $[\xi_{m_j}, \xi_{m_j+1})$ can be chosen in the basic interval. From the local support property (1.36) we know

$$s(x) = \sum_{i=m_j-p}^{m_j} c_i B_{i,p,\xi}(x) = 0, \quad x \in [\xi_{m_j}, \xi_{m_j+1}).$$

The local linear independence property (see Sect. 1.2.3) implies $c_{m_j-p} = \cdots = c_{m_j} = 0$, and in particular $c_j = 0$. $\square$

1.3.2 The Piecewise Polynomial Space $\mathbb{S}_p^r(\Delta)$

We now prove that the spline space $\mathbb{S}_{p,\xi}$ is nothing else than a space of piecewise polynomials of degree p defined by a given sequence of break points and by some prescribed smoothness. The set of knots ξ must be suitably selected according to the break points and the smoothness conditions. Therefore, the B-splines are a basis of such a space of piecewise polynomials.

Let Δ be a sequence of distinct real numbers,

$$\Delta := \{\eta_0 < \eta_1 < \cdots < \eta_{\ell+1}\}.$$

The elements in Δ are called **break points**. Moreover, let $r := (r_1, \ldots, r_\ell)$ be a vector of integers such that $-1 \leq r_i \leq p$ for $i = 1, \ldots, \ell$. The space $\mathbb{S}_p^r(\Delta)$ of piecewise polynomials of degree p with smoothness r over the partition Δ is defined by

$$\mathbb{S}_p^r(\Delta) := \big\{ s : [\eta_0, \eta_{\ell+1}] \to \mathbb{R} : s \in \mathbb{P}_p([\eta_i, \eta_{i+1})), \ i = 0, \ldots, \ell - 1,$$

$$s \in \mathbb{P}_p([\eta_\ell, \eta_{\ell+1}]), \ s \in C^{r_i}(\eta_i), \ i = 1, \ldots, \ell \big\}. \tag{1.42}$$

Suppose that $s^{\{i\}} \in \mathbb{P}_p$ is the polynomial equal to the restriction of a given function $s \in \mathbb{S}_p^r(\Delta)$ to the interval $[\eta_i, \eta_{i+1})$, $i = 0, \ldots, \ell$. Since $s \in C^{r_i}(\eta_i)$, we have

$$s^{\{i\}}(x) - s^{\{i-1\}}(x) = \sum_{j=r_i+1}^{p} c_{i,j}(x - \eta_i)^j,$$

for some coefficients $c_{i,j}$. It follows that $\mathbb{S}_p^r(\Delta)$ is spanned by the set of functions

$$\big\{ 1, x, \ldots, x^p, (x - \eta_1)_+^{r_1+1}, \ldots, (x - \eta_1)_+^p, \ldots, (x - \eta_\ell)_+^{r_\ell+1}, \ldots, (x - \eta_\ell)_+^p \big\}, \tag{1.43}$$

where the **truncated power** function $(\cdot)_+^p$ is defined by

$$(x)_+^p := \begin{cases} x^p, & x > 0, \\ 0, & x < 0, \end{cases} \tag{1.44}$$

and the value at zero is defined by taking the right limit.

It is easy to see that the functions in (1.43) are linearly independent. Indeed, let

$$s(x) := \sum_{j=0}^{p} c_{0,j}\, x^j + \sum_{i=1}^{\ell} \sum_{j=r_i+1}^{p} c_{i,j}(x - \eta_i)_+^j = 0, \quad x \in [\eta_0, \eta_{\ell+1}].$$

On $[\eta_0, \eta_1)$ we have $s(x) = \sum_{j=0}^{p} c_{0,j}\, x^j$ and it follows that $c_{0,0} = \cdots = c_{0,p} = 0$. Suppose for some $1 \le k \le \ell$ that $c_{i,j} = 0$ for $i < k$. Then, on $[\eta_k, \eta_{k+1})$ we have $s(x) = \sum_{j=r_k+1}^{p} c_{k,j}(x - \eta_k)^j = 0$ showing that all $c_{k,j} = 0$.

This implies that the set of functions in (1.43) is a basis for $\mathbb{S}_p^r(\Delta)$, the so-called **truncated power basis**. As a consequence,

$$\dim(\mathbb{S}_p^r(\Delta)) = p + 1 + \sum_{i=1}^{\ell}(p - r_i).$$

The next theorem shows that the set of B-splines in (1.31) defined over a specific knot sequence $\boldsymbol{\xi}$ forms an alternative basis for $\mathbb{S}_p^r(\Delta)$. This was first proved by Curry and Schoenberg in [16].

Theorem 6 (Characterization of Spline Space) *The piecewise polynomial space* $\mathbb{S}_p^r(\Delta)$ *is characterized in terms of B-splines by*

$$\mathbb{S}_p^r(\Delta) = \mathbb{S}_{p,\xi},$$

where the knot sequence $\boldsymbol{\xi} := \{\xi_i\}_{i=1}^{n+p+1}$ *with* $n := \dim(\mathbb{S}_p^r(\Delta))$ *is constructed such that*

$$\xi_1 \le \cdots \le \xi_{p+1} := \eta_0, \quad \eta_{\ell+1} =: \xi_{n+1} \le \cdots \le \xi_{n+p+1},$$

and

$$\xi_{p+2}, \ldots, \xi_n := \overbrace{\eta_1, \ldots, \eta_1}^{p-r_1}, \ldots, \overbrace{\eta_\ell, \ldots, \eta_\ell}^{p-r_\ell}.$$

Proof From the piecewise polynomial and smoothness properties of B-splines it follows that the B-spline space $\mathbb{S}_{p,\xi}$ is a subspace of $\mathbb{S}_p^r(\Delta)$. Moreover, the constructed knot sequence $\boldsymbol{\xi}$ is $(p + 1)$-basic, so $\dim(\mathbb{S}_{p,\xi}) = n$ by the linear independence property of B-splines. This implies that $\mathbb{S}_p^r(\Delta) = \mathbb{S}_{p,\xi}$. □

Example 6 Consider $\Delta := \{\eta_0 < \eta_1 < \eta_2 < \eta_3\}$ and the space $\mathbb{S}_3^r(\Delta)$ with $r = (r_1, r_2) = (2, 1)$. It follows from Theorem 6 that $\mathbb{S}_3^r(\Delta) = \mathbb{S}_{3,\xi}$, where

$$\xi = \{\xi_i\}_{i=1}^{7+3+1} = \{\eta_0 = \eta_0 = \eta_0 = \eta_0 < \eta_1 < \eta_2 = \eta_2 < \eta_3 = \eta_3 = \eta_3 = \eta_3\}.$$

This knot sequence is 4-open.

Finally, we give a characterization for the space spanned by the r-th derivatives of B-splines for $0 \leq r \leq p$, i.e.,

$$D_+^r \mathbb{S}_{p,\xi} := \left\{ s : [\xi_{p+1}, \xi_{n+1}] \to \mathbb{R} : s = D_+^r \left(\sum_{j=1}^n c_j B_{j,p,\xi} \right), \ c_j \in \mathbb{R} \right\}.$$

Theorem 7 (Characterization of Derivative Spline Space) *Given a knot sequence* $\xi := \{\xi_i\}_{i=1}^{n+p+1}$, *we have for* $0 \leq r \leq p$,

$$D_+^r \mathbb{S}_{p,\xi} = \mathbb{S}_{p-r,\xi_r},$$

where $\xi_r := \{\xi_i\}_{i=r+1}^{n+p+1-r}$.

Proof The result is obvious for $r = 0$. Let us now consider the case $r = 1$, for which we note that

$$\{B_{1,p-1,\xi_1}, \ldots, B_{n-1,p-1,\xi_1}\} = \{B_{2,p-1,\xi}, \ldots, B_{n,p-1,\xi}\}.$$

By the differentiation formula (1.40) it is clear that

$$D_+ \left(\sum_{j=1}^n c_j B_{j,p,\xi} \right) = p \sum_{j=2}^n \left(\frac{c_j - c_{j-1}}{\xi_{j+p} - \xi_j} \right) B_{j,p-1,\xi} \in \mathbb{S}_{p-1,\xi_1}.$$

On the other hand, suppose $s \in \mathbb{S}_{p-1,\xi_1}$, represented as $s = \sum_{j=2}^n d_j B_{j,p-1,\xi}$. Then, by using again the differentiation formula, we can write $s = D_+(\sum_{j=1}^n c_j B_{j,p,\xi})$, where c_1 can be any real number and

$$c_j = c_{j-1} + \frac{\xi_{j+p} - \xi_j}{p} d_j, \quad j = 2, \ldots, n.$$

For $r > 1$ we use the relation $D_+^r = D_+ D_+^{r-1}$. $\qquad\square$

By combining Theorems 6 and 7 it follows that for $0 \leq r \leq p$,

$$\mathbb{S}_{p-r}^{r-r}(\Delta) = D_+^r \mathbb{S}_{p,\xi},$$

where $r - r := \left(\max(r_1 - r, -1), \ldots, \max(r_\ell - r, -1) \right)$ and the knot sequence ξ is constructed as in Theorem 6.

1.3.3 B-Spline Representation of Polynomials

Polynomials can be represented in terms of B-splines of at least the same degree. We now derive an explicit expression for their B-spline coefficients by using the dual polynomials and the (local) Marsden identity.

Theorem 8 (Marsden Identity) *We have*

$$(y - x)^p = \sum_{j=1}^{n} \psi_{j,p,\xi}(y) B_{j,p,\xi}(x), \quad x \in [\xi_{p+1}, \xi_{n+1}], \quad y \in \mathbb{R}, \tag{1.45}$$

where $\psi_{j,p,\xi}(y) := (y - \xi_{j+1}) \cdots (y - \xi_{j+p})$ is the polynomial of degree p that is dual to $B_{j,p,\xi}$.

Proof This follows immediately from the local version (1.17). Indeed, if $x \in [\xi_{p+1}, \xi_{n+1})$ then $x \in [\xi_m, \xi_{m+1})$ for some $p + 1 \le m \le n$, and by the local support property (1.36) we get

$$(y - x)^p = \sum_{j=m-p}^{m} \psi_{j,p,\xi}(y) B_{j,p,\xi}(x) = \sum_{j=1}^{n} \psi_{j,p,\xi}(y) B_{j,p,\xi}(x).$$

Taking into account the left continuity of B-splines at the endpoint ξ_{n+1}, see (1.34), we arrive at the Marsden identity (1.45).   □

Differentiating $p - k$ times with respect to y in (1.45) results in the following formula.

Corollary 1 *For $k = 0, 1, \ldots, p$ we have*

$$\frac{(y - x)^k}{k!} = \sum_{j=1}^{n} \left(\frac{1}{p!} D^{p-k} \psi_{j,p,\xi}(y) \right) B_{j,p,\xi}(x), \quad x \in [\xi_{p+1}, \xi_{n+1}], \quad y \in \mathbb{R}. \tag{1.46}$$

Corollary 1 immediately leads to the following properties.

- **Representation of Monomials.** For $k = 0, 1, \ldots, p$ we have

$$x^k = \sum_{j=1}^{n} \xi_{j,p,\xi}^{*,k} B_{j,p,\xi}(x), \quad x \in [\xi_{p+1}, \xi_{n+1}], \tag{1.47}$$

where

$$\xi_{j,p,\xi}^{*,k} := (-1)^k \frac{k!}{p!} D^{p-k} \psi_{j,p,\xi}(0). \tag{1.48}$$

This follows from (1.46) with $y = 0$.

- **Partition of Unity.** Taking $k = 0$ in (1.47) gives

$$\sum_{j=1}^{n} B_{j,p,\xi}(x) = 1, \quad x \in [\xi_{p+1}, \xi_{n+1}].\tag{1.49}$$

Since the B-splines are nonnegative it follows that they form a **nonnegative partition of unity** on $[\xi_{p+1}, \xi_{n+1}]$.

- **Greville Points.** Taking $k = 1$ in (1.47) gives for $p \geq 1$,

$$x = \sum_{j=1}^{n} \xi_{j,p,\xi}^{*} B_{j,p,\xi}(x), \quad x \in [\xi_{p+1}, \xi_{n+1}],\tag{1.50}$$

where

$$\xi_{j,p,\xi}^{*} := \xi_{j,p,\xi}^{*,1} = \frac{\xi_{j+1} + \cdots + \xi_{j+p}}{p}.\tag{1.51}$$

The number $\xi_{j,p,\xi}^{*}$ is called **Greville point.**[3] It is also known as **knot average** or **node**.

Example 7 For $p = 3$ Eq. (1.47) gives

$$1 = \sum_{j=1}^{n} B_{j,3,\xi}(x),$$

$$x = \sum_{j=1}^{n} \frac{\xi_{j+1} + \xi_{j+2} + \xi_{j+3}}{3} B_{j,3,\xi}(x),$$

$$x^2 = \sum_{j=1}^{n} \frac{\xi_{j+1}\xi_{j+2} + \xi_{j+1}\xi_{j+3} + \xi_{j+2}\xi_{j+3}}{3} B_{j,3,\xi}(x),$$

$$x^3 = \sum_{j=1}^{n} \xi_{j+1}\xi_{j+2}\xi_{j+3} B_{j,3,\xi}(x).$$

We finally present an expression for the B-spline coefficients of a general polynomial.

---

[3]An explicit expression of (1.51) was given by Greville in [24]. According to Schoenberg [43], Greville reviewed the paper [43] introducing some elegant simplifications.

Proposition 3 (Representation of Polynomials) *Any polynomial g of degree p can be represented as*

$$g(x) = \sum_{j=1}^{n} \Lambda_{j,p,\xi}(g) B_{j,p,\xi}(x), \quad x \in [\xi_{p+1}, \xi_{n+1}], \tag{1.52}$$

where

$$\Lambda_{j,p,\xi}(g) := \frac{1}{p!} \sum_{r=0}^{p} (-1)^{p-r} D^r \psi_{j,p,\xi}(\tau_j) D^{p-r} g(\tau_j), \quad \tau_j \in \mathbb{R}. \tag{1.53}$$

Proof The polynomial g can be represented in Taylor form (1.95) as

$$g(x) = \sum_{r=0}^{p} \frac{(x - \tau_j)^{p-r}}{(p-r)!} D^{p-r} g(\tau_j), \quad \tau_j \in \mathbb{R}.$$

The result follows when we apply (1.46) with $k = p - r$.   □

Note that, if τ_j is a root of ψ_j of multiplicity μ_j then $D^r \psi_i(\tau_j) = 0$, $r = 0, 1, \ldots, \mu_j - 1$ and (1.53) becomes

$$\Lambda_{j,p,\xi}(g) = \frac{1}{p!} \sum_{r=\mu_j}^{p} (-1)^{p-r} D^r \psi_{j,p,\xi}(\tau_j) D^{p-r} g(\tau_j), \quad \tau_j \in \mathbb{R}. \tag{1.54}$$

Example 8 The polynomial $g(x) = ax^2 + bx + c$ can be represented in terms of quadratic B-splines:

$$ax^2 + bx + c = \sum_{j=1}^{n} c_j B_{j,2,\xi}(x).$$

From (1.52)–(1.53) with $\psi_{j,2,\xi}(y) := (y - \xi_{j+1})(y - \xi_{j+2})$, we obtain that

$$\begin{aligned}
c_j = \Lambda_{j,2,\xi}(g) &= \frac{1}{2} \Big[(\tau_j - \xi_{j+1})(\tau_j - \xi_{j+2}) 2a \\
&\quad - (2\tau_j - \xi_{j+1} - \xi_{j+2})(2a\tau_j + b) \\
&\quad + 2(a\tau_j^2 + b\tau_j + c) \Big] \\
&= a\,\xi_{j+1}\xi_{j+2} + b\,\frac{\xi_{j+1} + \xi_{j+2}}{2} + c.
\end{aligned}$$

1.3.4   B-Spline Representation of Splines

In the previous section we have derived an explicit expression for the B-spline coefficients of polynomials; see (1.52). The next theorem extends this result by providing an explicit expression for the B-spline coefficients of any spline in $\mathbb{S}_{p,\xi}$.

Theorem 9 (Representation of B-Spline Coefficients) *Any element s in the space $\mathbb{S}_{p,\xi}$ can be represented as*[4]

$$s(x) = \sum_{j=1}^{n} \Lambda_{j,p,\xi}(s) B_{j,p,\xi}(x), \quad x \in [\xi_{p+1}, \xi_{n+1}], \tag{1.55}$$

where

$$\Lambda_{j,p,\xi}(s) := \frac{1}{p!} \begin{cases} \sum_{r=\mu_j}^{p} (-1)^{p-r} D^r \psi_{j,p,\xi}(\tau_j) D_+^{p-r} s(\tau_j), & \text{if } \tau_j = \xi_j, \\ \sum_{r=\mu_j}^{p} (-1)^{p-r} D^r \psi_{j,p,\xi}(\tau_j) D^{p-r} s(\tau_j), & \text{if } \xi_j < \tau_j < \xi_{j+p+1}, \\ \sum_{r=\mu_j}^{p} (-1)^{p-r} D^r \psi_{j,p,\xi}(\tau_j) D_-^{p-r} s(\tau_j), & \text{if } \tau_j = \xi_{j+p+1}, \end{cases} \tag{1.56}$$

and where $\mu_j \geq 0$ is the number of times τ_j appears in $\xi_{j+1}, \ldots, \xi_{j+p}$.

Proof Suppose $\xi_j \leq \tau_j < \xi_{j+p+1}$ and let $I_j := [\xi_{m_j}, \xi_{m_j+1})$ be the interval containing τ_j. The restriction of s to I_j is a polynomial and so by Proposition 3 we find

$$s(x) = \sum_{i=m_j-p}^{m_j} \left(\frac{1}{p!} \sum_{r=0}^{p} (-1)^{p-r} D^r \psi_{i,p,\xi}(\tau_j) D_+^{p-r} s(\tau_j) \right) B_{i,p,\xi}(x), \quad x \in I_j. \tag{1.57}$$

Note that since $\xi_j \leq \tau_j < \xi_{j+p+1}$ we have $j \leq m_j \leq j + p$ which implies $m_j - p \leq j \leq m_j$. By taking $i = j$ in (1.57) and using the local linear independence of the B-splines, we obtain

$$\Lambda_{j,p,\xi}(s) := \frac{1}{p!} \sum_{r=0}^{p} (-1)^{p-r} D^r \psi_{j,p,\xi}(\tau_j) D_+^{p-r} s(\tau_j).$$

Since $D^r \psi_{j,p,\xi}(\tau_j) = 0$ for $r < \mu_j$ we obtain the top term in (1.56). In the middle term we can replace $D_+^{p-r} s(\tau_j)$ by $D^{p-r} s(\tau_j)$ since $s \in C^{p-\mu_j}(\tau_j)$. The proof of the last term is similar using D_- instead of D_+.   □

---

[4]The value $\Lambda_{j,p,\xi}(s)$ is known as the **de Boor–Fix functional** [7] applied to s.

Note that the operator $\Lambda_{j,p,\xi}$ in (1.54) is identical to $\Lambda_{j,p,\xi}$ in (1.56). However, in the spline case we need the restriction $\tau_j \in [\xi_j, \xi_{j+p+1}]$.

Because the set of B-splines $\{B_{j,p,\xi}\}_{j=1}^n$ is a basis for the space $\mathbb{S}_{p,\xi}$, the coefficients $\Lambda_{j,p,\xi}(s)$ are uniquely determined for any $s \in \mathbb{S}_{p,\xi}$. Thus, the right-hand side in (1.56) does not depend on the choice of τ_j. This is an astonishing property considering the complexity of the expression. For example, one could take the Greville point $\xi_{j,p,\xi}^*$ defined in (1.51) as a valid choice for the point τ_j. It is easy to verify that $\xi_{j,p,\xi}^* \in [\xi_j, \xi_{j+p+1}]$, and moreover, $\xi_{j,p,\xi}^* \in (\xi_j, \xi_{j+p+1})$ if $B_{j,p,\xi}$ is a continuous function.

Example 9 We consider the quadratic spline

$$s(x) = \sum_{j=1}^n c_j B_{j,2,\xi}(x),$$

and we illustrate that some derivative terms in the expression (1.56) can be canceled by specific choices of τ_j. Assume for simplicity $\xi_j < \xi_{j+1} < \xi_{j+2} < \xi_{j+3}$.

- If τ_j is the Greville point $\xi_{j,2,\xi}^* := (\xi_{j+1}+\xi_{j+2})/2$, then there is no first derivative term. Indeed, we have

$$c_j = \Lambda_{j,2,\xi}(s) = s(\xi_{j,2,\xi}^*) - \frac{(\xi_{j+2} - \xi_{j+1})^2}{8} D^2 s(\xi_{j,2,\xi}^*).$$

Moreover, since $s \in \mathbb{P}_2$ on $[\xi_{j+1}, \xi_{j+2}]$, we can replace $D^2 s(\xi_{j,2,\xi}^*)$ by a difference quotient

$$D^2 s(\xi_{j,2,\xi}^*) = \left(s(\xi_{j+2}) - 2s(\xi_{j,2,\xi}^*) + s(\xi_{j+1})\right) \Big/ \left(\frac{\xi_{j+2} - \xi_{j+1}}{2}\right)^2,$$

to obtain

$$c_j = -\frac{1}{2}s(\xi_{j+1}) + 2s(\xi_{j,2,\xi}^*) - \frac{1}{2}s(\xi_{j+2}). \tag{1.58}$$

- If τ_j is equal to ξ_{j+1} or ξ_{j+2}, then there is no second derivative term. Indeed, we have

$$c_j = \Lambda_{j,2,\xi}(s) = s(\tau_j) + \frac{\xi_{j,2,\xi}^* - \tau_j}{2} Ds(\tau_j), \quad \tau_j \in \{\xi_{j+1}, \xi_{j+2}\}.$$

A similar property holds for any p: if τ_j is chosen as one of the interior knots $\xi_{j+1}, \ldots, \xi_{j+p}$, then there is no p-th derivative term in the expression of $\Lambda_{j,p,\xi}(s)$.

1.3.5   Knot Insertion

In this section we are addressing the problem of representing a given spline on a refined knot sequence. In particular, we focus on the special case where only a single knot is inserted. Since any refined knot sequence can be reached by repeatedly inserting one knot at a time, it suffices to deal with this case.

Without loss of generality, we assume that the spline $s = \sum_{j=1}^{n} c_j B_{j,p,\boldsymbol{\xi}}$ is given on a $(p+1)$-basic knot sequence $\boldsymbol{\xi} := \{\xi_i\}_{i=1}^{n+p+1}$. We want to insert a knot ξ in some subinterval $[\xi_m, \xi_{m+1})$ of $[\xi_{p+1}, \xi_{n+1})$, resulting in a new $(p+1)$-basic knot sequence $\tilde{\boldsymbol{\xi}} := \{\tilde{\xi}_i\}_{i=1}^{n+p+2}$ defined by

$$\tilde{\xi}_i := \begin{cases} \xi_i, & \text{if } 1 \le i \le m, \\ \xi, & \text{if } i = m+1, \\ \xi_{i-1}, & \text{if } m+2 \le i \le n+p+2. \end{cases} \tag{1.59}$$

The B-spline form of s on the new knot sequence can be computed with the aid of the following procedure introduced by Böhm [3].

Theorem 10 (Knot Insertion) *Let the $(p+1)$-basic knot sequence $\tilde{\boldsymbol{\xi}} := \{\tilde{\xi}_i\}_{i=1}^{n+p+2}$ be obtained from the $(p+1)$-basic knot sequence $\boldsymbol{\xi} := \{\xi_i\}_{i=1}^{n+p+1}$ by inserting just one knot ξ, such that $\xi_m \le \xi < \xi_{m+1}$ as in (1.59). Then,*

$$s(x) = \sum_{j=1}^{n} c_j B_{j,p,\boldsymbol{\xi}}(x) = \sum_{i=1}^{n+1} \tilde{c}_i B_{i,p,\tilde{\boldsymbol{\xi}}}(x), \quad x \in [\xi_{p+1}, \xi_{n+1}], \tag{1.60}$$

where

$$\tilde{c}_i = \begin{cases} c_i, & \text{if } i \le m-p, \\ \dfrac{\xi - \xi_i}{\xi_{i+p} - \xi_i} c_i + \dfrac{\xi_{i+p} - \xi}{\xi_{i+p} - \xi_i} c_{i-1}, & \text{if } m-p < i \le m, \\ c_{i-1}, & \text{if } i > m. \end{cases} \tag{1.61}$$

Proof From Theorem 6 it follows that $\mathbb{S}_{p,\boldsymbol{\xi}}$ is a subspace of $\mathbb{S}_{p,\tilde{\boldsymbol{\xi}}}$, since we have reduced the continuity requirement at ξ_m if $\xi = \xi_m$ or introduced another segment otherwise. Hence, the B-splines in $\mathbb{S}_{p,\boldsymbol{\xi}}$ belong to $\mathbb{S}_{p,\tilde{\boldsymbol{\xi}}}$, and we can write

$$B_{j,p,\boldsymbol{\xi}} = \sum_{i=1}^{n+1} \alpha_{i,j,p} B_{i,p,\tilde{\boldsymbol{\xi}}}, \quad j = 1, \dots, n,$$

for some real numbers $\alpha_{i,j,p}$. Suppose $s \in \mathbb{S}_{p,\xi}$ is given by (1.60). Then,

$$\sum_{j=1}^{n} c_j B_{j,p,\xi} = \sum_{i=1}^{n+1} \left(\sum_{j=1}^{n} \alpha_{i,j,p} c_j \right) B_{i,p,\tilde{\xi}}.$$

By linear independence of the B-splines in $\mathbb{S}_{p,\tilde{\xi}}$ we obtain

$$\tilde{c}_i = \sum_{j=1}^{n} \alpha_{i,j,p} c_j, \quad i = 1, \ldots, n+1. \tag{1.62}$$

Note that each $\alpha_{i,j,p}$ is independent of the c's.

Now, consider the function $f_y(x) := (y - x)^p$ for fixed $y \in \mathbb{R}$. By the Marsden identity (1.45) we have

$$(y - x)^p = \sum_{j=1}^{n} c_j B_{j,p,\xi}(x) = \sum_{i=1}^{n+1} \tilde{c}_i B_{i,p,\tilde{\xi}}(x), \quad x \in [\xi_{p+1}, \xi_{n+1}], \quad y \in \mathbb{R},$$

where

$$c_j = \psi_{j,p,\xi}(y) = (y - \xi_{j+1}) \cdots (y - \xi_{j+p}),$$

and

$$\tilde{c}_i = \psi_{i,p,\tilde{\xi}}(y) = (y - \tilde{\xi}_{i+1}) \cdots (y - \tilde{\xi}_{i+p}).$$

Hence, for the function $f_y(x)$, the identity (1.62) takes the form

$$\psi_{i,p,\tilde{\xi}}(y) = \sum_{j=1}^{n} \alpha_{i,j,p} \psi_{j,p,\xi}(y), \quad i = 1, \ldots, n+1. \tag{1.63}$$

From the relation (1.59) between the knot sequences $\tilde{\xi}$ and ξ, we deduce that $\psi_{i,p,\tilde{\xi}} = \psi_{i,p,\xi}$ for $i \leq m - p$, and $\psi_{i,p,\tilde{\xi}} = \psi_{i-1,p,\xi}$ for $i > m$, and using the dual recurrence relation (1.11) that for $m - p < i \leq m$,

$$\psi_{i,p,\tilde{\xi}}(y) = (y - \xi)\psi_{i,p-1,\xi}(y) = \frac{\xi - \xi_i}{\xi_{i+p} - \xi_i} \psi_{i,p,\xi} + \frac{\xi_{i+p} - \xi}{\xi_{i+p} - \xi_i} \psi_{i-1,p,\xi}.$$

Then, (1.61) follows from (1.62) and (1.63). $\qquad\qquad\qquad\qquad\qquad\qquad \square$

When several knots have to be inserted simultaneously, alternative algorithms can be used instead of repeating the single knot insertion procedure given in Theorem 10. In Sect. 1.4.3 we provide such a simultaneous knot insertion algorithm in case of uniform knot sequences. A more general (but also more complex) knot insertion algorithm is known as the Oslo algorithm [11].

- **Convex Combination.** From relation (1.61) we see that the coefficients $\tilde{c}_i$ are a convex combination of the coefficients c_i. In general, the coefficients obtained after repeated knot insertion are a convex combination of the original coefficients.

- **Evaluation.** Repeated knot insertion gives rise to an evaluation process for spline functions in B-spline form. Indeed, the evaluation of a spline s at the point x can be achieved by the repeated insertion of x as a knot till it has multiplicity p. Then, assuming that for some m,

$$\xi_m < x = \xi_{m+1} = \cdots = \xi_{m+p} < \xi_{m+p+1},$$

we can conclude from (1.29) and (1.49) that

$$B_{j,p,\xi}(x) = \begin{cases} 1, & \text{if } j = m, \\ 0, & \text{otherwise,} \end{cases}$$

and

$$s(x) = \sum_{j=1}^{n} c_j B_{j,p,\xi}(x) = c_m B_{m,p,\xi}(x) = c_m.$$

When comparing (1.61) with (1.39), we observe that single knot insertion is nothing else than applying once the B-spline coefficient recurrence relation. This evaluation procedure is a fast and numerically stable algorithm introduced by de Boor [4].

1.3.6 Condition Number

A basis $\{B_j\}$ of a normed space is said to be **stable** with respect to a vector norm if there are positive constants K_L and K_U such that

$$K_L^{-1}\|c\| \le \left\|\sum_j c_j B_j\right\| \le K_U\|c\|, \tag{1.64}$$

for all coefficient vectors $c := (c_j)$. For simplicity we use the same symbol $\|\cdot\|$ for the norm in the space and the vector norm. The number

$$\kappa := \inf\{K_L K_U : K_L \text{ and } K_U \text{ satisfy } (1.64)\} \tag{1.65}$$

is called the **condition number** of the basis $\{B_j\}$ with respect to $\|\cdot\|$.

Such condition numbers give an upper bound for how much an error in coefficients can be magnified in function values and vice versa. Indeed, if

$f := \sum_j c_j B_j \neq 0$ and $g := \sum_j d_j B_j$ then it follows immediately from (1.64) that

$$\frac{1}{\kappa} \frac{\|c - d\|}{\|c\|} \leq \frac{\|f - g\|}{\|f\|} \leq \kappa \frac{\|c - d\|}{\|c\|},$$

where $c := (c_j)$ and $d := (d_j)$. Many other applications are given in [5] and it is interesting to have estimates for the size of κ.

We consider the L_q-norm for functions and the q-norm for vectors with $1 \leq q \leq \infty$. We focus on a scaled version of the B-spline basis defined on $[\xi_1, \xi_{n+p+1}]$,

$$\{N_{j,p,q,\xi}\}_{j=1}^n := \{\gamma_{j,p,\xi}^{-1/q} B_{j,p,\xi}\}_{j=1}^n, \tag{1.66}$$

where $\gamma_{j,p,\xi} := (\xi_{j+p+1} - \xi_j)/(p+1)$; see also (1.30). The knot sequence ξ is assumed to be $(p+1)$-basic in order to have linearly independent B-splines. This also ensures that $\gamma_{j,p,\xi} > 0$. The q-**norm condition number** of the basis in (1.66) will be denoted by $\kappa_{p,q,\xi}$, i.e.,

$$\kappa_{p,q,\xi} := \sup_{c \neq 0} \frac{\left\| \sum_{j=1}^n c_j N_{j,p,q,\xi} \right\|_{L_q([\xi_1, \xi_{n+p+1}])}}{\|c\|_q}$$

$$\times \sup_{c \neq 0} \frac{\|c\|_q}{\left\| \sum_{j=1}^n c_j N_{j,p,q,\xi} \right\|_{L_q([\xi_1, \xi_{n+p+1}])}}. \tag{1.67}$$

The next theorem shows that the scaled B-spline basis above is stable in any L_q-norm independently of the knot sequence ξ. It also provides an upper bound for the q-norm condition number which does not depend on ξ. To this end, we first state the **Hölder inequality for sums**:

$$\sum_{j=1}^n |x_j y_j| \leq \|x\|_q \|y\|_{q'}, \tag{1.68}$$

where q, q' are integers so that

$$\frac{1}{q} + \frac{1}{q'} = 1, \quad 1 \leq q \leq \infty. \tag{1.69}$$

In particular, $q' = \infty$ if $q = 1$ and $q' = 2$ if $q = 2$.

Theorem 11 *For any $p \geq 0$ there exists a positive constant K_p depending only on p, such that for any vector $c := (c_1, \ldots, c_n)$ and for any $1 \leq q \leq \infty$ we have*

$$K_p^{-1} \|c\|_q \leq \left\| \sum_{j=1}^n c_j N_{j,p,q,\xi} \right\|_{L_q([\xi_1, \xi_{n+p+1}])} \leq \|c\|_q. \tag{1.70}$$

Proof We first prove the upper inequality. By using the nonnegative partition of unity property of B-splines, the upper bound for $q = \infty$ is straightforward. For $q = 1$, we have

$$\int_{\xi_1}^{\xi_{n+p+1}} \left| \sum_{j=1}^{n} c_j N_{j,p,q,\xi}(x) \right| dx \leq \sum_{j=1}^{n} |c_j| \gamma_{j,p,\xi}^{-1} \int_{\xi_j}^{\xi_{j+p+1}} B_{j,p,\xi}(x) \, dx = \|c\|_1.$$

Finally, we consider $1 < q < \infty$. By applying the Hölder inequality (1.68) and again the nonnegative partition of unity property of B-splines, we obtain for $x \in [\xi_1, \xi_{n+p+1}]$,

$$\left| \sum_{j=1}^{n} c_j N_{j,p,q,\xi}(x) \right| \leq \sum_{j=1}^{n} |c_j| \gamma_{j,p,\xi}^{-1/q} \left(B_{j,p,\xi}(x) \right)^{1/q} \left| \left| B_{j,p,\xi}(x) \right| \right|^{1-1/q}$$

$$\leq \left(\sum_{j=1}^{n} |c_j|^q \gamma_{j,p,\xi}^{-1} B_{j,p,\xi}(x) \right)^{1/q} \left(\sum_{j=1}^{n} B_{j,p,\xi}(x) \right)^{1-1/q}$$

$$\leq \left(\sum_{j=1}^{n} |c_j|^q \gamma_{j,p,\xi}^{-1} B_{j,p,\xi}(x) \right)^{1/q}.$$

Raising both sides of this inequality to the q-th power and integrating gives the inequality

$$\int_{\xi_1}^{\xi_{n+p+1}} \left| \sum_{j=1}^{n} c_j N_{j,p,q,\xi}(x) \right|^q dx \leq \sum_{j=1}^{n} |c_j|^q \gamma_{j,p,\xi}^{-1} \int_{\xi_j}^{\xi_{j+p+1}} B_{j,p,\xi}(x) \, dx = \|c\|_q^q.$$

Taking the q-th roots on both sides proves the upper inequality in (1.70).

We now focus on the lower inequality. We extend ξ to a $(p + 1)$-open knot sequence $\hat{\xi}$ by possibly increasing the multiplicity of ξ_1 and ξ_{n+p+1} to $p + 1$. Clearly, the set of B-splines on ξ is a subset of the set of B-splines on $\hat{\xi}$, and any linear combination of the B-splines on ξ is a linear combination of the B-splines on $\hat{\xi}$ where the extra B-splines have coefficients zero. Therefore, without loss of generality, we can assume that the knot sequence is open with the basic interval $[\xi_1, \xi_{n+p+1}]$. The lower bound then follows from Lemma 5; see Sect. 1.5.3.1. □

Finally, we define a condition number that is independent of the knot sequence,

$$\kappa_{p,q} := \sup_{\xi} \kappa_{p,q,\xi}. \tag{1.71}$$

Theorem 11 shows that

$$\kappa_{p,q} \leq K_p < \infty.$$

It is known that $\kappa_{p,q}$ grows like 2^p for all $1 \leq q \leq \infty$; see [34, 40] where it has been proved that

$$\frac{1}{p+1}2^{p-1/2} \leq \kappa_{p,q} \leq (p+1)2^{p+1}, \quad 1 \leq q \leq \infty. \tag{1.72}$$

1.4 Cardinal B-Splines

A particularly interesting case of B-spline functions is obtained when the knot sequence is uniformly spaced. Without loss of generality, we can assume that the knot sequence is given by the set of integers $\mathbb{Z}$. It is natural to index the knots as $\xi_j = j$, $j \in \mathbb{Z}$. Due to the translation invariance property (1.9) we have

$$B_{j,p,\mathbb{Z}}(x) = B_{0,p,\mathbb{Z}}(x - j), \quad j \in \mathbb{Z}, \quad x \in \mathbb{R}. \tag{1.73}$$

Therefore, all the B-splines on the knot sequence $\mathbb{Z}$ are integer translates of a single function. This motivates the following definition.

Definition 3 The function $M_p := B[0, 1, \ldots, p + 1]$ is the **cardinal B-spline** of degree p.

Example 10 Figure 1.3 illustrates the cardinal B-splines M_p for $p = 1, \ldots, 5$.

1.4.1 Main Properties

Cardinal B-splines possess several interesting features. Of course, they inherit all the properties of general B-splines, and in particular the following ones.

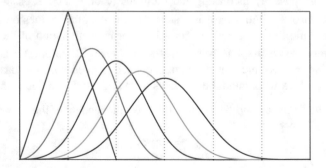

Fig. 1.3 The cardinal B-splines M_p for $p = 1, \ldots, 5$. The uniform knot positions are visualized by vertical dotted lines

- **Local Support.** From (1.6) it follows that the cardinal B-spline M_p is locally supported on the interval $[0, p+1]$.

- **Nonnegativity, Piecewise Structure and Smoothness.** From (1.7), (1.8) and (1.26) it follows that the cardinal B-spline M_p is a nonnegative, piecewise polynomial of degree p belonging to the class $C^{p-1}(\mathbb{R})$.

- **Differentiation and Integration.** The formulas (1.22) and (1.30) simplify in the case of cardinal B-splines to

$$D_+ M_p(x) = M_{p-1}(x) - M_{p-1}(x-1), \quad p \geq 1, \tag{1.74}$$

and

$$\int_{\mathbb{R}} M_p(x)\, dx = 1. \tag{1.75}$$

- **Recurrence Relation.** From Definition 2 we obtain the following recurrence relation for cardinal B-splines,

$$M_0(x) = \begin{cases} 1, & \text{if } x \in [0,1), \\ 0, & \text{otherwise,} \end{cases} \tag{1.76}$$

$$M_p(x) = \frac{x}{p} M_{p-1}(x) + \frac{p+1-x}{p} M_{p-1}(x-1), \quad p \geq 1. \tag{1.77}$$

The uniformity of the knot sequence endows the cardinal B-splines with several additional properties. A key feature is based on convolution.

- **Convolution.** The convolution of two functions f and g is defined by

$$(f * g)(x) := \int_{\mathbb{R}} f(x-y)g(y)\, dy.$$

The cardinal B-spline M_p can be characterized using convolution by

$$M_p(x) = (M_{p-1} * M_0)(x) = \int_0^1 M_{p-1}(x-y)\, dy, \quad p \geq 1, \tag{1.78}$$

and

$$M_p(x) = \big(\overbrace{M_0 * \cdots * M_0}^{p+1}\big)(x). \tag{1.79}$$

Proof From (1.74) we deduce

$$M_p(x) = \int_0^x (M_{p-1}(y) - M_{p-1}(y-1))\,dy$$

$$= \int_0^x M_{p-1}(y)\,dy - \int_{-1}^{x-1} M_{p-1}(y)\,dy$$

$$= \int_{x-1}^x M_{p-1}(y)\,dy = \int_0^1 M_{p-1}(x-y)\,dy.$$

Applying recursively (1.78) immediately gives (1.79).   □

- **Fourier Transform.** The Fourier transform of a function $f \in L_2(\mathbb{R})$ is defined by

$$\widehat{f}(\theta) := \int_{\mathbb{R}} f(x)\,e^{-i\theta x}\,dx,$$

where $i := \sqrt{-1}$ denotes the imaginary unit. The Fourier transform of the cardinal B-spline M_p is given by

$$\widehat{M_p}(\theta) = \left(\frac{1 - e^{-i\theta}}{i\theta}\right)^{p+1}. \qquad (1.80)$$

Proof From (1.76), a direct computation gives

$$\widehat{M_0}(\theta) = \frac{1 - e^{-i\theta}}{i\theta}.$$

An interesting property of the Fourier transform of a convolution is

$$\widehat{(f * g)}(\theta) = \widehat{f}(\theta)\widehat{g}(\theta), \quad \forall f, g \in L_2(\mathbb{R}); \qquad (1.81)$$

see, e.g., [39]. Hence, by combining (1.81) with (1.79) we deduce that $\widehat{M_p}(\theta) = \left(\widehat{M_0}(\theta)\right)^{p+1}$, which implies (1.80).   □

- **Symmetry.** The cardinal B-spline M_p is symmetric with respect to the midpoint of its support, namely $(p+1)/2$. More generally,

$$D^r M_p\left(\frac{p+1}{2} + x\right) = (-1)^r D^r M_p\left(\frac{p+1}{2} - x\right), \quad r = 0, \ldots, p-1, \qquad (1.82)$$

and

$$D_-^p M_p\left(\frac{p+1}{2} + x\right) = (-1)^p D_+^p M_p\left(\frac{p+1}{2} - x\right). \qquad (1.83)$$

Proof It suffices to prove that $M_p(p + 1 - x) = M_p(x)$. The general result then follows from repeated differentiations. We proceed by induction. It is easy to check that it is true for $p = 0$. Assuming the symmetry property holds for degree $p - 1$ and using (1.78), we get

$$M_p(p + 1 - x) = \int_0^1 M_{p-1}(p + 1 - x - t)\, dt = \int_0^1 M_{p-1}(x - 1 + t)\, dt$$

$$= -\int_1^0 M_{p-1}(x - t)\, dt = \int_0^1 M_{p-1}(x - t)\, dt = M_p(x). \qquad \square$$

We now focus on the set of integer translates of the cardinal B-spline M_p, i.e.,

$$\{ M_p(\cdot - j), \ j \in \mathbb{Z} \}. \tag{1.84}$$

They have the following properties.

- **Linear Independence.** From (1.73) it follows that the integer translates $M_p(\cdot - j)$, $j \in \mathbb{Z}$, are (locally) linearly independent on $\mathbb{R}$. They span the space of piecewise polynomials of degree p and smoothness $p - 1$ with integer break points; see (1.42).

- **Partition of Unity.** From (1.49) and (1.73) we get

$$\sum_{j \in \mathbb{Z}} M_p(x - j) = 1, \quad x \in \mathbb{R}.$$

Due to the local support of cardinal B-splines, the above series reduces to a finite sum for any x. More precisely, referring to (1.21), we have

$$\sum_{j=m-p}^{m} M_p(x - j) = 1, \quad x \in [m, m + 1).$$

- **Greville Points.** From (1.50)–(1.51) and (1.73) we have

$$x = \sum_{j \in \mathbb{Z}} \zeta_{j,p}^* M_p(x - j), \quad x \in \mathbb{R},$$

with

$$\zeta_{j,p}^* := \frac{(1 + j) + \cdots + (p + j)}{p} = \frac{p + 1}{2} + j. \tag{1.85}$$

1.4.2 Inner Products

Inner products of cardinal B-splines and their translates can be interpreted as evaluations of higher-degree cardinal B-splines; similar results also hold for their derivatives.[5]

Theorem 12 (Inner Product) *Given* $p_1, p_2 \geq 0$, *we have*

$$\int_{\mathbb{R}} M_{p_1}(y) M_{p_2}(y+x) \, dy = M_{p_1+p_2+1}(p_1+1+x) = M_{p_1+p_2+1}(p_2+1-x).$$

Proof From the symmetry property (1.82)–(1.83) and the convolution relation (1.78) of cardinal B-splines, we get

$$\int_{\mathbb{R}} M_{p_1}(y) M_{p_2}(y+x) \, dy = \int_{\mathbb{R}} M_{p_1}(y) M_{p_2}(p_2+1-y-x) \, dy$$

$$= (M_{p_1} * M_{p_2})(p_2+1-x)$$

$$= \big(\overbrace{M_0 * \cdots * M_0}^{p_1+1} * \overbrace{M_0 * \cdots * M_0}^{p_2+1}\big)(p_2+1-x)$$

$$= M_{p_1+p_2+1}(p_2+1-x).$$

Finally, again by symmetry of cardinal B-splines, we have

$$M_{p_1+p_2+1}(p_1+1+x) = M_{p_1+p_2+1}(p_2+1-x),$$

which completes the proof.   □

Theorem 13 (Inner Product of Derivatives) *Given* $p_1 \geq r_1 \geq 0$ *and* $p_2 \geq r_2 \geq 0$, *we have*

$$\int_{\mathbb{R}} D_+^{r_1} M_{p_1}(y) \, D_+^{r_2} M_{p_2}(y+x) \, dy = (-1)^{r_1} D^{r_1+r_2} M_{p_1+p_2+1}(p_1+1+x)$$

$$= (-1)^{r_2} D^{r_1+r_2} M_{p_1+p_2+1}(p_2+1-x).$$

Proof Because of the (anti-)symmetry of higher order derivatives of cardinal B-splines given in (1.82), we have

$$(-1)^{r_1} D^{r_1+r_2} M_{p_1+p_2+1}(p_1+1+x)$$

$$= (-1)^{r_1} (-1)^{r_1+r_2} D^{r_1+r_2} M_{p_1+p_2+1}(p_1+p_2+2-(p_1+1+x))$$

$$= (-1)^{r_2} D^{r_1+r_2} M_{p_1+p_2+1}(p_2+1-x).$$

---

[5]The inner product formula for cardinal B-splines traces back to [44]. The formula for derivatives of cardinal B-splines can be found in [21] and a generalization for multivariate box splines in [48].

So, we only have to show one of both equalities in the theorem. This can be proved by induction on the order of derivatives. The base case ($r_1 = r_2 = 0$) simply follows from Theorem 12. We consider two inductive steps: in the first inductive step we increase the order of derivative of M_{p_1} by one, i.e., $r_1 \rightarrow r_1 + 1$, and in the second inductive step we increase the order of derivative of M_{p_2} by one, i.e., $r_2 \rightarrow r_2 + 1$.

1. ($r_1 \rightarrow r_1 + 1$). Using (1.74) and the induction hypothesis, we have

$$
\int_{\mathbb{R}} D_+^{r_1+1} M_{p_1}(y)\, D_+^{r_2} M_{p_2}(y+x)\, dy
$$

$$
= \int_{\mathbb{R}} \left(D_+^{r_1} M_{p_1-1}(y) - D_+^{r_1} M_{p_1-1}(y-1) \right) D_+^{r_2} M_{p_2}(y+x)\, dy
$$

$$
= \int_{\mathbb{R}} D_+^{r_1} M_{p_1-1}(y) D_+^{r_2} M_{p_2}(y+x)\, dy
$$

$$
\quad - \int_{\mathbb{R}} D_+^{r_1} M_{p_1-1}(y-1) D_+^{r_2} M_{p_2}(y+x)\, dy
$$

$$
= (-1)^{r_1} \left(D^{r_1+r_2} M_{p_1+p_2}(p_1 + x) - D^{r_1+r_2} M_{p_1+p_2}(p_1 + 1 + x) \right)
$$

$$
= (-1)^{r_1+1}\, D^{r_1+r_2+1} M_{p_1+p_2+1}(p_1 + 1 + x).
$$

2. ($r_2 \rightarrow r_2 + 1$). This inductive step can be proved in a completely analogous way as the first inductive step.   □

Due to the relevance of the set (1.84), the results in Theorems 12 and 13 are of particular interest when we consider integer shifts, i.e., $x \in \mathbb{Z}$. In this case, the above inner products reduce to evaluations of cardinal B-splines and their derivatives at either integer or half-integer points. Moreover, there is a relation with the Greville points (1.85). Indeed, if $p_1 = p_2 = p$ and $x = i$ in Theorem 12, then

$$
\int_{\mathbb{R}} M_p(x) M_p(x+i)\, dx = M_{2p+1}(p+1+i) = M_{2p+1}(\zeta^*_{i,2p+1}).
$$

A similar relation holds for the inner products of derivatives in Theorem 13. Thanks to the recurrence relation for derivatives (1.74), the inner products of derivatives of cardinal B-splines and its integer translates reduce to evaluations of cardinal B-splines at either integer or half-integer points.

1.4.3   Uniform Knot Insertion

In Sect. 1.3.5 we have seen how to insert a (single) knot into an existing knot sequence without changing the shape of a given spline function defined on that knot sequence. For uniform knot sequences, we can provide a simple alternative algorithm for inserting simultaneously a knot in each knot interval.

Let us consider the B-splines of degree p over the uniform knot sequence given by $\mathbb{Z}/2$. In this case, it is natural to index the knots as

$$\xi_i = \begin{cases} k, & \text{if } i = 2k, \\ k+1/2, & \text{if } i = 2k+1, \end{cases} \quad i \in \mathbb{Z}.$$

From the definition we have $B_{i,p,\mathbb{Z}/2}(x) = M_p(2x - i)$ for $i \in \mathbb{Z}$. Since $\mathbb{S}_{p,\mathbb{Z}} \subset \mathbb{S}_{p,\mathbb{Z}/2}$, the cardinal B-spline M_p is a **refinable function**, i.e., it can be written as a linear combination of translated and dilated versions of itself:

$$M_p(x) = \sum_{i=0}^{p+1} \alpha_{i,p} M_p(2x - i). \tag{1.86}$$

We are now looking for a relation between the coefficients of a given spline function corresponding to knots in $\mathbb{Z}$ and the coefficients of the same function corresponding to knots in $\mathbb{Z}/2$. The following simultaneous knot insertion procedure was introduced by Lane and Riesenfeld [32].

Theorem 14 (Uniform Knot Insertion) *Consider the uniform knot sequences $\mathbb{Z}$ and $\mathbb{Z}/2$. Then,*

$$s(x) = \sum_{j \in \mathbb{Z}} c_j M_p(x - j) = \sum_{i \in \mathbb{Z}} \tilde{c}_i M_p(2x - i), \tag{1.87}$$

with $\tilde{c}_i = \tilde{c}_i^{[p]}$ defined recursively by

$$\tilde{c}_i^{[p]} := \frac{\tilde{c}_i^{[p-1]} + \tilde{c}_{i-1}^{[p-1]}}{2}, \tag{1.88}$$

starting from

$$\tilde{c}_i^{[0]} := \begin{cases} c_j, & \text{if } i = 2j, \\ c_j, & \text{if } i = 2j+1. \end{cases} \tag{1.89}$$

Proof For $p = 0$ we can directly check that

$$M_0(x) = M_0(2x) + M_0(2x - 1),$$

leading to (1.87) with (1.89). We proceed by induction on p. Assume the relation (1.87) with (1.88) holds for cardinal B-splines of degree $p - 1$. Then, by using the convolution property (1.78) we get

$$\sum_{j \in \mathbb{Z}} c_j M_p(x - j) = \int_0^1 \sum_{j \in \mathbb{Z}} c_j M_{p-1}(x - y - j) \, dy$$

$$= \int_0^1 \sum_{i \in \mathbb{Z}} \tilde{c}_i^{[p-1]} M_{p-1}(2x - 2y - i) \, dy$$

$$= \sum_{i \in \mathbb{Z}} \tilde{c}_i^{[p-1]} \left(\int_0^{1/2} M_{p-1}(2x - 2y - i) \, dy \right.$$

$$\left. + \int_{1/2}^1 M_{p-1}(2x - 2y - i) \, dy \right)$$

$$= \sum_{i \in \mathbb{Z}} \frac{\tilde{c}_i^{[p-1]}}{2} \left(M_p(2x - i) + M_p(2x - i - 1) \right)$$

$$= \sum_{i \in \mathbb{Z}} \frac{\tilde{c}_i^{[p-1]} + \tilde{c}_{i-1}^{[p-1]}}{2} M_p(2x - i),$$

which concludes the proof.   □

The knot insertion procedure in Theorem 14 can be geometrically described as follows. First, every coefficient is doubled. Second, a sequence of p sets of coefficients is constructed by taking averages of the previous set of coefficients.

The coefficients $\{\alpha_{i,p}\}$ in (1.86) can be directly computed from Theorem 14, and we obtain the explicit expression

$$\alpha_{i,p} = \frac{1}{2^p} \binom{p+1}{i}, \quad i = 0, \ldots, p+1. \tag{1.90}$$

They are called the **subdivision mask** of the (uniform) B-spline refinement scheme of degree p.

1.5 Spline Approximation

In this section we discuss how well a sufficiently smooth function can be approximated in the spline space spanned by a given set of B-splines. Exploiting the properties of the B-spline basis presented in the previous sections, we explicitly construct a spline which achieves optimal approximation accuracy for the function and its derivatives, and we determine the corresponding error estimates. The construction method we are going to present is local and linear.

1.5.1 Preliminaries

Let I be a finite interval of the real line. A function $f : I \to \mathbb{R}$ is a **piecewise continuous** function on I if it is bounded and continuous except at a finite number of points, where the value is obtained by taking the limit either from the left or the right. We denote the space of these functions by $C^{-1}(I)$.

For $r \in \mathbb{N}_0$ and $1 \le q \le \infty$ the one-dimensional **Sobolev spaces** are defined by

$$W_q^r(I) := \left\{ f : I \to \mathbb{R} : D^j f \in L_q(I), \ j = 0, \ldots, r \right\}. \tag{1.91}$$

They are normed spaces with norm

$$\|f\|_{W_q^r(I)}^2 := \sum_{j=0}^{r} \|D^j f\|_{L_q(I)}^2, \tag{1.92}$$

called **Sobolev norm**. It can be shown that for $r \in \mathbb{N}$ and $1 < q < \infty$,

$$C^r(I) \subset W_\infty^r(I) \subset W_q^r(I) \subset W_1^r(I) \subset C^{r-1}(I). \tag{1.93}$$

The **Hölder inequality for integrals** is given by

$$\int_a^b |f(x)g(x)| \, dx \le \|f\|_{L_q(I)} \|g\|_{L_{q'}(I)}, \tag{1.94}$$

where $I := [a, b]$ and q, q' are integers satisfying (1.69).

The **Taylor polynomial** of degree p at the point a to a function $f \in W_1^{p+1}([a, b])$ is defined by

$$\mathcal{T}_{a,p} f(x) := \sum_{j=0}^{p} \frac{(x - a)^j}{j!} D^j f(a), \tag{1.95}$$

and its approximation error can be expressed in integral form for $x \in [a, b]$ as

$$f(x) - \mathcal{T}_{a,p} f(x) = \frac{1}{p!} \int_a^b (x - y)_+^p D^{p+1} f(y) \, dy. \tag{1.96}$$

Every polynomial $g \in \mathbb{P}_p$ can be written in **Taylor form** as $g = \mathcal{T}_{a,p} g$.

Theorem 15 (Taylor Interpolation Error) *Let $f \in W_q^{p+1}([a, b])$ with $1 \le q \le \infty$, and let $\mathcal{T}_{a,p} f$ be the Taylor polynomial of degree p to f at the point a. Then, for any $x \in [a, b]$ and $0 \le r \le p$,*

$$|D^r(f - \mathcal{T}_{a,p} f)(x)| \le \frac{(b - a)^{p+1-r-1/q}}{(p - r)!} \|D^{p+1} f\|_{L_q([a,b])}, \tag{1.97}$$

and

$$\|D^r(f - \mathcal{T}_{a,p}f)\|_{L_q([a,b])} \leq \frac{(b-a)^{p+1-r}}{(p-r)!}\|D^{p+1}f\|_{L_q([a,b])}. \tag{1.98}$$

Proof By differentiating the integral form of the Taylor approximation error (1.96) and using the Hölder inequality (1.94), we obtain

$$|D^r(f - \mathcal{T}_{a,p}f)(x)| = \frac{1}{(p-r)!}\int_a^b (x-y)_+^{p-r} D^{p+1}f(y)\,dy$$

$$\leq \frac{1}{(p-r)!}\left[\int_a^b (x-y)_+^{(p-r)q'}\,dy\right]^{1/q'} \|D^{p+1}f\|_{L_q([a,b])}$$

$$\leq \frac{(b-a)^{p-r+1/q'}}{(p-r)!\,((p-r)q'+1)^{1/q'}}\|D^{p+1}f\|_{L_q([a,b])}.$$

Since $1/q + 1/q' = 1$ and $(p-r)q' \geq 0$, we obtain (1.97). Finally, taking the L_q-norm shows (1.98). □

For the sake of simplicity one can use the following weaker, but simpler upper bound,

$$\|D^r(f - \mathcal{T}_{a,p}f)\|_{L_q([a,b])} \leq (b-a)^{p+1-r}\|D^{p+1}f\|_{L_q([a,b])}. \tag{1.99}$$

1.5.2 Spline Quasi-Interpolation

In general, a spline approximating a function f can be written in terms of B-splines as

$$\mathcal{Q}f(x) := \sum_{j=1}^n \lambda_j(f)B_{j,p,\xi}(x) \tag{1.100}$$

for suitable coefficients $\lambda_j(f)$. The spline in (1.100) will be referred to as a **quasi-interpolant** to f whenever it provides a "reasonable" approximation to f.

Both interpolation and least squares are examples of quasi-interpolation methods. They are global methods since we have to solve an n by n system of linear equations to find their coefficients $\lambda_j(f)$. It follows that the value of the spline (1.100) at a point depends on all the data.

In this section we focus on **local linear methods**, i.e., methods where each λ_j is a linear functional only depending on the values of f in the support of $B_{j,p,\xi}$. In principle, it suffices to be "near" the support of $B_{j,p,\xi}$, but we want to keep the presentation as simple as possible. In order to deal with point evaluator functionals we assume here that $f \in C^{-1}([a, b])$, where $[a, b]$ is a bounded interval. We consider a spline space $\mathbb{S}_{p,\boldsymbol{\xi}}$, where the knot sequence $\boldsymbol{\xi}$ is $(p + 1)$-basic and the basic interval $[\xi_{p+1}, \xi_{n+1}]$ is equal to $[a, b]$.

With the aim of constructing a spline quasi-interpolant with optimal accuracy, we need to introduce some basic approximation properties of quasi-interpolants of the form (1.100). Since we are interested in local methods, we start with the following definition.

Definition 4 We say that a linear functional $\lambda : C^{-1}([a, b]) \to \mathbb{R}$ is **supported on a nonempty set** $\mathscr{S} \subset [a, b]$ if $\lambda(f) = 0$ for any $f \in C^{-1}([a, b])$ which vanishes on $\mathscr{S}$.

Note that the set $\mathscr{S}$ in this definition is not uniquely defined and is not necessary minimal.

To construct our quasi-interpolant, we first require linear functionals that are supported on intervals consisting of a few knot intervals. This will ensure that $\mathscr{Q}f$ only depends locally on f. To ensure a good approximation power, we also require polynomial reproduction up to a given degree. Finally, to bound the error, a boundedness assumption on the linear functionals is needed. This leads to the following definitions.

Definition 5 The quasi-interpolant $\mathscr{Q}$ given by (1.100) is called a **local quasi-interpolant** if

(i) each λ_j is supported on the interval I_j, where

$$I_j := [\xi_j, \xi_{j+p+1}] \cap [a, b], \tag{1.101}$$

such that I_j has nonempty interior;
(ii) the λ_j are chosen so that (1.100) reproduces $\mathbb{P}_l$, i.e.,

$$\mathscr{Q}g(x) = g(x) \text{ for all } x \in [a, b] \text{ and all } g \in \mathbb{P}_l, \tag{1.102}$$

for some l with $0 \leq l \leq p$.

Definition 6 A local quasi-interpolant $\mathscr{Q}$ is called **bounded** in an L_q-norm, $1 \leq q \leq \infty$, if there is a constant $C_{\mathscr{Q}}$ such that for each λ_j we have

$$|\lambda_j(f)| \leq C_{\mathscr{Q}} h_{j,p,\boldsymbol{\xi}}^{-1/q} \|f\|_{L_q(I_j)} \text{ for all } f \in C^{-1}(I_j), \tag{1.103}$$

where

$$h_{j,p,\boldsymbol{\xi}} := \max_{\max(j,p+1) \leq k \leq \min(j+p,n)} \xi_{k+1} - \xi_k. \tag{1.104}$$

Note that $h_{j,p,\boldsymbol{\xi}}$ is the largest length of a knot interval in the intersection of the basic interval with the support of $B_{j,p,\boldsymbol{\xi}}$. The requirement (1.101) ensures that the spline in (1.100) provides a local approximation to f. The polynomial reproduction as stated in (1.102) coupled with the boundedness of the linear functionals are the main ingredients to prove the approximation power of any bounded local quasi-interpolant.

We now give both a local and a global version of the approximation power of bounded local quasi-interpolants. To turn a local bound into a global bound we first state the following lemma.

Lemma 1 *Suppose that $f \in L_q([\xi_{p+1}, \xi_{n+1}])$ for some q, $1 \le q < \infty$, and that $m_{i_1}, \ldots, m_{i_2}$ are integers with $m_{i_1} < \cdots < m_{i_2}$, $\xi_{p+1} \le \xi_{m_{i_1}}$ and $\xi_{m_{i_2}+k} \le \xi_{n+1}$ for some positive integer k and integers $i_1 \le i_2$. Then,*

$$\left(\sum_{j=i_1}^{i_2} \|f\|_{L_q([\xi_{m_j}, \xi_{m_j+k}])}^q \right)^{1/q} \le k^{1/q} \|f\|_{L_q([\xi_{p+1}, \xi_{n+1}])}. \qquad (1.105)$$

Proof Under the stated assumptions, each knot interval in $[\xi_{p+1}, \xi_{n+1}]$ is counted at most k times and moreover all the local intervals $[\xi_{m_j}, \xi_{m_j+k}]$ are contained in $[\xi_{p+1}, \xi_{n+1}]$. The definition of the L_q-norm gives immediately (1.105). □

Theorem 16 (Quasi-Interpolation Error) *Let $\mathcal{Q}$ be a bounded local quasi-interpolant in an L_q-norm, $1 \le q \le \infty$, as in Definitions 5 and 6. Let l, p be integers with $0 \le l \le p$. Suppose $\xi_m < \xi_{m+1}$ for some $p + 1 \le m \le n$, and let $f \in W_q^{l+1}(J_m)$ with*

$$J_m := [\xi_{m-p}, \xi_{m+p+1}] \cap [a, b].$$

Then,

$$\|f - \mathcal{Q}f\|_{L_q([\xi_m, \xi_{m+1}])} \le \frac{(2p+1)^{l+1}}{l!} (1 + C_{\mathcal{Q}}) h_{m,\xi}^{l+1} \|D^{l+1} f\|_{L_q(J_m)}, \qquad (1.106)$$

where $h_{m,\xi}$ is the largest length of a knot interval in J_m. Moreover, if $f \in W_q^{l+1}([a, b])$ then

$$\|f - \mathcal{Q}f\|_{L_q([a,b])} \le \frac{(2p+1)^{l+1+1/q}}{l!} (1 + C_{\mathcal{Q}}) h_{\xi}^{l+1} \|D^{l+1} f\|_{L_q([a,b])}, \qquad (1.107)$$

where

$$h_{\xi} := \max_{p+1 \le j \le n} \xi_{j+1} - \xi_j.$$

Proof Note that f is continuous since $l \ge 0$. Suppose $x \in [\xi_m, \xi_{m+1})$. By the local partition of unity (1.21) and by (1.103) we have

$$|\mathcal{Q}f(x)| \le \max_{m-p \le j \le m} |\lambda_j(f)| \le C_{\mathcal{Q}} \max_{m-p \le j \le m} h_{j,p,\xi}^{-1/q} \|f\|_{L_q(I_j)}.$$

Since $\xi_{m+1} - \xi_m \leq \min_{m-p \leq j \leq m} h_{j,p,\xi}$ and $J_m = \cup_{m-p \leq j \leq m} I_j$ we find

$$\|\mathscr{Q}f\|_{L_q([\xi_m,\xi_{m+1}])} \leq C_{\mathscr{Q}}\|f\|_{L_q(J_m)}. \tag{1.108}$$

From (1.102) we know that $\mathscr{Q}$ reproduces any polynomial $g \in \mathbb{P}_l$, and so the triangle inequality gives

$$\|f - \mathscr{Q}f\|_{L_q([\xi_m,\xi_{m+1}])} \leq \|f - g\|_{L_q([\xi_m,\xi_{m+1}])} + \|\mathscr{Q}(f - g)\|_{L_q([\xi_m,\xi_{m+1}])}.$$

Since $[\xi_m, \xi_{m+1}] \subset J_m$ and by (1.108) for any $g \in \mathbb{P}_l$, we have

$$\|f - \mathscr{Q}f\|_{L_q([\xi_m,\xi_{m+1}])} \leq (1 + C_{\mathscr{Q}})\|f - g\|_{L_q(J_m)}. \tag{1.109}$$

Let $a_m := \max(\xi_{m-p}, a)$, and choose $g := \mathscr{T}_{a_m,l}f$, where $\mathscr{T}_{a_m,l}f$ is the Taylor polynomial of degree l defined in (1.95) with $a = a_m$. Then, by (1.98) with $r = 0$ we have

$$\|f - g\|_{L_q(J_m)} \leq \frac{(2p + 1)^{l+1}}{l!} h_{m,\xi}^{l+1} \|D^{l+1}f\|_{L_q(J_m)}. \tag{1.110}$$

Combining the inequalities (1.109) and (1.110) gives the local bound.

Since each J_m is contained in the basic interval $[a, b]$ the global bound follows immediately from the local one and Lemma 1. $\square$

Example 11 Let $\boldsymbol{\xi}$ be a $(p + 1)$-open knot sequence for $p \geq 1$, and consider the operator

$$\mathscr{V}_{p,\xi}f(x) := \sum_{j=1}^{n} f(\xi_{j,p,\xi}^*)B_{j,p,\xi}(x), \tag{1.111}$$

where $\xi_{j,p,\xi}^*$ is the j-th Greville point of degree p; see (1.51). This operator is known as the **Schoenberg operator**, and was introduced in [43, Section 10]. It is a bounded local quasi-interpolant in the L_∞-norm with $l = 1$ and $C_{\mathscr{Q}} = 1$. Note that $\xi_{j,p,\xi}^*$ belongs to $[\xi_{j+1}, \xi_{j+p}]$. Therefore, Theorem 16 implies for any $f \in W_\infty^2([a, b])$,

$$\|f - \mathscr{V}_{p,\xi}f\|_{L_\infty([a,b])} \leq 2(2p + 1)^2 h_\xi^2 \|D^2f\|_{L_\infty([a,b])}. \tag{1.112}$$

The next proposition can be used to find the degree l of polynomials reproduced by a linear quasi-interpolant.

Proposition 4 *Let*

$$\{\varphi_{j,0}, \ldots, \varphi_{j,l}\}, \quad j = 1, \ldots, n, \quad 0 \leq l \leq p \tag{1.113}$$

be n sets of basis functions for $\mathbb{P}_l$, and let

$$\varphi_{j,i} = \sum_{k=1}^{n} c_{j,i,k} B_{k,p,\xi} \qquad (1.114)$$

be their B-spline representations. The linear quasi-interpolant (1.100) *reproduces $\mathbb{P}_l$ provided the corresponding linear functionals satisfy*

$$\lambda_j(\varphi_{j,i}) = c_{j,i,j}, \quad j = 1, \ldots, n, \quad i = 0, \ldots, l. \qquad (1.115)$$

Proof On the basic interval, any $g \in \mathbb{P}_l$ can be written both in terms of the φ's and the B-splines, say

$$g = \sum_{i=0}^{l} b_{j,i} \varphi_{j,i} = \sum_{k=1}^{n} b_k B_{k,p,\xi}, \quad j = 1, \ldots, n. \qquad (1.116)$$

By (1.114) and (1.116) for $j = 1, \ldots, n$,

$$g = \sum_{i=0}^{l} b_{j,i} \left(\sum_{k=1}^{n} c_{j,i,k} B_{k,p,\xi} \right) = \sum_{k=1}^{n} \left(\sum_{i=0}^{l} b_{j,i} c_{j,i,k} \right) B_{k,p,\xi} = \sum_{k=1}^{n} b_k B_{k,p,\xi}.$$

By linear independence of the B-splines and choosing $j = k$ we obtain

$$b_k = \sum_{i=0}^{l} b_{k,i} c_{k,i,k}. \qquad (1.117)$$

Similarly, for $\mathcal{Q}g$ using (1.116) with $j = k$,

$$\mathcal{Q}g := \sum_{k=1}^{n} \lambda_k(g) B_{k,p,\xi} = \sum_{k=1}^{n} \lambda_k \left(\sum_{i=0}^{l} b_{k,i} \varphi_{k,i} \right) B_{k,p,\xi}.$$

From the linearity of λ_k and (1.115), (1.117) and finally (1.116) again we obtain

$$\mathcal{Q}g = \sum_{k=1}^{n} \sum_{i=0}^{l} b_{k,i} \lambda_k(\varphi_{k,i}) B_{k,p,\xi} = \sum_{k=1}^{n} \sum_{i=0}^{l} b_{k,i} c_{k,i,k} B_{k,p,\xi} = \sum_{k=1}^{n} b_k B_{k,p,\xi} = g.$$

□

The next proposition gives a sufficient condition for a quasi-interpolant to reproduce the whole spline space, i.e., to be a projector onto $\mathbb{S}_{p,\xi}$.

Proposition 5 *The linear quasi-interpolant* (1.100) *reproduces the whole spline space, i.e.,*

$$\mathcal{Q}s(x) = s(x), \quad s \in \mathbb{S}_{p,\xi}, \quad x \in [\xi_{p+1}, \xi_{n+1}], \qquad (1.118)$$

if $\mathcal{Q}$ reproduces $\mathbb{P}_p$ and each linear functional λ_j is supported on one knot interval[6]

$$[\xi_{m_j}^+, \xi_{m_j+1}^-] \subset [\xi_j, \xi_{j+p+1}], \text{ with } \xi_{m_j} < \xi_{m_j+1}. \tag{1.119}$$

Proof Let j with $1 \leq j \leq n$ be fixed. By the linearity it suffices to prove that

$$\lambda_j(B_{i,p,\xi}) = \delta_{i,j}, \quad i = 1, \dots, n,$$

where $\delta_{i,j}$ stands for the classical Kronecker delta. On the interval $[\xi_{m_j}^+, \xi_{m_j+1}^-]$ the local support property implies that $\lambda_j(B_{i,p,\xi}) = 0$ for $i \notin \{m_j - p, \dots, m_j\}$. This follows because we use the left limit at ξ_{m_j+1} if necessary. Since $B_{i,p,\xi} \in \mathbb{P}_p$ on this interval, we have

$$B_{i,p,\xi}(x) = \mathcal{Q}(B_{i,p,\xi})(x) = \sum_{k=m_j-p}^{m_j} \lambda_k(B_{i,p,\xi}) B_{k,p,\xi}(x), \quad x \in [\xi_{m_j}, \xi_{m_j+1}),$$

and by local linear independence of the B-splines we obtain $\lambda_k(B_{i,p,\xi}) = \delta_{i,k}$ for $k = m_j - p, \dots, m_j$. In particular, it holds for $k = j$ since the condition (1.119) implies that $m_j - p \leq j \leq m_j$. $\qquad\square$

Example 12 Let $p = 2$, and let ξ be a 3-open knot sequence with at most double knots in the interior. We consider the operator

$$\mathcal{Q}_{2,\xi} f(x) := \sum_{j=1}^{n} \left(a_{2,0} f(\xi_{j+1}) + a_{2,1} f(\xi_{j,2,\xi}^*) + a_{2,2} f(\xi_{j+2}) \right) B_{j,2,\xi}(x),$$

where $\xi_{j,2,\xi}^* = (\xi_{j+1} + \xi_{j+2})/2$ is the j-th Greville point of degree 2. It can be checked (see also Example 9) that if we choose $a_{2,0} = a_{2,2} = -1/2$ and $a_{2,1} = 2$ then $\mathcal{Q}_{2,\xi}$ reproduces $\mathbb{P}_2$, i.e., $l = 2$. Proposition 5 says that it is even a projector on the spline space $\mathbb{S}_{2,\xi}$. Moreover,

$$\left| -\frac{1}{2} f(\xi_{j+1}) + 2 f(\xi_{j,2,\xi}^*) - \frac{1}{2} f(\xi_{j+2}) \right| \leq 3 \|f\|_{L_\infty([\xi_j, \xi_{j+3}])}.$$

It follows that $\mathcal{Q}_{2,\xi}$ is a bounded local quasi-interpolant in the L_∞-norm with $l = 2$ and $C_\mathcal{Q} = 3$. In this case, Theorem 16 implies for any $f \in W_\infty^3([a, b])$,

$$\|f - \mathcal{Q}_{2,\xi} f\|_{L_\infty([a,b])} \leq 4 \frac{5^3}{2!} h_\xi^3 \|D^3 f\|_{L_\infty([a,b])},$$

showing that the error is $O(h_\xi^3)$.

---

[6]This notation means that if $\lambda_j(f)$ uses the value of f or one of its derivatives at ξ_{m_j} (or ξ_{m_j+1}) then this value is obtained by taking the one sided limit from the right (or the left).

1.5.3 Approximation Power of Splines

In this section we want to understand how well a function can be approximated by a spline. In order words, we want to investigate the **distance** between a general function f and the piecewise polynomial space $\mathbb{S}_p^r(\Delta)$ defined in (1.42). From Theorem 6 we know that $\mathbb{S}_p^r(\Delta) = \mathbb{S}_{p,\xi}$ for a suitable choice of the knot sequence $\xi := \{\xi_i\}_{i=1}^{n+p+1}$. In particular, ξ can be chosen to be $(p+1)$-open. Therefore, without loss of generality, we consider the distance between a general function f and the spline space $\mathbb{S}_{p,\xi}$ of degree p over the $(p+1)$-open knot sequence ξ. For a given $f \in L_q([\xi_{p+1}, \xi_{n+1}])$ with $1 \le q \le \infty$, we define

$$\mathrm{dist}_q(f, \mathbb{S}_{p,\xi}) := \inf_{s \in \mathbb{S}_{p,\xi}} \|f - s\|_{L_q([\xi_{p+1}, \xi_{n+1}])}. \tag{1.120}$$

We are also interested in estimates for the distance between derivatives of f and derivative spline spaces. To this end, in this section we use the simplified notation $D^r s := D_+^r s$ for the derivatives of a spline $s \in \mathbb{S}_{p,\xi}$ with the usual convention of left continuity at the right endpoint of the basic interval. Note that with such a notation we ensure that $D^r s(x)$ exists for all x. In the same spirit, we use the notation $D^r \mathbb{S}_{p,\xi} := D_+^r \mathbb{S}_{p,\xi}$ for the r-th derivative spline space. We recall from Sect. 1.3.2 that this derivative space is a piecewise polynomial space of degree $p - r$ with a certain smoothness, i.e.,

$$\mathbb{S}_{p-r}^{r-r}(\Delta) = D^r \mathbb{S}_{p,\xi},$$

where the partition Δ consists of the distinct break points in the knot sequence ξ and the smoothness r is related to the multiplicity of the knots, according to the rule in (1.35). This leads to the following more general definition of distance. For a given $f \in W_q^r([\xi_{p+1}, \xi_{n+1}])$ with $1 \le q \le \infty$ and $0 \le r \le p$, we define

$$\mathrm{dist}_q(D^r f, D^r \mathbb{S}_{p,\xi}) := \inf_{s \in \mathbb{S}_{p,\xi}} \|D^r(f - s)\|_{L_q([\xi_{p+1}, \xi_{n+1}])}. \tag{1.121}$$

We will derive the following upper bound for $\mathrm{dist}_q(D^r f, D^r \mathbb{S}_{p,\xi})$.

Theorem 17 (Distance to a Function) *For any* $0 \le r \le l \le p$ *and* $f \in W_q^{l+1}([\xi_{p+1}, \xi_{n+1}])$ *with* $1 \le q \le \infty$ *we have*

$$\mathrm{dist}_q(D^r f, D^r \mathbb{S}_{p,\xi}) \le K(h_\xi)^{l+1-r} \|D^{l+1} f\|_{L_q([\xi_{p+1}, \xi_{n+1}])},$$

where $h_\xi := \max_{p+1 \le j \le n}(\xi_{j+1} - \xi_j)$ *and* K *is a constant depending only on* p.

The distance result will be shown by explicitly constructing a suitable spline quasi-interpolant which achieves this order of approximation; see Theorem 18. For sufficiently smooth f, the upper bound behaves like $(h_\xi)^{p+1-r}$.

1.5.3.1   A Spline Quasi-Interpolant

Given an integer $p \geq 0$ and a $(p + 1)$-open knot sequence $\boldsymbol{\xi}$, we define a specific spline approximant of degree p over $\boldsymbol{\xi}$ to a given function f. Let $[\xi_{m_{j,p}}, \xi_{m_{j,p}+1}]$ be a knot interval of largest length in $[\xi_j, \xi_{j+p+1}]$ for any $j = 1, \ldots, n$ and $h_{j,p,\boldsymbol{\xi}} := \xi_{m_{j,p}+1} - \xi_{m_{j,p}} > 0$. The spline approximant to f is constructed as

$$\mathcal{Q}_{p,\boldsymbol{\xi}} f(x) := \sum_{j=1}^{n} \mathcal{L}_{j,p,\boldsymbol{\xi}}(f) B_{j,p,\boldsymbol{\xi}}(x), \tag{1.122}$$

where

$$\mathcal{L}_{j,p,\boldsymbol{\xi}}(f) := \frac{1}{h_{j,p,\boldsymbol{\xi}}} \int_{\xi_{m_{j,p}}}^{\xi_{m_{j,p}+1}} \left(\sum_{i=0}^{p} a_{j,i} \left(\frac{x - \xi_{m_{j,p}}}{h_{j,p,\boldsymbol{\xi}}} \right)^i \right) f(x)\, dx, \tag{1.123}$$

and the coefficients $a_{j,i}$, $i = 0, \ldots, p$ are such that

$$\mathcal{L}_{j,p,\boldsymbol{\xi}} \left(\left(\frac{x - \xi_{m_{j,p}}}{h_{j,p,\boldsymbol{\xi}}} \right)^i \right) = c_{j,i,j}, \quad i = 0, \ldots, p, \tag{1.124}$$

where

$$\left(\frac{x - \xi_{m_{j,p}}}{h_{j,p,\boldsymbol{\xi}}} \right)^i = \sum_{k=m_{j,p}-p}^{m_{j,p}} c_{j,i,k} B_{k,p,\boldsymbol{\xi}}(x), \quad x \in [\xi_{m_{j,p}}, \xi_{m_{j,p}+1}), \quad i = 0, \ldots, p. \tag{1.125}$$

In the next lemmas we collect some properties for the spline approximation (1.122).

Lemma 2 *The above spline approximation is well defined and reproduces polynomials, i.e., for any polynomial $g \in \mathbb{P}_p$ we have*

$$\mathcal{Q}_{p,\boldsymbol{\xi}} g(x) = g(x), \quad x \in [\xi_{p+1}, \xi_{n+1}]. \tag{1.126}$$

Moreover, it is a projector onto the spline space $\mathbb{S}_{p,\boldsymbol{\xi}}$, i.e., for any spline $s \in \mathbb{S}_{p,\boldsymbol{\xi}}$ we have

$$\mathcal{Q}_{p,\boldsymbol{\xi}} s(x) = s(x), \quad x \in [\xi_{p+1}, \xi_{n+1}], \tag{1.127}$$

and, in particular,

$$s(x) = \sum_{j=1}^{n} \mathcal{L}_{j,p,\boldsymbol{\xi}}(s) B_{j,p,\boldsymbol{\xi}}(x), \quad x \in [\xi_{p+1}, \xi_{n+1}]. \tag{1.128}$$

Proof By applying $\mathscr{L}_{j,p,\xi}$ to the polynomials $\left(\frac{x-\xi_{m_{j,p}}}{h_{j,p,\xi}}\right)^r$, $r = 0, \ldots, p$, the coefficients $a_{j,i}$ are given by the solution of the linear system

$$H_{p+1}\boldsymbol{a}_j = \boldsymbol{c}_j, \tag{1.129}$$

where $\boldsymbol{a}_j := (a_{j,0}, \ldots, a_{j,p})^T$, $\boldsymbol{c}_j := (c_{j,0,j}, \ldots, c_{j,p,j})^T$, and H_{p+1} is a $(p+1) \times (p+1)$ matrix with elements

$$(H_{p+1})_{i+1,r+1} := \frac{1}{h_{j,p,\xi}} \int_{\xi_{m_{j,p}}}^{\xi_{m_{j,p}+1}} \left(\frac{x - \xi_{m_{j,p}}}{h_{j,p,\xi}}\right)^{r+i} \mathrm{d}x = \frac{1}{i+r+1},$$

for $i, r = 0, \ldots, p$. This is the well-known Hilbert matrix which is nonsingular and it follows that the spline approximation (1.122) is well defined. From Proposition 4 we deduce that (1.126) holds.

Since we only integrate over one subinterval when we define $\mathscr{L}_{j,p,\xi}$, we conclude that it reproduces not only polynomials but also splines, and (1.127) follows from Proposition 5. $\qquad\square$

Lemma 3 *For $p \geq 0$ and $1 \leq q \leq \infty$ we have for any $f \in L_q([\xi_{m_{j,p}}, \xi_{m_{j,p}+1}])$,*

$$|\mathscr{L}_{j,p,\xi}(f)| \leq Ch_{j,p,\xi}^{-1/q}\|f\|_{L_q([\xi_{m_{j,p}},\xi_{m_{j,p}+1}])}, \quad j = 1, \ldots, n, \tag{1.130}$$

where C is a constant depending only on p.

Proof By (1.20), (1.10) and (1.13) we have

$$|c_{j,i,j}| = \frac{i!}{p!}\left|\frac{D^{p-i}\psi_{j,p,\xi}(\xi_{m_{j,p}})}{h_{j,p,\xi}^i}\right| \leq \left(\frac{\xi_{j+p+1} - \xi_j}{h_{j,p,\xi}}\right)^i \leq (p+1)^i, \quad i = 0, \ldots, p.$$

Here we used that $[\xi_{m_{j,p}}, \xi_{m_{j,p}+1}]$ is a knot interval of largest length in $[\xi_j, \xi_{j+p+1}]$. Since $0 \leq \frac{x-\xi_{m_{j,p}}}{h_{j,p,\xi}} \leq 1$ for $x \in [\xi_{m_{j,p}}, \xi_{m_{j,p}+1}]$, we get from (1.123),

$$|\mathscr{L}_{j,p,\xi}(f)| \leq (p+1)h_{j,p,\xi}^{-1}\|\boldsymbol{a}_j\|_\infty \|f\|_{L_1([\xi_{m_{j,p}},\xi_{m_{j,p}+1}])}$$

$$\leq (p+1)h_{j,p,\xi}^{-1}\|H_{p+1}^{-1}\|_\infty\|\boldsymbol{c}_j\|_\infty\|f\|_{L_1([\xi_{m_{j,p}},\xi_{m_{j,p}+1}])}.$$

This gives $|\mathscr{L}_{j,p,\xi}(f)| \leq Ch_{j,p,\xi}^{-1}\|f\|_{L_1([\xi_{m_{j,p}},\xi_{m_{j,p}+1}])}$, where $C := \|H_{p+1}^{-1}\|_\infty(p+1)^{p+1}$ only depends on p. By the Hölder inequality (1.94) we arrive at (1.130). $\qquad\square$

We now give a bound for the derivative of $\mathcal{Q}_{p,\boldsymbol{\xi}} f$. To this end, we recall from (1.25) that

$$\Delta_{m,k} := \min_{m-k+1 \leq i \leq m} h_{i,k}, \quad h_{i,k} := \xi_{i+k} - \xi_i, \quad k = 1, \ldots, p,$$

and that $\Delta_{m,k} > 0$ for all k if $\xi_m < \xi_{m+1}$.

Lemma 4 *Suppose $\xi_m < \xi_{m+1}$ for some $p + 1 \leq m \leq n$, and let $f \in L_q([\xi_{m-p}, \xi_{m+p+1}])$ with $1 \leq q \leq \infty$. Then, we have for $0 \leq r \leq p$,*

$$\|D^r(\mathcal{Q}_{p,\boldsymbol{\xi}} f)\|_{L_q([\xi_m,\xi_{m+1}])} \leq C\left(\prod_{k=p-r+1}^{p} \frac{1}{\Delta_{m,k}} \right) \|f\|_{L_q([\xi_{m-p},\xi_{m+p+1}])}, \quad (1.131)$$

where $\Delta_{m,k}$ is defined in (1.25) and C is a constant depending only on p.

Proof From the quasi-interpolant definition (1.122), the local support property (1.36) and Lemma 3, we have for $x \in [\xi_m, \xi_{m+1})$,

$$|D^r(\mathcal{Q}_{p,\boldsymbol{\xi}} f)(x)| = \left| \sum_{j=m-p}^{m} \mathcal{L}_{j,p,\boldsymbol{\xi}}(f) D^r B_{j,p,\boldsymbol{\xi}}(x) \right|$$

$$\leq \max_{m-p \leq j \leq m} |D^r B_{j,p,\boldsymbol{\xi}}(x)| \sum_{j=m-p}^{m} |\mathcal{L}_{j,p,\boldsymbol{\xi}}(f)|$$

$$\leq (p+1) \max_{m-p \leq j \leq m} |D^r B_{j,p,\boldsymbol{\xi}}(x)|$$

$$\times \max_{m-p \leq j \leq m} h_{j,p,\boldsymbol{\xi}}^{-1/q} \|f\|_{L_q([\xi_{m-p},\xi_{m+p+1}])}.$$

Note that $[\xi_m, \xi_{m+1}] \subset [\xi_j, \xi_{j+p+1}]$ for $j = m - p, \ldots, m$. Since $h_{j,p,\boldsymbol{\xi}}$ is the length of the largest knot interval in $[\xi_j, \xi_{j+p+1}]$, we have $\xi_{m+1} - \xi_m \leq h_{j,p,\boldsymbol{\xi}}$ for $j = m - p, \ldots, m$. Replacing $|D^r B_{j,p,\boldsymbol{\xi}}(x)|$ by the upper bound given in Proposition 2 and taking the L_q-norm results in (1.131). $\quad\square$

The next lemma will complete the proof of Theorem 11 related to the condition number. Note that $[\xi_{p+1}, \xi_{n+1}] = [\xi_1, \xi_{n+p+1}]$ because the knot sequence $\boldsymbol{\xi}$ is open.

Lemma 5 *For any $p \geq 0$, there exists a positive constant K_p depending only on p, such that for any vector $\boldsymbol{c} := (c_1, \ldots, c_n)$ and for any $1 \leq q \leq \infty$ we have*

$$\|\boldsymbol{c}\|_q \leq K_p \left\| \sum_{j=1}^{n} c_j N_{j,p,q,\boldsymbol{\xi}} \right\|_{L_q([\xi_{p+1},\xi_{n+1}])}, \quad (1.132)$$

where $N_{j,p,q,\boldsymbol{\xi}} := \gamma_{j,p,\boldsymbol{\xi}}^{-1/q} B_{j,p,\boldsymbol{\xi}}$ and $\gamma_{j,p,\boldsymbol{\xi}} := (\xi_{j+p+1} - \xi_j)/(p + 1)$.

Proof Let $s := \sum_{j=1}^{n} \gamma_{j,p,\boldsymbol{\xi}}^{-1/q} c_j B_{j,p,\boldsymbol{\xi}}$. Observe that (1.128) and (1.130) imply

$$|\gamma_{j,p,\boldsymbol{\xi}}^{-1/q} c_j| = |\mathscr{L}_{j,p,\boldsymbol{\xi}}(s)| \leq C h_{j,p,\boldsymbol{\xi}}^{-1/q} \|s\|_{L_q([\xi_{m_{j,p}}, \xi_{m_{j,p}+1}])}.$$

Since $\gamma_{j,p,\boldsymbol{\xi}}/h_{j,p,\boldsymbol{\xi}} \leq 1$ we obtain

$$|c_j| \leq C\|s\|_{L_q([\xi_{m_{j,p}}, \xi_{m_{j,p}+1}])} \leq C\|s\|_{L_q([\xi_j, \xi_{j+p+1}])}.$$

Raising both sides to the q-th power and summing over j gives

$$\sum_{j=1}^{n} |c_j|^q \leq C^q \sum_{j=1}^{n} \int_{\xi_j}^{\xi_{j+p+1}} |s(x)|^q \, dx \leq (p+1)C^q \|s\|_{L_q([\xi_{p+1}, \xi_{n+1}])}^q.$$

When taking the q-th roots on both sides, we arrive at the inequality in (1.132) with $K_p := (p+1)C \geq (p+1)^{1/q}C$, which only depends on p. $\qquad\square$

1.5.3.2 Distance to a Function

The quasi-interpolant $\mathscr{Q}_{p,\boldsymbol{\xi}} f$ described in the previous section can be used to obtain an upper bound for the distance between a given function f and the spline space $\mathbb{S}_{p,\boldsymbol{\xi}}$ for $p \geq 0$, $n > p+1$ and $\boldsymbol{\xi} := \{\xi_j\}_{j=1}^{n+p+1}$; see Theorem 18. We recall that the knot sequence $\boldsymbol{\xi}$ is $(p+1)$-open. We start by giving a local and global upper bound for (the derivatives of) the difference between f and $\mathscr{Q}_{p,\boldsymbol{\xi}} f$.

Proposition 6 *Suppose* $\xi_m < \xi_{m+1}$ *for some* $p+1 \leq m \leq n$, *and let* $f \in W_q^{l+1}([\xi_{m-p}, \xi_{m+p+1}])$ *with* $0 \leq l \leq p$ *and* $1 \leq q \leq \infty$. *If* $\mathscr{Q}_{p,\boldsymbol{\xi}} f$ *is defined as in* (1.122), *then we have for any* $0 \leq r \leq l$,

$$\|D^r(f - \mathscr{Q}_{p,\boldsymbol{\xi}} f)\|_{L_q([\xi_m, \xi_{m+1}])} \leq K_m (\xi_{m+p+1} - \xi_{m-p})^{l+1-r} \|D^{l+1} f\|_{L_q([\xi_{m-p}, \xi_{m+p+1}])}.$$

Here,

$$K_m := 1 + C \prod_{k=p-r+1}^{p} \frac{\xi_{m+p+1} - \xi_{m-p}}{\Delta_{m,k}},$$

$\Delta_{m,k}$ *is defined in* (1.25) *and* C *is a constant depending only on* p.

Proof From Lemma 2 we know that $\mathscr{Q}_{p,\boldsymbol{\xi}}$ reproduces any polynomial in $\mathbb{P}_l$, and so the triangle inequality gives

$$\|D^r(f - \mathscr{Q}_{p,\boldsymbol{\xi}} f)\|_{L_q([\xi_m, \xi_{m+1}])}$$
$$\leq \|D^r(f - g)\|_{L_q([\xi_m, \xi_{m+1}])} + \|D^r \mathscr{Q}_{p,\boldsymbol{\xi}}(f - g)\|_{L_q([\xi_m, \xi_{m+1}])},$$

for any $g \in \mathbb{P}_l$. Let us now set $g := \mathscr{T}_{\xi_m, l} f$, where $\mathscr{T}_{\xi_m, l} f$ is the Taylor polynomial of degree l defined in (1.95) with $a = \xi_m$, $b = \xi_{m+1}$. Then, Eq. (1.99) implies

$$\|D^r(f - g)\|_{L_q([\xi_m, \xi_{m+1}])} \leq (\xi_{m+1} - \xi_m)^{l+1-r} \|D^{l+1} f\|_{L_q([\xi_m, \xi_{m+1}])}.$$

On the other hand, since $f - g \in L_q([\xi_{m-p}, \xi_{m+p+1}])$, it follows from Lemma 4 that

$$\|D^r \mathscr{Q}_{p,\xi}(f - g)\|_{L_q([\xi_m, \xi_{m+1}])} \leq C \left(\prod_{k=p-r+1}^{p} \frac{1}{\Delta_{m,k}} \right) \|f - g\|_{L_q([\xi_{m-p}, \xi_{m+p+1}])},$$

where C is a constant depending only on p. Combining the above three inequalities gives the result. $\qquad \square$

We know that the ratio $\frac{\xi_{m+p+1} - \xi_{m-p}}{\Delta_{m,k}}$ is well defined because $\Delta_{m,k} > 0$. For a uniform knot sequence

$$\frac{\xi_{m+p+1} - \xi_{m-p}}{\Delta_{m,k}} = \frac{2p+1}{k}.$$

For a general knot sequence it is related to the "local mesh ratio", i.e., the ratio between the lengths of the largest and smallest knot intervals in a neighborhood of ξ_m.

The local error bound in Proposition 6 can be turned into a global one as in the following proposition.

Proposition 7 *Let $f \in W_q^{l+1}([\xi_{p+1}, \xi_{n+1}])$ with $0 \leq l \leq p$ and $1 \leq q \leq \infty$. If $\mathscr{Q}_{p,\xi} f$ is defined as in (1.122) then, for any $0 \leq r \leq l$,*

$$\|D^r(f - \mathscr{Q}_{p,\xi} f)\|_{L_q([\xi_{p+1}, \xi_{n+1}])} \leq K h_{\xi}^{l+1-r} \|D^{l+1} f\|_{L_q([\xi_{p+1}, \xi_{n+1}])}, \qquad (1.133)$$

where $h_{\xi} := \max_{p+1 \leq j \leq n}(\xi_{j+1} - \xi_j)$, and

$$K := (2p+1)^{l+2-r} \left[1 + C \max_{p+1 \leq m \leq n} \prod_{k=p-r+1}^{p} \frac{\xi_{m+p+1} - \xi_{m-p}}{\Delta_{m,k}} \right],$$

where $\Delta_{m,k}$ is defined in (1.25) and C is a constant depending only on p.

Proof For $q = \infty$ the result follows immediately from Proposition 6 by taking into account that ξ is $(p+1)$-open. We now assume $1 \leq q < \infty$. Since

$$\max_{p+1 \leq m \leq n}(\xi_{m+p+1} - \xi_{m-p}) \leq (2p+1)h_{\xi},$$

the result follows from Lemma 1 and the local error bound in Proposition 6. $\qquad \square$

The expression K in the upper bound in Proposition 7 depends on the position of the knots for $r > 0$. However, for any knot sequence $\boldsymbol{\xi}$, it is possible to construct a coarser knot sequence $\boldsymbol{\xi}^\sharp$ such that the corresponding K only depends on p. This can be obtained by a clever thinning process. The idea of thinning out a knot sequence to get a quasi-uniform sequence is credited to [47]; see [45, Section 6.4] for details. Since $\boldsymbol{\xi}^\sharp$ is a subsequence of $\boldsymbol{\xi}$, we have that $\mathbb{S}_{p,\boldsymbol{\xi}^\sharp}$ is a subspace of $\mathbb{S}_{p,\boldsymbol{\xi}}$. In particular, for any $f \in L_q([\xi_{p+1}, \xi_{n+1}])$ the spline approximation

$$s_p := \mathscr{Q}_{p,\boldsymbol{\xi}^\sharp} f$$

as defined in (1.122) belongs to the spline space $\mathbb{S}_{p,\boldsymbol{\xi}}$. This spline quasi-interpolant leads to the following important result.

Theorem 18 (Approximation Error) *Let $f \in W_q^{l+1}([\xi_{p+1}, \xi_{n+1}])$ with $1 \le q \le \infty$ and $0 \le l \le p$. Then, there exists $s_p \in \mathbb{S}_{p,\boldsymbol{\xi}}$ such that*

$$\|D^r(f - s_p)\|_{L_q([\xi_{p+1},\xi_{n+1}])} \le K h_{\boldsymbol{\xi}}^{l+1-r} \|D^{l+1} f\|_{L_q([\xi_{p+1},\xi_{n+1}])}, \quad 0 \le r \le l,$$
(1.134)

where $h_{\boldsymbol{\xi}} := \max_{p+1 \le j \le n}(\xi_{j+1} - \xi_j)$ and K is a constant depending only on p.

The constant K in Theorem 18 grows exponentially with p. However, this dependency on p can be removed in some cases; see [1, Theorem 2] and [52, Theorem 7] for details. Theorem 18 immediately leads to the distance result in Theorem 17.

1.6 Hierarchical Splines and the Truncation Mechanism

The hierarchical spline model is a simple strategy to mix locally spline spaces of different resolution (different mesh size and/or different degree). Hierarchical spline representations are defined in terms of a sequence of nested B-spline bases and a hierarchy of locally refined domains. In this section we define such hierarchical splines and focus on a set of basis functions with properties similar to B-splines.

1.6.1 Hierarchical B-Splines

Let I be a closed interval of the real line, and consider a sequence of strictly nested spline spaces defined on I, say

$$\mathbb{S}_{p_1,\boldsymbol{\xi}_1} \subset \mathbb{S}_{p_2,\boldsymbol{\xi}_2} \subset \cdots \subset \mathbb{S}_{p_L,\boldsymbol{\xi}_L}.$$
(1.135)

We assume that each knot sequence involved in (1.135),

$$\boldsymbol{\xi}_\ell := \{\xi_{1,\ell} \le \xi_{2,\ell} \le \cdots \le \xi_{n_\ell + p_\ell + 1, \ell}\}, \quad \ell = 1, \ldots, L,$$

is $(p+1)$-basic with basic interval I. Nestedness of the spaces is ensured if and only if

$$0 \le p_{\ell+1} - p_\ell \le \mu_{\boldsymbol{\xi}_{\ell+1}}(\xi) - \mu_{\boldsymbol{\xi}_\ell}(\xi), \quad \xi \in \boldsymbol{\xi}_\ell \cap I, \quad \ell = 1, \ldots, L-1. \quad (1.136)$$

Note that (1.136) implies that $(\boldsymbol{\xi}_\ell \cap I) \subseteq (\boldsymbol{\xi}_{\ell+1} \cap I)$. The assumption of dealing with $(p+1)$-basic knot sequences ensures that the corresponding n_ℓ B-splines are linearly independent on I. We denote the B-spline basis of the space $\mathbb{S}_{p_\ell, \boldsymbol{\xi}_\ell}$ by

$$\mathscr{B}_\ell := \{B_{j,\ell} := B_{j,p_\ell,\boldsymbol{\xi}_\ell}, \ j = 1, \ldots, n_\ell\}. \quad (1.137)$$

Next, consider a sequence of nested, closed subsets of I,

$$I \supseteq \Omega_1 \supseteq \Omega_2 \supseteq \cdots \supseteq \Omega_L, \quad (1.138)$$

where Ω_ℓ is the union of some closed knot intervals related to the knot sequence $\boldsymbol{\xi}_\ell$. Note that each Ω_ℓ may consist of disjoint intervals. We assume that each connected component of Ω_1 has nonempty interior. The collection of those subsets in (1.138) is denoted by

$$\boldsymbol{\Omega} := \{\Omega_1, \Omega_2, \ldots, \Omega_L\}, \quad (1.139)$$

and will be simply referred to as the **domain hierarchy** in I. We also set $\Omega_{L+1} := \emptyset$. Finally, for a given function f on I, we define its support on $\boldsymbol{\Omega}$ as

$$\mathrm{supp}_{\boldsymbol{\Omega}}(f) := \mathrm{supp}(f) \cap \Omega_1.$$

Given a sequence of spline spaces and bases as in (1.135)–(1.137) and a domain hierarchy as in (1.138)–(1.139), we construct the corresponding set of hierarchical B-splines (in short, HB-splines) as follows.[7]

Definition 7 Given a domain hierarchy $\boldsymbol{\Omega}$, the corresponding set of **HB-splines** is denoted by $\mathscr{H}_{\boldsymbol{\Omega}}$ and defined recursively as follows:

(i) $\mathscr{H}_1 := \{B_{j,1} \in \mathscr{B}_1 : \mathrm{supp}_{\boldsymbol{\Omega}}(B_{j,1}) \ne \emptyset\}$;
(ii) for $\ell = 2, \ldots, L$:

$$\mathscr{H}_\ell := \mathscr{H}_\ell^C \cup \mathscr{H}_\ell^F,$$

---

[7]The HB-splines in Definition 7 were introduced by Kraft [28, 29] and further elaborated in [53]. However, the concept of hierarchical splines has a long history; for example, it was used in preconditioning [18, 54], adaptive modeling [19, 20] and adaptive finite elements [25, 30].

where

$$\mathscr{H}_\ell^C := \{B_{j,k} \in \mathscr{H}_{\ell-1} : \mathrm{supp}_{\boldsymbol{\Omega}}(B_{j,k}) \nsubseteq \Omega_\ell\},$$

$$\mathscr{H}_\ell^F := \{B_{j,\ell} \in \mathscr{B}_\ell : \mathrm{supp}_{\boldsymbol{\Omega}}(B_{j,\ell}) \subseteq \Omega_\ell\};$$

(iii) $\mathscr{H}_{\boldsymbol{\Omega}} := \mathscr{H}_L$.

To obtain the set of HB-splines, we first take all the B-splines in $\mathscr{B}_1$ whose support overlaps Ω_1. Then, we apply a recursive procedure which selects at each level ℓ all the B-splines obtained in the previous step whose support is not entirely contained in Ω_ℓ and all the B-splines in $\mathscr{B}_\ell$ whose support is entirely contained in Ω_ℓ.

Example 13 An example of the recursive definition of HB-splines is illustrated in Fig. 1.4. We consider three nested knot sequences, with knots of multiplicity 4 at the two extrema of the intervals and single knots elsewhere, as in Fig. 1.4a. This

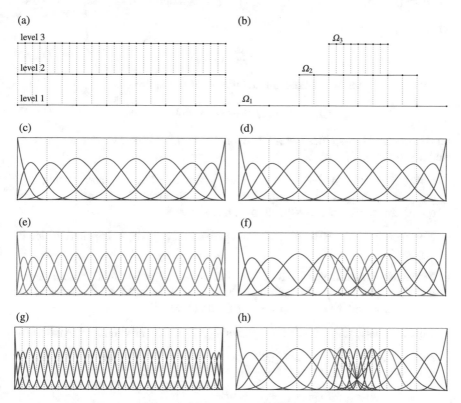

Fig. 1.4 An example of cubic HB-splines where the domain hierarchy consists of three levels. The knot positions are visualized by vertical dotted lines in (**c**)–(**h**). (**a**) Knot sequences. (**b**) Domain hierarchy. (**c**) $\mathscr{B}_1$. (**d**) $\mathscr{H}_1$. (**e**) $\mathscr{B}_2$. (**f**) $\mathscr{H}_2$. (**g**) $\mathscr{B}_3$. (**h**) $\mathscr{H}_3 = \mathscr{H}_{\boldsymbol{\Omega}}$

allows us to construct the three sets of cubic B-splines shown in Fig. 1.4c, e, g, whose dimensions are $n_1 = 10$, $n_2 = 17$ and $n_3 = 31$, respectively. The domain hierarchy is defined by the subsets $\Omega_1 = [\xi_{4,1}, \xi_{11,1}]$, $\Omega_2 = [\xi_{8,2}, \xi_{16,2}]$ and $\Omega_3 = [\xi_{16,3}, \xi_{24,3}]$, and is shown in Fig. 1.4b. Obviously, $\mathcal{H}_1$ coincides with $\mathcal{B}_1$. Furthermore, $\mathcal{H}_2^C$ is obtained from $\mathcal{H}_1$ by removing $B_{6,1}$, and $\mathcal{H}_2^F = \{B_{8,2}, \ldots, B_{12,2}\}$. Hence, $\mathcal{H}_2 = \mathcal{H}_2^C \cup \mathcal{H}_2^F$ consists of $9 + 5 = 14$ elements. Finally, $\mathcal{H}_3^C$ is obtained from $\mathcal{H}_2$ by removing $B_{10,2}$, and $\mathcal{H}_3^F = \{B_{16,3}, \ldots, B_{20,3}\}$. Hence, $\mathcal{H}_3 = \mathcal{H}_3^C \cup \mathcal{H}_3^F$ consists of $13 + 5 = 18$ elements. The sets $\mathcal{H}_1$, $\mathcal{H}_2$ and $\mathcal{H}_3$ are shown in Fig. 1.4d, f, h.

For each $\ell \in \{1, \ldots, L\}$, let $J_{\ell,\Omega}$ be the set of indices of the B-splines in $\mathcal{B}_\ell$ belonging to $\mathcal{H}_\Omega$, i.e.,

$$J_{\ell,\Omega} := \{j : B_{j,\ell} \in \mathcal{B}_\ell \cap \mathcal{H}_\Omega\}. \tag{1.140}$$

From Definition 7 it follows that

$$J_{\ell,\Omega} = \{j : B_{j,\ell} \in \mathcal{B}_\ell, \; \mathrm{supp}_\Omega(B_{j,\ell}) \cap \Gamma_\ell \neq \emptyset, \; \mathrm{supp}_\Omega(B_{j,\ell}) \subseteq \Omega_\ell\}, \tag{1.141}$$

where

$$\Gamma_\ell := \Omega_\ell \setminus \Omega_{\ell+1}. \tag{1.142}$$

Given this index set, we can reconstruct the set of HB-splines as

$$\mathcal{H}_\Omega = \{B_{j,\ell}, \; j \in J_{\ell,\Omega}, \; \ell = 1, \ldots, L\}. \tag{1.143}$$

Since the set of HB-splines is a mixture of standard B-splines, we deduce immediately the following properties.

- **Local Support.** An HB-spline is locally supported on an interval that only depends on the level it was introduced in the hierarchical construction and not on the choice of subsets in the domain hierarchy.

- **Nonnegativity.** An HB-spline is nonnegative everywhere, and positive inside its support.

- **Piecewise Structure.** An HB-spline is a piecewise polynomial, whose degree and smoothness depends on the level it was introduced in the hierarchical construction and the spline space used on that level.

- **Linear Independence.** The HB-splines in $\mathcal{H}_\Omega$ are linearly independent on Ω_1.

Proof We first note that if $J_{\ell,\Omega}$ is nonempty then Γ_ℓ has nonempty interior for any ℓ; see (1.141) and (1.142). We must prove that if

$$s(x) = \sum_{\ell=1}^{L} \sum_{j \in J_{\ell,\Omega}} c_{j,\ell} B_{j,\ell}(x) = 0, \quad x \in \Omega_1, \tag{1.144}$$

then $c_{j,\ell} = 0$ for all j and ℓ in (1.144). We know from the local linear independence property that the B-splines $B_{j,1}$, $j \in J_{1,\Omega}$ are linearly independent on Γ_1. Moreover, from (1.141) it follows that only those functions are nonzero on Γ_1. Hence, we conclude that $c_{j,1} = 0$ for $j \in J_{1,\Omega}$ in (1.144). We can repeat the same argument for the remaining terms in (1.144) going level by level in the hierarchy. Indeed, for $\ell = 2, \ldots, L$, the B-splines $B_{j,\ell}$, $j \in J_{\ell,\Omega}$ are linearly independent on Γ_ℓ, and only those functions are nonzero on Γ_ℓ except for functions already considered before at previous levels. This implies that $c_{j,\ell} = 0$ for $j \in J_{\ell,\Omega}$ with $\ell = 2, \ldots, L$. □

The space spanned by the HB-splines in $\mathscr{H}_\Omega$ is called the **hierarchical spline space** on Ω and is denoted by

$$\mathbb{S}_\Omega := \left\{ s : \dot{\Omega}_1 \to \mathbb{R} : s = \sum_{\ell=1}^{L} \sum_{j \in J_{\ell,\Omega}} c_{j,\ell} B_{j,\ell}, \ c_{j,\ell} \in \mathbb{R} \right\}. \quad (1.145)$$

Such hierarchical space has some interesting properties.

- **Dimension.** By the linear independence of the HB-splines, the space $\mathbb{S}_\Omega$ is a vector space of dimension $\sum_{\ell=1}^{L} |J_{\ell,\Omega}|$.

- **Nestedness.** Let the domain hierarchy $\tilde{\Omega}$ be obtained from another domain hierarchy Ω such that $\Omega_1 = \tilde{\Omega}_1$ and $\Omega_\ell \subseteq \tilde{\Omega}_\ell$ for $\ell = 2, \ldots, L$. Then, $\mathbb{S}_\Omega \subseteq \mathbb{S}_{\tilde{\Omega}}$.

Proof We first note that any B-spline $B_{j,\ell-1} \in \mathscr{B}_{\ell-1}$ whose support is entirely contained in Ω_ℓ can be represented exactly in terms of B-splines $B_{i,\ell} \in \mathscr{B}_\ell$ whose support is also contained in Ω_ℓ. Consider the intermediate spaces $\mathscr{H}_\ell$ and $\tilde{\mathscr{H}}_\ell$ arising in Definition 7. From their construction it directly follows

$$\text{span}(\mathscr{H}_{\ell-1}) \subseteq \text{span}(\mathscr{H}_\ell) \quad \text{and} \quad \text{span}(\tilde{\mathscr{H}}_{\ell-1}) \subseteq \text{span}(\tilde{\mathscr{H}}_\ell). \quad (1.146)$$

We now show that $\text{span}(\mathscr{H}_\ell) \subseteq \text{span}(\tilde{\mathscr{H}}_\ell)$ for all $\ell = 1, \ldots, L$. This clearly holds for $\ell = 1$ since $\Omega_1 = \tilde{\Omega}_1$ and hence $\mathscr{H}_1 = \tilde{\mathscr{H}}_1$. We proceed by induction on ℓ, and assume that the statement is true for $\ell - 1$. Then, we have

$$\text{span}(\mathscr{H}_\ell^C) \subseteq \text{span}(\mathscr{H}_{\ell-1}) \subseteq \text{span}(\tilde{\mathscr{H}}_{\ell-1}) \subseteq \text{span}(\tilde{\mathscr{H}}_\ell),$$

and

$$\text{span}(\mathscr{H}_\ell^F) \subseteq \text{span}(\tilde{\mathscr{H}}_\ell^F) \subseteq \text{span}(\tilde{\mathscr{H}}_\ell).$$

This implies

$$\text{span}(\mathscr{H}_\ell) = \text{span}(\mathscr{H}_\ell^C) \cup \text{span}(\mathscr{H}_\ell^F) \subseteq \text{span}(\tilde{\mathscr{H}}_\ell).$$

As a consequence, $\mathbb{S}_\Omega = \text{span}(\mathscr{H}_L) \subseteq \text{span}(\tilde{\mathscr{H}}_L) = \mathbb{S}_{\tilde{\Omega}}$. □

- **Polynomial Embedding.** The space $\mathbb{S}_{\boldsymbol{\Omega}}$ contains (at least) all polynomials of degree less than or equal to p_1.

  Proof Let g be a polynomial in $\mathbb{P}_{p_1}$. From Sect. 1.3.3 we know that g belongs to the coarsest spline space $\mathbb{S}_{p_1,\boldsymbol{\xi}_1}$ in the sequence (1.135). Hence, taking into account (1.146), we conclude that $g \in \mathrm{span}(\mathscr{H}_1) \subseteq \mathrm{span}(\mathscr{H}_L) = \mathbb{S}_{\boldsymbol{\Omega}}$. $\qquad\square$

1.6.2 Truncated Hierarchical B-Splines

HB-splines do not satisfy the partition of unity property. In addition, the number of overlapping basis functions associated with different hierarchical levels easily increases. This motivates the construction of another basis for the hierarchical spline space. The construction is based on the following truncation mechanism [22].

Definition 8 Given $\ell \in \{2, \ldots, L\}$, let $s \in \mathbb{S}_{p_\ell,\boldsymbol{\xi}_\ell}$ be represented in the B-spline basis $\mathscr{B}_\ell$, i.e.,

$$s = \sum_{j=1}^{n_\ell} c_{j,\ell}\, B_{j,\ell}. \tag{1.147}$$

The **truncation** of s at level ℓ is defined as the sum of the terms appearing in (1.147) related to the B-splines whose support is not a subset of Ω_ℓ, i.e.,

$$\mathrm{trunc}_{\ell,\boldsymbol{\Omega}}(s) := \sum_{j\,:\,\mathrm{supp}_{\boldsymbol{\Omega}}(B_{j,\ell}) \not\subseteq \Omega_\ell} c_{j,\ell}\, B_{j,\ell}. \tag{1.148}$$

By successively truncating the functions constructed in Definition 7, we obtain the truncated hierarchical B-splines (in short, THB-splines).[8]

Definition 9 Given a domain hierarchy $\boldsymbol{\Omega}$, the corresponding set of **THB-splines** is denoted by $\mathscr{T}_{\boldsymbol{\Omega}}$ and defined recursively as follows:

(i) $\mathscr{T}_1 := \{B_{j,1} \in \mathscr{B}_1 : \mathrm{supp}_{\boldsymbol{\Omega}}(B_{j,1}) \neq \emptyset\}$;
(ii) for $\ell = 2, \ldots, L$:

$$\mathscr{T}_\ell := \mathscr{T}_\ell^C \cup \mathscr{T}_\ell^F,$$

where

$$\mathscr{T}_\ell^C := \{\mathrm{trunc}_{\ell,\boldsymbol{\Omega}}(B_{j,k,\Omega_{\ell-1}}^t) : B_{j,k,\Omega_{\ell-1}}^t \in \mathscr{T}_{\ell-1},\ \mathrm{supp}_{\boldsymbol{\Omega}}(B_{j,k,\Omega_{\ell-1}}^t) \not\subseteq \Omega_\ell\},$$

$$\mathscr{T}_\ell^F := \{B_{j,\ell} \in \mathscr{B}_\ell : \mathrm{supp}_{\boldsymbol{\Omega}}(B_{j,\ell}) \subseteq \Omega_\ell\};$$

---

[8]The truncation approach was introduced in [22] for hierarchical tensor-product splines, but was already developed before in the context of hierarchical Powell–Sabin splines [50]. A generalization towards a broad class of hierarchical spaces can be found in [23].

(iii) $\mathscr{T}_{\boldsymbol{\Omega}} := \mathscr{T}_L$.

To obtain the THB-splines, we apply a recursive procedure building a set $\mathscr{T}_\ell$ at level ℓ. This set consists of two subsets, the coarse set $\mathscr{T}_\ell^C$ and the fine set $\mathscr{T}_\ell^F$. To construct the elements B_{j,k,Ω_ℓ}^t of $\mathscr{T}_\ell^C$, we first express any function $B_{j,k,\Omega_{\ell-1}}^t \in \mathscr{T}_{\ell-1}$ with respect to the B-spline basis $\mathscr{B}_\ell$, and then we apply the truncation as in (1.148) with $s = B_{j,k,\Omega_{\ell-1}}^t$. The fine set $\mathscr{T}_\ell^F$ consists of all B-splines in $\mathscr{B}_\ell$ whose support is entirely contained in Ω_ℓ, exactly as in the HB-spline case; see Definition 7.

When comparing Definition 9 with Definition 7, we see that the number of THB-splines in the set $\mathscr{T}_{\boldsymbol{\Omega}}$ is equal to the number of HB-splines in the set $\mathscr{H}_{\boldsymbol{\Omega}}$. In the following, the THB-splines in $\mathscr{T}_{\boldsymbol{\Omega}}$ are denoted by $B_{j,\ell,\boldsymbol{\Omega}}^T$ for $j \in J_{\ell,\boldsymbol{\Omega}}$ and $\ell = 1, \dots, L$.

Example 14 When unrolling the recursive definition of THB-splines for $L = 3$, we get

$$B_{j,1,\boldsymbol{\Omega}}^T = \mathrm{trunc}_{3,\boldsymbol{\Omega}}(\mathrm{trunc}_{2,\boldsymbol{\Omega}}(B_{j,1})), \quad j \in J_{1,\boldsymbol{\Omega}},$$

$$B_{j,2,\boldsymbol{\Omega}}^T = \mathrm{trunc}_{3,\boldsymbol{\Omega}}(B_{j,2}), \quad j \in J_{2,\boldsymbol{\Omega}},$$

$$B_{j,3,\boldsymbol{\Omega}}^T = B_{j,3}, \quad j \in J_{3,\boldsymbol{\Omega}}.$$

Example 15 Figure 1.5 illustrates the truncation mechanism applied to the set of HB-splines depicted in Fig. 1.4 (Example 13). Obviously, $\mathscr{T}_1$ coincides with $\mathscr{H}_1$.

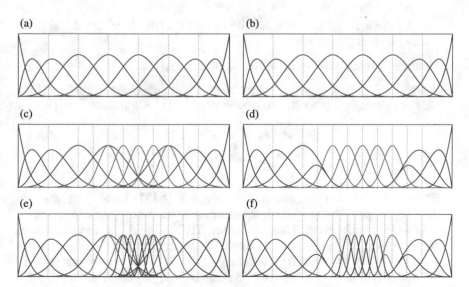

Fig. 1.5 HB-splines and THB-splines with respect to the same domain hierarchy as in Fig. 1.4b. (a) $\mathscr{H}_1$. (b) $\mathscr{T}_1$. (c) $\mathscr{H}_2$. (d) $\mathscr{T}_2$. (e) $\mathscr{H}_3$. (f) $\mathscr{T}_3$

Furthermore, $\mathscr{T}_2^C$ is obtained from $\mathscr{H}_2^C$ by applying the truncation mechanism to its elements; this only results in a modification of the elements $B_{4,1}$, $B_{5,1}$, $B_{7,1}$ and $B_{8,1}$. On the other hand, we have $\mathscr{T}_2^F = \mathscr{H}_2^F$. Finally, $\mathscr{T}_3^C$ is obtained from $\mathscr{H}_3^C$ by modifying $B_{4,1}$, $B_{5,1}$, $B_{7,1}$, $B_{8,1}$ (truncated at level 2) and $B_{8,2}$, $B_{9,2}$, $B_{11,2}$, $B_{12,2}$ (truncated at level 3), while $\mathscr{T}_3^F = \mathscr{H}_3^F$. It is clear that $\mathscr{T}_\ell = \mathscr{T}_\ell^C \cup \mathscr{T}_\ell^F$ and $\mathscr{H}_\ell = \mathscr{H}_\ell^C \cup \mathscr{H}_\ell^F$ have the same number of elements for $\ell = 2, 3$.

The next properties can be easily deduced from the definition of THB-splines.

- **Relation to HB-Splines.** Each THB-spline in $\mathscr{T}_\mathbf{\Omega}$ is uniquely related to a single HB-spline in $\mathscr{H}_\mathbf{\Omega}$ possibly by successive truncations, i.e.,

$$B_{j,\ell,\mathbf{\Omega}}^T = \mathrm{Trunc}_{\ell,\mathbf{\Omega}}(B_{j,\ell}), \qquad (1.149)$$

where for any $s \in \mathbb{S}_{p_\ell,\mathbf{\xi}_\ell}$ with $\ell = 1, \ldots, L-1$,

$$\mathrm{Trunc}_{\ell,\mathbf{\Omega}}(s) := \mathrm{trunc}_{L,\mathbf{\Omega}}(\mathrm{trunc}_{L-1,\mathbf{\Omega}}(\cdots(\mathrm{trunc}_{\ell+1,\mathbf{\Omega}}(s))\cdots)),$$

and for any $s \in \mathbb{S}_{p_L,\mathbf{\xi}_L}$,

$$\mathrm{Trunc}_{L,\mathbf{\Omega}}(s) := s.$$

From (1.149) in combination with (1.147)–(1.148), it is clear that

$$B_{j,\ell,\mathbf{\Omega}}^T(x) = B_{j,\ell}(x), \quad x \in \Gamma_\ell. \qquad (1.150)$$

- **Local Support.** From (1.149) it follows that a THB-spline has the same or smaller support than its related HB-spline.

- **Nonnegativity.** A THB-spline is nonnegative on Ω_1.

Proof Fix $1 \le \ell_1 < \ell_2 \le L$. Because of the nestedness of the spaces in (1.135), we can write the B-spline $B_{j,\ell_1} \in \mathscr{B}_{\ell_1}$ in terms of the B-splines in $\mathscr{B}_{\ell_2}$, i.e.,

$$B_{j,\ell_1}(x) = \sum_{i=1}^{n_{\ell_2}} c_{i,\ell_2}^{j,\ell_1} B_{i,\ell_2}(x), \quad x \in \Omega_1. \qquad (1.151)$$

From Sect. 1.3.5 we know that the coefficients in (1.151) are all nonnegative in case $p_{\ell_1} = p_{\ell_2}$. This property holds in general, also when $p_{\ell_1} < p_{\ell_2}$, and we refer to [12] for its proof. Then, since each THB-spline $B_{j,\ell,\mathbf{\Omega}}^T$ can be deduced from the B-spline $B_{j,\ell}$ possibly by successive truncations, see (1.149), it follows from (1.147)–(1.148) that $B_{j,\ell,\mathbf{\Omega}}^T$ can be written as a linear combination of B-splines of the finest level L with nonnegative coefficients. This implies that $B_{j,\ell,\mathbf{\Omega}}^T$ is nonnegative. $\qquad\qquad\square$

- **Linear Independence.** The THB-splines in $\mathscr{T}_\mathbf{\Omega}$ are linearly independent on Ω_1.

Proof We must prove that if

$$s(x) = \sum_{\ell=1}^{L} \sum_{j \in J_{\ell,\Omega}} c_{j,\ell} B_{j,\ell}^{T}(x) = 0, \quad x \in \Omega_1, \tag{1.152}$$

then $c_{j,\ell} = 0$ for all j and ℓ in (1.152). This can be shown using exactly the same line of arguments as in the case of HB-splines (see (1.144)), taking into account relation (1.150). □

The next theorem shows that the THB-splines in $\mathcal{T}_{\Omega}$ form an alternative basis for the hierarchical spline space $\mathbb{S}_{\Omega}$ in (1.145).

Theorem 19 (Hierarchical Spline Space) *The THB-splines in $\mathcal{T}_{\Omega}$ span the same space as the HB-splines in $\mathcal{H}_{\Omega}$, i.e.,*

$$\mathbb{S}_{\Omega} = \mathrm{span}(\mathcal{H}_{\Omega}) = \mathrm{span}(\mathcal{T}_{\Omega}). \tag{1.153}$$

Proof Consider the intermediate spaces $\mathcal{H}_{\ell}$ and $\mathcal{T}_{\ell}$ in Definitions 7 and 9, respectively. From their construction it directly follows

$$\mathrm{span}(\mathcal{H}_{\ell-1}) \subseteq \mathrm{span}(\mathcal{H}_{\ell}) \quad \text{and} \quad \mathrm{span}(\mathcal{T}_{\ell-1}) \subseteq \mathrm{span}(\mathcal{T}_{\ell}).$$

We now show that $\mathrm{span}(\mathcal{H}_{\ell}) = \mathrm{span}(\mathcal{T}_{\ell})$ for all $\ell = 1, \dots, L$. This clearly holds for $\ell = 1$ since $\mathcal{H}_1 = \mathcal{T}_1$. We proceed by induction on ℓ, and assume that the statement is true for $\ell - 1$. Then, we have

$$\mathrm{span}(\mathcal{H}_{\ell}^{C}) \subseteq \mathrm{span}(\mathcal{H}_{\ell-1}) = \mathrm{span}(\mathcal{T}_{\ell-1}) \subseteq \mathrm{span}(\mathcal{T}_{\ell}),$$

and

$$\mathrm{span}(\mathcal{H}_{\ell}^{F}) = \mathrm{span}(\mathcal{T}_{\ell}^{F}) \subseteq \mathrm{span}(\mathcal{T}_{\ell}).$$

This implies

$$\mathrm{span}(\mathcal{H}_{\ell}) = \mathrm{span}(\mathcal{H}_{\ell}^{C}) \cup \mathrm{span}(\mathcal{H}_{\ell}^{F}) \subseteq \mathrm{span}(\mathcal{T}_{\ell}).$$

Finally, since both sets $\mathcal{H}_{\ell}$ and $\mathcal{T}_{\ell}$ have the same number of elements and these elements are all linearly independent, it follows that $\mathrm{span}(\mathcal{H}_{\ell}) = \mathrm{span}(\mathcal{T}_{\ell})$. As a consequence, $\mathrm{span}(\mathcal{H}_{\Omega}) = \mathrm{span}(\mathcal{H}_{L}) = \mathrm{span}(\mathcal{T}_{L}) = \mathrm{span}(\mathcal{T}_{\Omega})$. □

The correspondence in (1.149) between the THB-spline $B_{j,\ell,\Omega}^{T}$ and a particular B-spline $B_{j,\ell} \in \mathcal{B}_{\ell}$ has an important consequence, namely the so-called property of **preservation of coefficients** [23]. This means that the THB-spline representation preserves certain coefficients of functions represented with respect to one of the B-spline bases $\mathcal{B}^{\ell}$.

Theorem 20 (Preservation of Coefficients) *Given $\ell \in \{1, \ldots, L\}$, let the restriction of $s \in \mathbb{S}_{\boldsymbol{\Omega}}$ to $\Gamma_\ell := \Omega_\ell \setminus \Omega_{\ell+1}$ be represented in the bases $\mathscr{T}_{\boldsymbol{\Omega}}$ and $\mathscr{B}_\ell$, i.e.,*

$$s(x) = \sum_{k=1}^{L} \sum_{j \in J_{k,\boldsymbol{\Omega}}} c_{j,k}^{T} B_{j,k,\boldsymbol{\Omega}}^{T}(x) = \sum_{i=1}^{n_\ell} c_{i,\ell} B_{i,\ell}(x), \quad x \in \Gamma_\ell. \tag{1.154}$$

Then,

$$c_{i,\ell}^{T} = c_{i,\ell}, \quad i \in J_{\ell,\boldsymbol{\Omega}}. \tag{1.155}$$

Proof We first note that if $J_{\ell,\boldsymbol{\Omega}}$ is nonempty then Γ_ℓ has nonempty interior. Assume now that Γ_ℓ has nonempty interior. Since $s \in \mathbb{S}_{\boldsymbol{\Omega}}$ and the spline spaces in (1.135) are nested, it is clear that the restriction of s to Γ_ℓ can be expressed as a linear combination of the B-splines in $\mathscr{B}_\ell$ restricted to Γ_ℓ as in (1.154). Let us focus on the sum

$$\sum_{j \in J_{k,\boldsymbol{\Omega}}} c_{j,k}^{T} B_{j,k,\boldsymbol{\Omega}}^{T}(x), \quad x \in \Gamma_\ell, \tag{1.156}$$

and consider three cases.

- If $k > \ell$, then the sum in (1.156) equals zero. Indeed, Definition 9 and (1.149) imply that

$$\mathrm{supp}_{\boldsymbol{\Omega}}(B_{j,k,\boldsymbol{\Omega}}^{T}) \subseteq \mathrm{supp}_{\boldsymbol{\Omega}}(B_{j,k}) \subseteq \Omega_k \subseteq \Omega_{\ell+1},$$

and consequently, we have $\mathrm{supp}_{\boldsymbol{\Omega}}(B_{j,k,\boldsymbol{\Omega}}^{T}) \cap \Gamma_\ell = \emptyset$.
- We now consider the case $k = \ell$. From (1.150) it immediately follows

$$\sum_{j \in J_{\ell,\boldsymbol{\Omega}}} c_{j,\ell}^{T} B_{j,\ell,\boldsymbol{\Omega}}^{T}(x) = \sum_{j \in J_{\ell,\boldsymbol{\Omega}}} c_{j,\ell}^{T} B_{j,\ell}(x), \quad x \in \Gamma_\ell.$$

- Finally, let $k < \ell$. In view of the truncation mechanism, we prove that THB-splines introduced at levels less than ℓ in the hierarchy can only contribute in terms of B-splines $B_{i,\ell}$ with $i \notin J_{\ell,\boldsymbol{\Omega}}$. To this end, let us rewrite the corresponding THB-splines $B_{j,k,\boldsymbol{\Omega}}^{T}$ in terms of the B-spline basis $\mathscr{B}_\ell$,

$$B_{j,k,\boldsymbol{\Omega}}^{T}(x) = \sum_{i=1}^{n_\ell} c_{i,\ell}^{j,k} B_{i,\ell}(x), \quad x \in \Gamma_\ell.$$

Due to the definition of $B_{j,k,\boldsymbol{\Omega}}^{T}$ and the truncation operation (1.148), we have

$$c_{i,\ell}^{j,k} = 0, \quad \text{if } i \in J_{\ell,\boldsymbol{\Omega}}.$$

Hence, for $k < \ell$ we arrive at

$$\sum_{j \in J_{k,\boldsymbol{\Omega}}} c_{j,k}^T B_{j,k,\boldsymbol{\Omega}}^T(x) = \sum_{i \notin J_{\ell,\boldsymbol{\Omega}}} \left(\sum_{j \in J_{k,\boldsymbol{\Omega}}} c_{j,k}^T c_{i,\ell}^{j,k} \right) B_{i,\ell}(x), \quad x \in \Gamma_\ell.$$

By combining the above three cases and taking into account the local linear independence of B-splines, we obtain the identity (1.154) where

$$c_{i,\ell} = \begin{cases} c_{i,\ell}^T, & \text{if } i \in J_{\ell,\boldsymbol{\Omega}}, \\ \sum_{k=1}^{\ell-1} \sum_{j \in J_{k,\boldsymbol{\Omega}}} c_{j,k}^T c_{i,\ell}^{j,k}, & \text{otherwise}, \end{cases}$$

which in particular gives (1.155). $\qquad\square$

Thanks to Theorem 20, many interesting features of B-spline representations can be transferred to THB-spline representations.

- **Representation of Polynomials.** Any polynomial g of degree p_1 can be represented as

$$g(x) = \sum_{\ell=1}^{L} \sum_{j \in J_{\ell,\boldsymbol{\Omega}}} \Lambda_{j,p_\ell,\boldsymbol{\xi}_\ell}(g) B_{j,\ell,\boldsymbol{\Omega}}^T(x), \quad x \in \Omega_1, \tag{1.157}$$

where $\Lambda_{j,p_\ell,\boldsymbol{\xi}_\ell}$ is defined in (1.53) with $p = p_\ell$ and $\boldsymbol{\xi} = \boldsymbol{\xi}_\ell$.

Proof Using the nestedness of the spaces (1.135), it is clear that $g \in \mathbb{S}_{p_\ell,\boldsymbol{\xi}_\ell}$ for $\ell = 1, \ldots, L$ and also that $g \in \mathbb{S}_{\boldsymbol{\Omega}}$. Then, consider its representation with respect to $\mathscr{T}_{\boldsymbol{\Omega}}$ and $\mathscr{B}_\ell$ for $\ell = 1, \ldots, L$. Theorem 20 in combination with Proposition 3 concludes the proof. $\qquad\square$

- **Partition of Unity.** By (1.49) we have

$$\sum_{\ell=1}^{L} \sum_{j \in J_{\ell,\boldsymbol{\Omega}}} B_{j,\ell,\boldsymbol{\Omega}}^T(x) = 1, \quad x \in \Omega_1. \tag{1.158}$$

Since the THB-splines are nonnegative it follows that they form a **nonnegative partition of unity** on Ω_1.

- **Greville Points.** By (1.50) we have

$$x = \sum_{\ell=1}^{L} \sum_{j \in J_{\ell,\boldsymbol{\Omega}}} \xi_{j,p_\ell,\boldsymbol{\xi}_\ell}^* B_{j,\ell,\boldsymbol{\Omega}}^T(x), \quad x \in \Omega_1, \tag{1.159}$$

where $\xi_{j,p_\ell,\boldsymbol{\xi}_\ell}^*$ are the Greville points defined in (1.51) with $p = p_\ell$ and $\boldsymbol{\xi} = \boldsymbol{\xi}_\ell$. Note that the Greville points are not necessarily distinct here.

- **Strong Stability.** The THB-spline basis is strongly stable with respect to the supremum norm, under mild assumptions on the underlying knot sequences required in the hierarchical construction. We refer the reader to [23] for a proof based on the property of preservation of coefficients. Strong stability in the hierarchical context means that the constants to be considered in the stability relation (1.64) of the basis do not depend on the number of hierarchical levels.

Example 16 The polynomial $g(x) = ax^2 + bx + c$ can be represented in terms of quadratic THB-splines:

$$ax^2 + bx + c = \sum_{\ell=1}^{L} \sum_{j \in J_{\ell,\Omega}} c_{j,\ell} B_{j,\ell,\Omega}^T(x).$$

From Theorem 20 and Example 8 we obtain that

$$c_{j,\ell} = \Lambda_{j,2,\xi_\ell}(g) = a\,\xi_{j+1,\ell}\xi_{j+2,\ell} + b\,\frac{\xi_{j+1,\ell} + \xi_{j+2,\ell}}{2} + c.$$

1.6.3 Quasi-Interpolation in Hierarchical Spaces

The above properties of THB-splines can be exploited to develop a general and very simple procedure for the construction of quasi-interpolants in hierarchical spline spaces [51].

Definition 10 Given for each spline space in (1.135) a quasi-interpolant in B-spline form, i.e.,

$$\mathcal{Q}_\ell f(x) := \sum_{j=1}^{n_\ell} \lambda_{j,\ell}(f) B_{j,\ell}(x), \quad x \in \Omega_1, \quad \ell = 1, \dots, L, \tag{1.160}$$

the corresponding **hierarchical quasi-interpolant** in $\mathbb{S}_\Omega$ is defined by

$$\mathcal{Q}_\Omega f(x) := \sum_{\ell=1}^{L} \sum_{j \in J_{\ell,\Omega}} \lambda_{j,\ell}(f) B_{j,\ell,\Omega}^T(x), \quad x \in \Omega_1. \tag{1.161}$$

According to Definition 10, in order to construct a quasi-interpolant in $\mathbb{S}_\Omega$, it suffices to consider first a quasi-interpolant in each space associated with a particular level in the hierarchy. Then, the coefficients of the proposed hierarchical quasi-interpolant are nothing else than a proper subset of the coefficients of the one-level quasi-interpolants.

We now show how to build hierarchical quasi-interpolants reproducing polynomials of a certain degree $p \leq p_1$. As described in Sect. 1.5.2, this is a crucial property feature to ensure good approximation properties.

Theorem 21 (Polynomial Reproduction) *Let $\mathcal{Q}_\ell$ be a given sequence of quasi-interpolants as in* (1.160), *let $\mathcal{Q}_{\boldsymbol{\Omega}}$ be the corresponding hierarchical quasi-interpolant as in* (1.161), *and let $p \leq p_1$. If*

$$\mathcal{Q}_\ell g = g, \quad \forall g \in \mathbb{P}_p, \quad \ell = 1, \ldots, L, \tag{1.162}$$

then

$$\mathcal{Q}_{\boldsymbol{\Omega}} g = g, \quad \forall g \in \mathbb{P}_p.$$

Proof Since the spaces in (1.135) are nested, we have $p_\ell \geq p$ for all ℓ. Let $g \in \mathbb{P}_p \subseteq \mathbb{P}_{p_\ell} \subseteq \mathbb{S}_{p_\ell, \boldsymbol{\xi}_\ell}$. Then, this polynomial can be uniquely represented as a linear combination of the B-splines in $\mathcal{B}_\ell$,

$$g(x) = \sum_{j=1}^{n_\ell} c_{j,\ell} B_{j,\ell}(x),$$

and since $\mathcal{Q}_\ell g = g$ we have $\lambda_{j,\ell}(g) = c_{j,\ell}$. On the other hand, $g \in \mathbb{S}_{\boldsymbol{\Omega}}$, so

$$g(x) = \sum_{\ell=1}^{L} \sum_{j \in J_{\ell,\boldsymbol{\Omega}}} c_{j,\ell}^T B_{j,\ell,\boldsymbol{\Omega}}^T(x).$$

From Theorem 20 it follows

$$c_{j,\ell}^T = c_{j,\ell} = \lambda_{j,\ell}(g), \quad j \in J_{\ell,\boldsymbol{\Omega}}, \quad \ell = 1, \ldots, L,$$

implying that $\mathcal{Q}_{\boldsymbol{\Omega}} g = g$. □

In the next theorem we present a sufficient condition for constructing quasi-interpolants that are projectors onto $\mathbb{S}_{\boldsymbol{\Omega}}$.

Theorem 22 (Spline Reproduction) *Let $\mathcal{Q}_\ell$ be a given sequence of quasi-interpolants as in* (1.160), *and let $\mathcal{Q}_{\boldsymbol{\Omega}}$ be the corresponding hierarchical quasi-interpolant as in* (1.161). *Assume*

$$\mathcal{Q}_\ell s = s, \quad \forall s \in \mathbb{S}_{p_\ell, \boldsymbol{\xi}_\ell}, \quad \ell = 1, \ldots, L,$$

and each $\lambda_{j,\ell}$ used in (1.161) *is supported on $\Gamma_\ell := \Omega_\ell \setminus \Omega_{\ell+1}$. Then,*

$$\mathcal{Q}_{\boldsymbol{\Omega}} s = s, \quad \forall s \in \mathbb{S}_{\boldsymbol{\Omega}}.$$

Proof Due to the linearity of the quasi-interpolant, it suffices to prove that

$$\lambda_{j,\ell}(B^T_{i,k,\boldsymbol{\Omega}}) = \delta_{i,j}\delta_{k,\ell}, \quad i \in J_{k,\boldsymbol{\Omega}}, \quad j \in J_{\ell,\boldsymbol{\Omega}}, \quad k,\ell = 1,\ldots,L, \quad (1.163)$$

where $\delta_{r,s}$ stands for the classical Kronecker delta. Let j and ℓ be fixed. To prove (1.163) we consider three cases.

- If $k > \ell$, then $B^T_{i,k,\boldsymbol{\Omega}}(x) = 0$ for $x \in \Gamma_\ell$; see Definition 9. Since $\lambda_{j,\ell}$ is only supported on Γ_ℓ, it follows from Definition 4 that $\lambda_{j,\ell}(B^T_{i,k,\boldsymbol{\Omega}}) = 0$.
- We now consider the case $k = \ell$. Since $\mathscr{Q}_\ell$ is a projector onto $\mathbb{S}_\ell$, we have that $\lambda_{j,\ell}(B_{i,\ell}) = \delta_{i,j}$. From (1.150) and the support restriction of $\lambda_{j,\ell}$, we obtain

$$\lambda_{j,\ell}(B^T_{i,\ell,\boldsymbol{\Omega}}) = \delta_{i,j}, \quad i,j \in J_{\ell,\boldsymbol{\Omega}}.$$

- Finally, let $k < \ell$. Any $B^T_{i,k,\boldsymbol{\Omega}}$ restricted to Γ_ℓ can then be expressed as a linear combination of the B-splines in $\mathscr{B}_\ell$ restricted to Γ_ℓ, i.e.,

$$B^T_{i,k,\boldsymbol{\Omega}}(x) = \sum_{r=1}^{n_\ell} c^{i,k}_{r,\ell} B_{r,\ell}(x), \quad x \in \Gamma_\ell,$$

where

$$c^{i,k}_{r,\ell} = 0, \quad \text{if } r \in J_{\ell,\boldsymbol{\Omega}},$$

as explained in the third case of the proof of Theorem 20. Thus, by the support restriction of $\lambda_{j,\ell}$, we have for $j \in J_{\ell,\boldsymbol{\Omega}}$,

$$\lambda_{j,\ell}(B^T_{i,k,\boldsymbol{\Omega}}) = \sum_{r=1}^{n_\ell} c^{i,k}_{r,\ell}\lambda_{j,\ell}(B_{r,\ell}) = \sum_{r=1}^{n_\ell} c^{i,k}_{r,\ell}\delta_{j,r} = c^{i,k}_{j,\ell} = 0.$$

The above three cases complete the proof. $\qquad\qquad\qquad\qquad\qquad\qquad\qquad\square$

Some remarks are in order here.

- **Constraints on** (1.160). The sequence of quasi-interpolants (1.160) considered in Theorem 22 needs to satisfy constraints more restrictive than those in Theorem 21: For each level ℓ, $\mathscr{Q}_\ell$ must be a projector onto $\mathbb{S}_{p_\ell,\boldsymbol{\xi}_\ell}$ and each $\lambda_{j,\ell}$, $j \in J_{\ell,\boldsymbol{\Omega}}$, must be supported on Γ_ℓ. The former constraint connects the sequence of quasi-interpolants $\mathscr{Q}_1, \ldots, \mathscr{Q}_L$ with the sequence of spaces $\mathbb{S}_{p_1,\boldsymbol{\xi}_1}, \ldots, \mathbb{S}_{p_L,\boldsymbol{\xi}_L}$ and has a similar counterpart in Theorem 21. The latter constraint links the same sequence of quasi-interpolants with the domain hierarchy $\boldsymbol{\Omega}$. Nevertheless, once a sequence of quasi-interpolants as in (1.160) satisfying the hypotheses of Theorem 22 is available, the construction of a hierarchical quasi-interpolant that is a projector onto $\mathbb{S}_{\boldsymbol{\Omega}}$ does not require additional efforts compared to a hierarchical quasi-interpolant that just reproduces polynomials.

- **Dual Basis.** Let $\{\lambda_{j,\ell}\}$ be a set of linear functionals as in (1.161) that provides a projector onto $\mathbb{S}_{\boldsymbol{\Omega}}$. Then, because of (1.163), it is a dual basis for the THB-spline basis $\mathscr{T}_{\boldsymbol{\Omega}}$.

- **Approximation Power.** Polynomial reproduction is one of the key ingredients to show the approximation power of spline quasi-interpolants; see Sect. 1.5.2. Boundedness of a hierarchical quasi-interpolation operator and optimal approximation accuracy can be achieved on domain hierarchies that are nicely graded (i.e., the boundaries of the different Ω_ℓ are sufficiently separated). Local error estimates for hierarchical quasi-interpolants of the form (1.161) can be found in [51] with respect to the L_∞-norm, and in [49] with respect to the general L_q-norm, $1 \leq q \leq \infty$.

Example 17 Let $p_\ell = 2$, and let ξ_ℓ be a 3-open knot sequence with at most double knots in the interior for each $\ell = 1, \ldots, L$. Then, we can choose the quasi-interpolants in (1.160) as in Example 12. This leads to the hierarchical quasi-interpolant

$$\mathscr{Q}_{\boldsymbol{\Omega}} f(x) = \sum_{\ell=1}^{L} \sum_{j \in J_{\ell,\boldsymbol{\Omega}}} \lambda_{j,\ell}(f) B_{j,\ell,\boldsymbol{\Omega}}^{T}(x), \quad x \in \Omega_1,$$

where

$$\lambda_{j,\ell}(f) = -\frac{1}{2} f(\xi_{j+1,\ell}) + 2f(\xi_{j,2,\xi_\ell}^*) - \frac{1}{2} f(\xi_{j+2,\ell}).$$

From Example 12 and Theorem 21 we deduce that this hierarchical quasi-interpolant reproduces the polynomial space $\mathbb{P}_2$. If $[\xi_{j+1,\ell}, \xi_{j+2,\ell}] \subseteq \Gamma_\ell$ for each $j \in J_{\ell,\boldsymbol{\Omega}}$, then it actually reproduces the entire hierarchical spline space $\mathbb{S}_{\boldsymbol{\Omega}}$, according to Theorem 22.

Example 18 Consider the quasi-interpolant constructed in Sect. 1.5.3.1 for each space $\mathbb{S}_{p_\ell,\xi_\ell}$ of level $\ell = 1, \ldots, L$. This leads to the hierarchical quasi-interpolant

$$\mathscr{Q}_{\boldsymbol{\Omega}} f(x) = \sum_{\ell=1}^{L} \sum_{j \in J_{\ell,\boldsymbol{\Omega}}} \mathscr{L}_{j,p_\ell,\xi_\ell}(f) B_{j,\ell,\boldsymbol{\Omega}}^{T}(x), \quad x \in \Omega_1,$$

where $\mathscr{L}_{j,p_\ell,\xi_\ell}$ is defined in (1.123) with $p = p_\ell$ and $\xi = \xi_\ell$; it is supported on a single knot interval $[\xi_{m_{j,p_\ell},\ell}, \xi_{m_{j,p_\ell}+1,\ell}]$. From Lemma 2 and Theorem 21 we deduce that this hierarchical quasi-interpolant reproduces the polynomial space $\mathbb{P}_{p_1}$. Theorem 22 says that if $[\xi_{m_{j,p_\ell},\ell}, \xi_{m_{j,p_\ell}+1,\ell}] \subseteq \Gamma_\ell$ for each $j \in J_{\ell,\boldsymbol{\Omega}}$, then the hierarchical quasi-interpolant reproduces the entire hierarchical spline space $\mathbb{S}_{\boldsymbol{\Omega}}$.

The hierarchical quasi-interpolant in Definition 10 can be interpreted as a telescopic approximant, where for each level an approximant of the residual is added.[9] To show this, we define the following set of indices

$$K_{\ell,\boldsymbol{\Omega}} := \{j \ : \ B_{j,\ell} \in \mathscr{B}_\ell, \ \operatorname{supp}_{\boldsymbol{\Omega}}(B_{j,\ell}) \subseteq \Omega_\ell\}.$$

Referring to (1.141), it is easy to see that $J_{\ell,\boldsymbol{\Omega}} \subseteq K_{\ell,\boldsymbol{\Omega}}$, and moreover $J_{L,\boldsymbol{\Omega}} = K_{L,\boldsymbol{\Omega}}$.

Theorem 23 (Telescopic Representation) *Let $\mathcal{Q}_\ell$ be a given sequence of quasi-interpolants as in (1.160), and let $\mathcal{Q}_{\boldsymbol{\Omega}}$ be the corresponding hierarchical quasi-interpolant as in (1.161). Assume*

$$\mathcal{Q}_\ell s = s, \quad \forall s \in \mathbb{S}_{p_\ell,\boldsymbol{\xi}_\ell}, \quad \ell = 1,\ldots,L, \tag{1.164}$$

then

$$\mathcal{Q}_{\boldsymbol{\Omega}} f = \sum_{\ell=1}^{L} f^{(\ell)}, \tag{1.165}$$

where

$$
\begin{aligned}
f^{(1)} &:= \sum_{j \in K_{1,\boldsymbol{\Omega}}} \lambda_{j,1}(f) B_{j,1}, \\
f^{(\ell)} &:= \sum_{j \in K_{\ell,\boldsymbol{\Omega}}} \lambda_{j,\ell}\big(f - f^{(1)} - \ldots - f^{(\ell-1)}\big) B_{j,\ell}, \quad \ell = 2,\ldots,L.
\end{aligned}
\tag{1.166}
$$

Proof Each quasi-interpolant $\mathcal{Q}_\ell$, $\ell = 1,\ldots,L$, is assumed to be a projector onto the space $\mathbb{S}_{p_\ell,\boldsymbol{\xi}_\ell}$, and because of the nestedness of the spaces $\mathbb{S}_{p_\ell,\boldsymbol{\xi}_\ell} \subset \mathbb{S}_{p_{\ell+1},\boldsymbol{\xi}_{\ell+1}}$, we know that every basis function $B_{j,\ell}$ can be represented as

$$B_{j,\ell} = \sum_{k=1}^{n_{\ell+1}} \lambda_{k,\ell+1}(B_{j,\ell})\, B_{k,\ell+1}, \tag{1.167}$$

where $\lambda_{k,\ell+1}(B_{j,\ell}) = 0$ if the support of $B_{k,\ell+1}$ is not contained in the support of $B_{j,\ell}$. By exploiting the definition of the truncated basis (1.149) and (1.167), we obtain

$$
\begin{aligned}
f^{(1)} &= \sum_{j \in K_{1,\boldsymbol{\Omega}}} \lambda_{j,1}(f) B_{j,1} \\
&= \sum_{j \in J_{1,\boldsymbol{\Omega}}} \lambda_{j,1}(f) B_{j,1,\boldsymbol{\Omega}}^T + \sum_{j \in K_{1,\boldsymbol{\Omega}}} \lambda_{j,1}(f) \bigg(\sum_{k \in K_{2,\boldsymbol{\Omega}}} \lambda_{k,2}(B_{j,1})\, B_{k,2} \bigg).
\end{aligned}
$$

---

[9]The general telescopic expression for the hierarchical quasi-interpolant was presented in [51]. A special telescopic approximation in the hierarchical setting was already considered in [29].

Moreover,

$$f^{(2)} = \sum_{j \in K_{2,\varOmega}} \lambda_{j,2}(f) B_{j,2} - \sum_{k \in K_{2,\varOmega}} \lambda_{k,2}(f^{(1)}) B_{k,2}$$

$$= \sum_{j \in K_{2,\varOmega}} \lambda_{j,2}(f) B_{j,2} - \sum_{k \in K_{2,\varOmega}} \left(\sum_{j \in K_{1,\varOmega}} \lambda_{j,1}(f) \lambda_{k,2}(B_{j,1}) \right) B_{k,2}.$$

Hence,

$$f^{(1)} + f^{(2)} = \sum_{j \in J_{1,\varOmega}} \lambda_{j,1}(f) B_{j,1,\varOmega}^T + \sum_{j \in K_{2,\varOmega}} \lambda_{j,2}(f) B_{j,2}. \tag{1.168}$$

We now remark that from the truncation definition (1.148)–(1.149) it follows that $\lambda_{k,3}(B_{j,1,\varOmega}^T) = 0$ for any $k \in K_{3,\varOmega}$ and $j \in J_{1,\varOmega}$, and so

$$\sum_{k \in K_{3,\varOmega}} \lambda_{k,3}(B_{j,1,\varOmega}^T) B_{k,3} = 0, \quad \forall j \in J_{1,\varOmega}. \tag{1.169}$$

By using similar arguments as before, we can write (1.168) as

$$f^{(1)} + f^{(2)} = \sum_{j \in J_{1,\varOmega}} \lambda_{j,1}(f) B_{j,1,\varOmega}^T + \sum_{j \in J_{2,\varOmega}} \lambda_{j,2}(f) B_{j,2,\varOmega}^T$$

$$+ \sum_{j \in K_{2,\varOmega}} \lambda_{j,2}(f) \left(\sum_{k \in K_{3,\varOmega}} \lambda_{k,3}(B_{j,2}) B_{k,3} \right),$$

and by means of (1.168) and (1.169) we obtain

$$f^{(3)} = \sum_{j \in K_{3,\varOmega}} \lambda_{j,3}(f) B_{j,3} - \sum_{k \in K_{3,\varOmega}} \lambda_{k,3}(f^{(1)} + f^{(2)}) B_{k,3}$$

$$= \sum_{j \in K_{3,\varOmega}} \lambda_{j,3}(f) B_{j,3} - \sum_{k \in K_{3,\varOmega}} \left(\sum_{j \in K_{2,\varOmega}} \lambda_{j,2}(f) \lambda_{k,3}(B_{j,2}) \right) B_{k,3},$$

resulting in

$$f^{(1)} + f^{(2)} + f^{(3)} = \sum_{j \in J_{1,\varOmega}} \lambda_{j,1}(f) B_{j,1,\varOmega}^T + \sum_{j \in J_{2,\varOmega}} \lambda_{j,2}(f) B_{j,2,\varOmega}^T$$

$$+ \sum_{j \in K_{3,\varOmega}} \lambda_{j,3}(f) B_{j,3}.$$

By iterating over all levels in the hierarchy and repeating the same arguments, we get the relation (1.165). □

The telescopic representation in Theorem 23 directly leads to the representation of the hierarchical quasi-interpolant in terms of the HB-spline basis, instead of in terms of the THB-spline basis (see Definition 10), under assumption (1.164). Indeed, as observed in [51], thanks to property (1.167), one can simply replace the index sets $K_{\ell,\Omega}$ by $J_{\ell,\Omega}$ in (1.166) and the relation (1.165) still remains true. This implies that (1.165) can be rewritten as

$$\mathscr{Q}_\Omega f = \sum_{\ell=1}^{L} \sum_{j \in J_{\ell,\Omega}} \lambda_{j,\ell}(f - \mathscr{Q}_{\ell-1,\Omega} f) B_{j,\ell}, \tag{1.170}$$

where

$$\mathscr{Q}_{0,\Omega} f := 0, \quad \mathscr{Q}_{r,\Omega} f := \sum_{k=1}^{r} \sum_{j \in J_{k,\Omega}} \lambda_{j,k}(f - \mathscr{Q}_{k-1,\Omega} f) B_{j,k}, \ r \geq 1. \tag{1.171}$$

1.7 Tensor-Product Structures and Adaptive Extensions

The most easy way to extend many of the previous results to the multivariate setting is to consider a tensor-product structure. For the sake of simplicity, we briefly focus here on the bivariate setting. The extension to higher dimensions is straightforward; it only requires a more involved indexing notation.

1.7.1 Tensor-Product B-Splines

Given two knot sequences

$$\boldsymbol{\xi}_k := \{\xi_{1,k} \leq \xi_{2,k} \leq \cdots \leq \xi_{n_k+p_k+1,k}\}, \quad k = 1, 2,$$

we define the **basic rectangle** as

$$R := [\xi_{p_1+1,1}, \xi_{n_1+1,1}] \times [\xi_{p_2+1,2}, \xi_{n_2+1,2}].$$

The **tensor-product B-splines** can be simply constructed as the product of univariate B-splines in each variable, i.e.,

$$B_{j_1,j_2,p_1,p_2,\boldsymbol{\xi}_1,\boldsymbol{\xi}_2}(x_1, x_2) := B_{j_1,p_1,\boldsymbol{\xi}_1}(x_1) B_{j_2,p_2,\boldsymbol{\xi}_2}(x_2), \tag{1.172}$$

for $j_k = 1, \ldots, n_k$ and $k = 1, 2$.

Example 19 Figure 1.6 shows a schematic representation of a tensor-product B-spline basis of bidegree $(p_1, p_2) = (3, 3)$. A $(p_k + 1)$-open knot sequence is chosen in each direction x_k, where the interior knots are all simple, and the corresponding univariate B-splines are depicted. Then, the set of tensor-product B-splines is obtained by computing the tensor product of the sets of univariate B-splines in each direction. Contour plots of some bicubic tensor-product B-splines are depicted in Fig. 1.7.

It is clear that tensor-product B-splines inherit all the nice features of univariate B-splines discussed in Sects. 1.2 and 1.3. In particular, they enjoy the following properties.

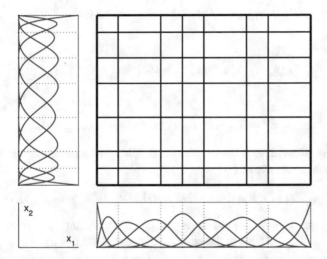

Fig. 1.6 Schematic representation of the (bivariate) tensor-product B-spline basis of bidegree $(p_1, p_2) = (3, 3)$ using a 4-open knot sequence in each direction. The knot lines are visualized by solid lines in the rectangular domain (this is the basic rectangle), and the sets of univariate B-splines are depicted for both directions

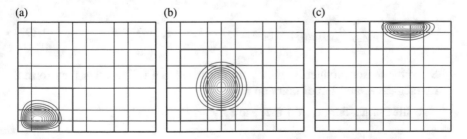

Fig. 1.7 Contour plots of some bicubic tensor-product B-splines $B_{j_1, j_2, 3, 3, \xi_1, \xi_2}$ defined on the tensor-product mesh given in Fig. 1.6. The bounding box of the support of each B-spline is visualized by solid blue lines. (**a**) $(j_1, j_2) = (3, 3)$. (**b**) $(j_1, j_2) = (5, 5)$. (**c**) $(j_1, j_2) = (7, 9)$

- **Local Support.** A tensor-product B-spline is locally supported on the rectangle given by the extreme knots used in the definition of its univariate B-splines in each direction. More precisely,

$$B_{j_1,j_2,p_1,p_2,\boldsymbol{\xi}_1,\boldsymbol{\xi}_2}(x_1,x_2) = 0, \quad (x_1,x_2) \notin S, \tag{1.173}$$

where

$$S := [\xi_{j_1,1}, \xi_{j_1+p_1+1,1}) \times [\xi_{j_2,2}, \xi_{j_2+p_2+1,2}).$$

- **Nonnegativity.** A tensor-product B-spline is nonnegative everywhere, and positive inside its support, i.e.,

$$B_{j_1,j_2,p_1,p_2,\boldsymbol{\xi}_1,\boldsymbol{\xi}_2}(x_1,x_2) \geq 0, \quad x_1, x_2 \in \mathbb{R}, \tag{1.174}$$

and

$$B_{j_1,j_2,p_1,p_2,\boldsymbol{\xi}_1,\boldsymbol{\xi}_2}(x_1,x_2) > 0, \quad (x_1,x_2) \in \mathring{S}, \tag{1.175}$$

where

$$\mathring{S} := (\xi_{j_1,1}, \xi_{j_1+p_1+1,1}) \times (\xi_{j_2,2}, \xi_{j_2+p_2+1,2}).$$

- **Piecewise Structure.** A tensor-product B-spline has a piecewise tensor-product polynomial structure, i.e.,

$$B_{j_1,j_2,p_1,p_2,\boldsymbol{\xi}_1,\boldsymbol{\xi}_2} \in \mathbb{P}_{p_1}([\xi_{m_1,1}, \xi_{m_1+1,1})) \otimes \mathbb{P}_{p_2}([\xi_{m_2,2}, \xi_{m_2+1,2})). \tag{1.176}$$

- **Smoothness.** If ξ is a knot of $B_{j_k,p_k,\boldsymbol{\xi}_k}$ of multiplicity $\mu \leq p_k + 1$ then $B_{j_1,j_2,p_1,p_2,\boldsymbol{\xi}_1,\boldsymbol{\xi}_2}$ belongs to the class $C^{p_k-\mu}$ across the line $x_k = \xi$ for $k = 1, 2$.

- **Linear Independence.** If each $\boldsymbol{\xi}_k$ is $(p_k + 1)$-basic for $k = 1, 2$, then the tensor-product B-splines $\{B_{j_1,j_2,p_1,p_2,\boldsymbol{\xi}_1,\boldsymbol{\xi}_2} : j_k = 1, \ldots, n_k, \ k = 1, 2\}$ are (locally) linearly independent on R.

- **Partition of Unity.** We have

$$\sum_{j_1=1}^{n_1} \sum_{j_2=1}^{n_2} B_{j_1,j_2,p_1,p_2,\boldsymbol{\xi}_1,\boldsymbol{\xi}_2}(x_1,x_2) = 1, \quad (x_1,x_2) \in R. \tag{1.177}$$

Since the tensor-product B-splines are nonnegative it follows that they form a **nonnegative partition of unity** on R.

- **Greville Points.** For $(x_1,x_2) \in R$ and $\ell_1, \ell_2 \in \{0, 1\}$, we have

$$x_1^{\ell_1} x_2^{\ell_2} = \sum_{j_1=1}^{n_1} \sum_{j_2=1}^{n_2} (\xi_{j_1,p_1,\boldsymbol{\xi}_1}^*)^{\ell_1} (\xi_{j_2,p_2,\boldsymbol{\xi}_2}^*)^{\ell_2} B_{j_1,j_2,p_1,p_2,\boldsymbol{\xi}_1,\boldsymbol{\xi}_2}(x_1,x_2), \tag{1.178}$$

where $\xi_{j_k,p_k,\boldsymbol{\xi}_k}^*$ is the Greville point defined in (1.51) for the knot sequence $\boldsymbol{\xi}_k$, $k = 1, 2$.

A tensor-product spline function is defined as

$$s(x_1, x_2) = \sum_{j_1=1}^{n_1} \sum_{j_2=1}^{n_2} c_{j_1,j_2} B_{j_1,j_2,p_1,p_2,\xi_1,\xi_2}(x_1, x_2), \quad c_{j_1,j_2} \in \mathbb{R}. \qquad (1.179)$$

Since the tensor-product B-splines are linearly independent, the space of spline functions has dimension $n_1 n_2$.

A main advantage of the representation in (1.179) is that its evaluation can be reduced to a sequence of evaluations of univariate spline functions:

$$s(x_1, x_2) = \sum_{j_1=1}^{n_1} d_{j_1,x_2} B_{j_1,p_1,\xi_1}(x_1), \quad d_{j_1,x_2} := \sum_{j_2=1}^{n_2} c_{j_1,j_2} B_{j_2,p_2,\xi_2}(x_2), \qquad (1.180)$$

or, equivalently,

$$s(x_1, x_2) = \sum_{j_2=1}^{n_2} d_{j_2,x_1} B_{j_2,p_2,\xi_2}(x_2), \quad d_{j_2,x_1} := \sum_{j_1=1}^{n_1} c_{j_1,j_2} B_{j_1,p_1,\xi_1}(x_1). \qquad (1.181)$$

Note that (1.180) requires n_1 univariate spline evaluations of degree p_2 and one univariate spline evaluation of degree p_1. On the other hand, (1.181) requires n_2 univariate spline evaluations of degree p_1 and one univariate spline evaluation of degree p_2. Thus, it is better to choose one of the two forms according to the minimal computational cost.

Other algorithms in the univariate B-spline setting (like knot insertion) can be extended in a similar way to the tensor-product B-spline setting.

1.7.2   Local Refinement

Despite their simple and elegant formulation, tensor-product B-spline structures have a main drawback. Any refinement of a knot sequence in one direction has a global effect in the other direction, and this prevents doing local refinement as illustrated in Fig. 1.8.

The hierarchical spline model provides a natural strategy to guarantee the locality of the refinement. As explained in Sect. 1.6, hierarchical spline spaces are a mixture of spline spaces of different resolution, localized by the domain hierarchy. Even though the concept of hierarchical splines was detailed in the univariate setting, it can be straightforwardly extended towards the bivariate (and multivariate) setting.

When selecting a sequence of nested tensor-product spline spaces on a common basic rectangle R in place of (1.135) and considering the corresponding tensor-product B-spline bases in place of (1.137), the definitions of tensor-product HB-splines and THB-splines follow verbatim Definitions 7 and 9, respectively. The properties (and their proofs) described in Sect. 1.6 also hold in the tensor-product extension. We refer the reader to [22, 23] for more details on tensor-product

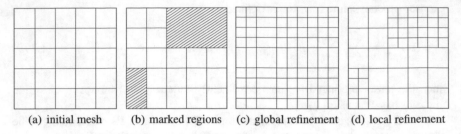

(a) initial mesh   (b) marked regions   (c) global refinement   (d) local refinement

Fig. 1.8 Given an initial tensor-product representation (**a**), an error estimator indicates regions of the mesh which require further refinement (**b**). The tensor-product structure necessarily implies a propagation of the refinement (**c**). Adaptive splines, instead, should provide a proper local control of the refinement procedure (**d**)

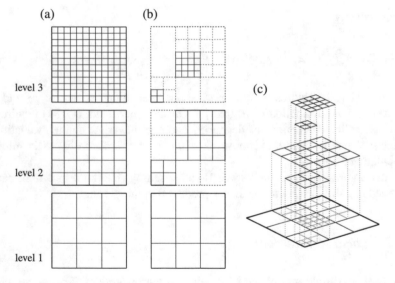

Fig. 1.9 An example of a two-dimensional domain hierarchy consisting of three levels. The knot lines are visualized by solid lines in the domain. (**a**) Global meshes. (**b**) Local meshes. (**c**) Domain hierarchy

THB-splines and their properties. A full treatment of the construction of related hierarchical quasi-interpolants and their approximation properties can be found in [49, 51].

Example 20 An example of a two-dimensional domain hierarchy together with its knot lines is illustrated in Fig. 1.9. We consider a nested sequence of three tensor-product spline spaces defined on a (uniform) knot mesh with open knots along the boundary (Fig. 1.9a). Assume the corresponding basic rectangle is denoted by R. Then, we select the subsets $R =: \Omega_1 \supseteq \Omega_2 \supseteq \Omega_3$ as a union of mesh elements at each level (Fig. 1.9b), and together they form the domain hierarchy Ω (Fig. 1.9c). On such domain hierarchy, we can define the corresponding HB-splines

(a)

(b)

(c)

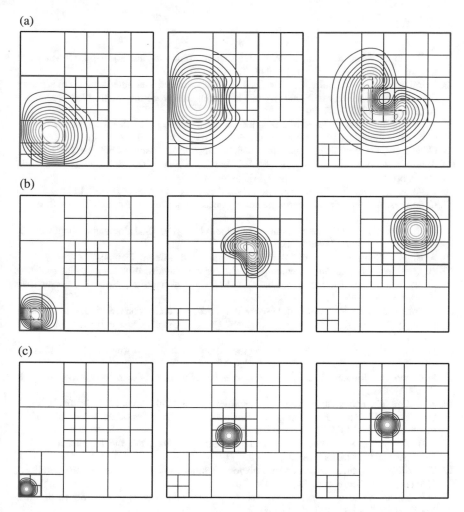

Fig. 1.10 Contour plots of some biquadratic tensor-product THB-splines of different levels defined on the domain hierarchy given in Fig. 1.9. The bounding box of the support of the untruncated version of each THB-spline is visualized by solid blue lines. (**a**) Level 1. (**b**) Level 2. (**c**) Level 3

and THB-splines according to Definitions 7 and 9, respectively. Contour plots of some biquadratic tensor-product THB-splines are depicted in Fig. 1.10. The shape of THB-splines related to coarser levels adapts nicely to the locally refined regions in Ω, as illustrated in Fig. 1.10a, b. THB-splines related to the finest level are nothing else than standard tensor-product B-splines, as illustrated in Fig. 1.10c.

Finally, we remark that there exist also other adaptive spline models based on local tensor-product structures, like (analysis-suitable) T-splines [2, 46] and LR-splines [9, 17].

References

1. L. Beirão da Veiga, A. Buffa, J. Rivas, G. Sangalli, Some estimates for h-p-k-refinement in isogeometric analysis. Numer. Math. **118**, 271–305 (2011)
2. L. Beirão da Veiga, A. Buffa, G. Sangalli, R. Vázquez, Analysis-suitable T-splines of arbitrary degree: definition, linear independence and approximation properties. Math. Models Methods Appl. Sci. **23**, 1979–2003 (2013)
3. W. Böhm, Inserting new knots into B-spline curves. Comput. Aided Des. **12**, 199–201 (1980)
4. C. de Boor, On calculating with B-splines. J. Approx. Theory **6**, 50–62 (1972)
5. C. de Boor, On local linear functionals which vanish at all B-splines but one, in *Theory of Approximation with Applications*, ed. by A.G. Law, N.B. Sahney (Academic Press, New York, 1976), pp. 120–145
6. C. de Boor, *A Practical Guide to Splines*, revised edn. (Springer, New York, 2001)
7. C. de Boor, G.J. Fix, Spline approximation by quasiinterpolants. J. Approx. Theory **8**, 19–45 (1973)
8. C. de Boor, A. Pinkus, The B-spline recurrence relations of Chakalov and of Popoviciu. J. Approx. Theory **124**, 115–123 (2003)
9. A. Bressan, Some properties of LR-splines. Comput. Aided Geom. Des. **30**, 778–794 (2013)
10. L. Chakalov, On a certain presentation of the Newton divided differences in interpolation theory and its applications. Godishnik na Sofijskiya Universitet, Fiziko-Matematicheski Fakultet **34**, 353–394 (1938)
11. E. Cohen, T. Lyche, R. Riesenfeld, Discrete B-splines and subdivision techniques in computer-aided geometric design and computer graphics. Comput. Graphics Image Process. **14**, 87–111 (1980)
12. E. Cohen, T. Lyche, L.L. Schumaker, Degree raising for splines. J. Approx. Theory **46**, 170–181 (1986)
13. E. Cohen, R.F. Riesenfeld, G. Elber, *Geometric Modeling with Splines: An Introduction* (A K Peters, Wellesley, 2001)
14. M.G. Cox, The numerical evaluation of B-splines. J. Inst. Math. Appl. **10**, 134–149 (1972)
15. H.B. Curry, I.J. Schoenberg, On spline distributions and their limits: the Pólya distribution functions. Bull. AMS **53**, 1114, Abstract 380t (1947)
16. H.B. Curry, I.J. Schoenberg, On Pólya frequency functions IV: the fundamental spline functions and their limits. J. Anal. Math. **17**, 71–107 (1966)
17. T. Dokken, T. Lyche, K.F. Pettersen, Polynomial splines over locally refined box-partitions. Comput. Aided Geom. Des. **30**, 331–356 (2013)
18. H.C. Elman, X. Zhang, Algebraic analysis of the hierarchical basis preconditioner. SIAM J. Matrix Anal. Appl. **16**, 192–206 (1995)
19. D.R. Forsey, R.H. Bartels, Hierarchical B-spline refinement. ACM SIGGRAPH Comput. Graph. **22**, 205–212 (1988)
20. D.R. Forsey, R.H. Bartels, Surface fitting with hierarchical splines. ACM Trans. Graph. **14**, 134–161 (1995)
21. C. Garoni, C. Manni, F. Pelosi, S. Serra-Capizzano, H. Speleers, On the spectrum of stiffness matrices arising from isogeometric analysis. Numer. Math. **127**, 751–799 (2014)
22. C. Giannelli, B. Jüttler, H. Speleers, THB-splines: the truncated basis for hierarchical splines. Comput. Aided Geom. Des. **29**, 485–498 (2012)
23. C. Giannelli, B. Jüttler, H. Speleers, Strongly stable bases for adaptively refined multilevel spline spaces. Adv. Comput. Math. **40**, 459–490 (2014)
24. T.N.E. Greville, On the normalisation of the B-splines and the location of the nodes for the case of unequally spaced knots, in *Inequalities*, ed. by O. Shisha (Academic Press, New York, 1967), pp. 286–290
25. E. Grinspun, P. Krysl, P. Schröder, CHARMS: a simple framework for adaptive simulation. ACM Trans. Graph. **21**, 281–290 (2002)

26. K. Höllig, J. Hörner, *Approximation and Modeling with B-Splines* (Society for Industrial and Applied Mathematics, Philadelphia, 2013)
27. J. Hoschek, D. Lasser, *Fundamentals of Computer Aided Geometric Design* (A K Peters, Wellesley, 1993)
28. R. Kraft, Adaptive and linearly independent multilevel B-splines, in *Surface Fitting and Multiresolution Methods*, ed. by A. Le Méhauté, C. Rabut, L.L. Schumaker (Vanderbilt University Press, Nashville, 1997), pp. 209–218
29. R. Kraft, Adaptive und linear unabhängige Multilevel B-Splines und ihre Anwendungen, Ph.D. thesis, University of Stuttgart, 1998
30. P. Krysl, E. Grinspun, P. Schröder, Natural hierarchical refinement for finite element methods. Int. J. Numer. Methods Eng. **56**, 1109–1124 (2003)
31. M.J. Lai, L.L. Schumaker, *Spline Functions on Triangulations* (Cambridge University Press, Cambridge, 2007)
32. J.M. Lane, R.F. Riesenfeld, A theoretical development for the computer generation and display of piecewise polynomial surfaces. IEEE Trans. Pattern Anal. Mach. Intell. **2**, 35–46 (1980)
33. P.S. de Laplace, *Théorie Analytique des Probabilités*, 3rd edn. (Courcier, Paris, 1820)
34. T. Lyche, A note on the condition numbers of the B-spline bases. J. Approx. Theory **22**, 202–205 (1978)
35. C. Manni, H. Speleers, Standard and non-standard CAGD tools for isogeometric analysis: a tutorial, in *IsoGeometric Analysis: A New Paradigm in the Numerical Approximation of PDEs*, ed. by A. Buffa, G. Sangalli. Lecture Notes in Mathematics, vol. 2161 (Springer, Cham, 2016), pp. 1–69
36. M. Marsden, An identity for spline functions and its application to variation diminishing spline approximation. J. Approx. Theory **3**, 7–49 (1970)
37. T. Popoviciu, Sur quelques propriétés des fonctions d'une ou de deux variables réelles, Ph.D. thesis, Presented to the Faculté des Sciences de Paris. Published by Institutul de Arte Grafice Ardealul, Cluj, 1933
38. H. Prautzsch, W. Böhm, M. Paluszny, *Bézier and B-Spline Techniques* (Springer, Berlin, 2002)
39. W. Rudin, *Real and Complex Analysis*, 3rd edn. (McGraw-Hill, Singapore, 1987)
40. K. Scherer, A.Y. Shadrin, New upper bound for the B-spline basis condition number: II. A proof of de Boor's 2^k-conjecture. J. Approx. Theory **99**, 217–229 (1999)
41. I.J. Schoenberg, Contributions to the problem of approximation of equidistant data by analytic functions. Part A.–On the problem of smoothing or graduation. A first class of analytic approximation formulae. Q. Appl. Math. **4**, 45–99 (1946)
42. I.J. Schoenberg, Contributions to the problem of approximation of equidistant data by analytic functions. Part B.–On the problem of osculatory interpolation. A second class of analytic approximation formulae. Q. Appl. Math. **4**, 112–141 (1946)
43. I.J. Schoenberg, On spline functions, in *Inequalities*, ed. by O. Shisha (Academic Press, New York, 1967), pp. 255–286
44. I.J. Schoenberg, Cardinal interpolation and spline functions. J. Approx. Theory **2**, 167–206 (1969)
45. L.L. Schumaker, *Spline Functions: Basic Theory*, 3rd edn. (Cambridge University Press, Cambridge, 2007)
46. T.W. Sederberg, J. Zheng, A. Bakenov, A. Nasri, T-splines and T-NURCCs. ACM Trans. Graph. **22**, 477–484 (2003)
47. A. Sharma, A. Meir, Degree of approximation of spline interpolation. J. Math. Mech. **15**, 759–768 (1966)
48. H. Speleers, Inner products of box splines and their derivatives. BIT Numer. Math. **55**, 559–567 (2015)
49. H. Speleers, Hierarchical spline spaces: quasi-interpolants and local approximation estimates. Adv. Comput. Math. **43**, 235–255 (2017)
50. H. Speleers, P. Dierckx, S. Vandewalle, Quasi-hierarchical Powell–Sabin B-splines. Comput. Aided Geom. Des. **26**, 174–191 (2009)

51. H. Speleers, C. Manni, Effortless quasi-interpolation in hierarchical spaces. Numer. Math. **132**, 155–184 (2016)
52. S. Takacs, T. Takacs, Approximation error estimates and inverse inequalities for B-splines of maximum smoothness. Math. Models Methods Appl. Sci. **26**, 1411–1445 (2016)
53. A.V. Vuong, C. Giannelli, B. Jüttler, B. Simeon, A hierarchical approach to adaptive local refinement in isogeometric analysis. Comput. Methods Appl. Mech. Eng. **200**, 3554–3567 (2011)
54. H. Yserentant, On the multilevel splitting of finite element spaces. Numer. Math. **49**, 379–412 (1986)

Chapter 2
Adaptive Multiscale Methods for the Numerical Treatment of Systems of PDEs

Angela Kunoth

Abstract These notes are concerned with numerical analysis issues arising in the solution of certain systems involving stationary and instationary linear variational problems. Standard examples are second order elliptic boundary value problems, where particular emphasis is placed on the treatment of essential boundary conditions, and linear parabolic equations. These operator equations serve as a core ingredient for control problems where in addition to the state, the solution of the PDE, a control is to be determined which together with the state minimizes a certain tracking-type objective functional. Having assured that the variational problems are well-posed, we discuss numerical schemes based on B-splines and B-spline-type wavelets as a particular multiresolution discretization methodology. The guiding principle is to devise fast and efficient solution schemes which are optimal in the number of arithmetic unknowns. We discuss optimal conditioning of the system matrices, numerical stability of discrete formulations, and adaptive approximations.

2.1 Introduction

Multilevel ingredients have for a variety of partial differential equations (PDEs) proved to achieve more efficient solution schemes than methods based on approximating the solution with respect to a fixed fine grid. The latter simple approach leads to the problem to solve a large ill-conditioned system of linear equations. The success of multilevel methods is due to the fact that solutions often exhibit a multiscale behaviour which one naturally wants to exploit. Among the first such schemes were multigrid methods. The basic idea of multigrid schemes is to successively solve smaller versions of the linear system which can be interpreted as discretizations with respect to coarser grids. Here 'efficiency of the scheme' means that one can solve the problem with respect to the finest grid with an amount

A. Kunoth (✉)
Mathematical Institute, University of Cologne, Cologne, Germany
e-mail: kunoth@math.uni-koeln.de

© Springer Nature Switzerland AG 2018
T. Lyche et al. (eds.), *Splines and PDEs: From Approximation Theory to Numerical Linear Algebra*, Lecture Notes in Mathematics 2219,
https://doi.org/10.1007/978-3-319-94911-6_2

of arithmetic operations which is proportional to the number of unknowns on this finest grid. Multigrid schemes provide an asymptotically optimal *preconditioner* for the original system on the finest grid. The search for such optimal preconditioners was one of the major topics in the solution of elliptic boundary value problems for many years. Another multiscale preconditioner which has this property is the BPX-preconditioner proposed first in [8]. It was proved to be asymptotically optimal with techniques from Approximation Theory in [27, 61]. In the context of isogeometric analysis, the BPX-preconditioner was further substantially optimized in [10]; this will be detailed in Sect. 2.2.

Wavelets as a particular example of a multiscale basis were constructed with compact support in the 1980s [36]. While mainly used for signal analysis and image compression, they were discovered to also provide optimal preconditioners in the above sense for elliptic boundary value problems [27, 47]. It was soon realized that biorthogonal spline-wavelets are better suited for the numerical solution of elliptic PDEs since they allow to work with piecewise polynomials instead of the only implicitly defined original wavelets [36], in addition to the fact that orthogonality of the Daubechies wavelets with respect to L_2 cannot really be exploited for elliptic PDEs. The principal ingredient that allows to prove optimality of the preconditioner are *norm equivalences* between Sobolev norms and sequence norms of weighted wavelet expansion coefficients. Optimal conditioning of the resulting linear system of equations can be achieved by applying the Fast Wavelet Transform together with a weighting in terms of an appropriate diagonal matrix. The terminology 'wavelets' here and in the sequel is to mean that these are classes of multiscale bases with three main properties: (R) Riesz basis property for the underlying function spaces, (L) locality of the basis functions, (CP) cancellation properties, all of which are detailed in Sect. 2.4.1.

After these initial results, research on using wavelets for numerically solving elliptic PDEs has gone into different directions. The original constructions in [18, 36] and many others are based on using the Fourier transform. Thus, these constructions provide bases for function spaces only on all of $\mathbb{R}$ or $\mathbb{R}^n$. In order for these tools to be applicable for the solution of PDEs which naturally live on a bounded domain $\Omega \subset \mathbb{R}^n$, there arose the need for having available constructions on bounded intervals without, of course, loosing the above mentioned properties (R), (L) and (CP). The first such systematic construction of biorthogonal spline-wavelets on $[0, 1]$ (and, by tensor products, on $[0, 1]^n$) was provided in [34].

Aside from the investigations to provide appropriate bases, the built-in potential of *adaptivity* for wavelets has played a prominent role when solving PDEs, on account of the fact that wavelets provide a locally supported Riesz basis for a whole range of function spaces. The key issue is to approximate the solution of the variational problem on an infinite-dimensional function space by the least amount of degrees of freedom up to a certain prescribed accuracy. Many approaches use wavelet coefficients in a heuristic way, i.e., judging approximation quality by the size of the wavelet coefficients together with thresholding. In contrast, *convergence* of wavelet-based adaptive methods for stationary variational problems was investigated systematically in [19–21]. These schemes are particularly designed to provide *optimal complexity* of the schemes: they provide the solution in a total amount of arithmetic operations which is comparable to the wavelet-best N-term approximation of the solution. This means that, given a prescribed tolerance, to find a sparse representation of the solution by extracting the N largest expansion coefficients of the solution during the solution process.

As soon as one aims at numerically solving a variational problem which can no longer be formulated in terms of a single elliptic operator equation such as a saddle point problem, one is faced with the problem of numerical stability. This means that finite approximations of the continuous well-posed problem may be ill-posed, obstructing its efficient numerical solution. This issue will also be addressed below.

In these notes, I also would like to discuss the potential proposed by wavelet methods for the following classes of problems. First, we will be concerned with second order elliptic PDEs with a particular emphasis placed on treating essential boundary conditions. Another interesting class that will be covered are linear parabolic PDEs which are formulated in full weak space-time from [66]. Then *PDE-constrained control problems* guided by elliptic boundary value problems are considered, leading to a *system* of elliptic PDEs. The starting point for designing efficient solution schemes are wavelet representations of continuous well-posed problems in their variational form. Viewing the numerical solution of such a discretized, yet still infinite-dimensional operator equation as an approximation helps to discover multilevel preconditioners for elliptic PDEs which yield *uniformly bounded condition numbers*. *Stability issues* like the LBB condition for saddle point problems are also discussed in this context. In addition, the compact support of the wavelets allows for sparse representations of the implicit information contained in systems of PDEs, the *adaptive approximation* of their solution.

More information and extensive literature on applying wavelets for more general PDEs addressing, among other things, the connection between adaptivity and nonlinear approximation and the evaluation of nonlinearities may be found in [16, 24, 25].

These notes are structured as follows. In Sect. 2.2, we begin with a simple elliptic PDE in variational form in the context of isogeometric analysis. For this problem, we address additive, BPX-type preconditioners and provide the main ingredients for showing optimality of the scheme with respect to the grid spacing. In Sect. 2.3, several well-posed variational problem classes are compiled to which later several aspects of the wavelet methodology are applied. The simplest example is a linear elliptic boundary value problem for which we derive two forms of an operator equation, the simplest one consisting just of one equation for homogeneous boundary conditions and a more complicated one in form of a saddle point problem where nonhomogeneous boundary conditions are treated by means of Lagrange multipliers. In Sect. 2.3.4, we consider a full weak space-time form of a linear parabolic PDE. These three formulations are then employed for the following classes of PDE-constrained control problems. In the *distributed control problems* in Sect. 2.3.5 the control is exerted through the right hand side of the PDE, while in *Dirichlet boundary control problems* in Sect. 2.3.6 the Dirichlet boundary condition

serves this purpose. The most potential for adaptive methods to be discussed below
are control problems constrained by parabolic PDEs as formulated in Sect. 2.3.7.

Section 2.4 is devoted to assembling necessary ingredients and basic properties
of wavelets which are required in the sequel. In particular, Sect. 2.4.4 collects the
essential construction principles for wavelets on bounded domains which do not
rely on Fourier techniques, namely, multiresolution analyses of function spaces and
the concept of stable completions. In Sect. 2.5 we formulate the problem classes
introduced in Sect. 2.3 in wavelet coordinates and derive in particular for the control
problems the resulting systems of linear equations arising from the optimality
conditions. Section 2.6 is devoted to the iterative solution of these systems. We
investigate fully iterative schemes on uniform grids and show that the resulting
systems can be solved in the wavelet framework together with a nested iteration
strategy with an amount of arithmetic operations which is proportional to the
total number of unknowns on the finest grid. Finally, in Sect. 2.6.2 a wavelet-
based adaptive scheme for the distributed control problem constrained by elliptic or
parabolic PDEs as in [29, 44] will be derived together with convergence results and
complexity estimates, relying on techniques from Nonlinear Approximation Theory.

Throughout these notes we will employ the following notational convention: the
relation $a \sim b$ will always stand for $a \lesssim b$ and $b \lesssim a$ where the latter inequality
means that b can be bounded by some constant times a uniformly in all parameters
on which a and b may depend. Norms and inner products are always indexed by
the corresponding function space. $L_p(\Omega)$ are for $1 \le p \le \infty$ the usual Lebesgue
spaces on a domain Ω, and $W_p^k(\Omega) \subset L_p(\Omega)$ denote for $k \in \mathbb{N}$ the Sobolev spaces
of functions whose weak derivatives up to order k are bounded in $L_p(\Omega)$. For $p = 2$,
we write as usual $H^k(\Omega) = W_2^k(\Omega)$.

2.2 BPX Preconditioning for Isogeometric Analysis

For a start, we consider linear elliptic PDEs in the framework of isogeometric
analysis, combining modern techniques from computer aided design with higher
order approximations of the solution. In this context, one exploits that the solution
exhibits a certain *smoothness*. We treat the physical domain by means of a regular
B-spline mapping from the parametric domain $\hat{\Omega} = (0, 1)^n$, $n \ge 2$, to the physical
domain Ω. The numerical solution of the PDE is computed by means of tensor
product B-splines mapped onto the physical domain. We will construct additive
BPX-type multilevel preconditioners and show that they are asymptotically optimal.
This means that the spectral condition number of the resulting preconditioned
stiffness matrix is independent of the grid spacing h. Together with a nested iteration
scheme, this enables an iterative solution scheme of optimal linear complexity. The
theoretical results are substantiated by numerical examples in two and three space
dimensions. The results of this section are essentially contained in [10].

We consider linear elliptic partial differential operators of order $2r = 2, 4$ on the domain Ω in variational form: for given $f \in H^{-r}(\Omega)$, find $u \in H_0^r(\Omega)$ such that

$$a(u, v) = \langle f, v \rangle \quad \text{for all } v \in H_0^r(\Omega) \tag{2.1}$$

holds. Here the energy space is $H_0^r(\Omega)$, a subset of the Sobolev space $H^r(\Omega)$, the space of square integrable functions with square integrable derivatives up to order r, containing homogeneous Dirichlet boundary conditions for $r = 1$ and homogeneous Dirichlet and Neumann derivatives for $r = 2$. The bilinear form $a(\cdot, \cdot)$ is derived from the linear elliptic PDE operator in a standard fashion, see, e.g., [7]. For example, the Laplacian is represented as $a(v, w) = \int_\Omega \nabla v \cdot \nabla w \, dx$. In order for the problem to be well-posed, we require the bilinear form $a(\cdot, \cdot)$: $H_0^r(\Omega) \times H_0^r(\Omega) \to \mathbb{R}$ to be symmetric, continuous and coercive on $H_0^r(\Omega)$. With $\langle \cdot, \cdot \rangle$, we denote on the right hand side of (2.1) the dual form between $H^{-r}(\Omega)$ and $H_0^r(\Omega)$. Our model problem (2.1) covers the second order Laplacian with homogeneous boundary conditions

$$-\Delta u = f \quad \text{on } \Omega, \qquad u|_{\partial\Omega} = 0, \tag{2.2}$$

as well as fourth order problems with corresponding homogeneous Dirichlet boundary conditions,

$$\Delta^2 u = f \quad \text{on } \Omega, \qquad u|_{\partial\Omega} = \mathbf{n} \cdot \nabla u|_{\partial\Omega} = 0 \tag{2.3}$$

where $\partial\Omega$ denotes the boundary of Ω and $\mathbf{n}$ the outward normal derivative at $\partial\Omega$. These PDEs serve as prototypes for more involved PDEs like Maxwell's equation or PDEs for linear and nonlinear elasticity. The reason we formulate these model problems of order $2r$ involving the parameter r is that this exhibits more clearly the order of the operator and the scaling in the subsequently used characterization of Sobolev spaces $H^r(\Omega)$. Thus, for the remainder of this section, the parameter $2r$ denoting the order of the PDE operator is fixed.

The assumptions on the bilinear form $a(\cdot, \cdot)$ entail that there exist constants $0 < c_A \le C_A < \infty$ such that the induced self-adjoint operator $\langle Av, w \rangle := a(v, w)$ satisfies the isomorphism relation

$$c_A \|v\|_{H^r(\Omega)} \le \|Av\|_{H^{-r}(\Omega)} \le C_A \|v\|_{H^r(\Omega)}, \quad v \in H_0^r(\Omega). \tag{2.4}$$

If the precise format of the constants in (2.4) does not matter, we abbreviate this relation as $\|v\|_{H^r(\Omega)} \lesssim \|Av\|_{H^{-r}(\Omega)} \lesssim \|v\|_{H^r(\Omega)}$, or shortly as

$$\|Av\|_{H^{-r}(\Omega)} \sim \|v\|_{H^r(\Omega)}. \tag{2.5}$$

Under these conditions, Lax-Milgram's theorem guarantees that, for any given $f \in H^{-r}(\Omega)$, the operator equation derived from (2.1)

$$Au = f \quad \text{in } H^{-r}(\Omega) \tag{2.6}$$

has a unique solution $u \in H_0^r(\Omega)$, see, e.g., [7].

In order to approximate the solution of (2.1) or (2.6), we choose a finite-dimensional subspace of $H_0^r(\Omega)$. We will construct these approximation spaces by using tensor products of B-splines as specified next.

2.2.1 B-Spline Discretizations

Our construction of optimal multilevel preconditioners will rely on tensor products so that principally any space dimension $n \in \mathbb{N}$ is permissible as long as storage permits; the examples cover the cases $n = 2, 3$. As discretization space, we choose in each spatial direction B-splines of the same degree p on uniform grids and with maximal smoothness. We begin with the univariate case and define B-splines on the interval $[0, 1]$ recursively with respect to their degree p. Given this positive integer p and some $m \in \mathbb{N}$, we call $\Xi := \{\xi_1, \ldots, \xi_{m+p+1}\}$ a p-open knot vector if the knots are chosen such that

$$0 = \xi_1 = \ldots = \xi_{p+1} < \xi_{p+2} < \ldots < \xi_m < \xi_{m+1} = \ldots = \xi_{m+p+1} = 1, \tag{2.7}$$

i.e., the boundary knots 0 and 1 have multiplicity $p + 1$ and the interior knots are single. For Ξ, B-spline functions of degree p are defined following the well-known Cox-de Boor recursive formula, see [38]. Starting point are the piecewise constants for $p = 0$ (or characteristic functions)

$$N_{i,0}(\zeta) = \begin{cases} 1, & \text{if } 0 \le \xi_i \le \zeta < \xi_{i+1} < 1, \\ 0, & \text{otherwise,} \end{cases} \tag{2.8}$$

with the modification that the last B-spline $N_{m,0}$ is defined also for $\zeta = 1$. For $p \ge 1$ the B-splines are defined as

$$N_{i,p}(\zeta) = \frac{\zeta - \xi_i}{\xi_{i+p} - \xi_i} N_{i,p-1}(\zeta) + \frac{\xi_{i+p+1} - \zeta}{\xi_{i+p+1} - \xi_{i+1}} N_{i+1,p-1}(\zeta), \quad \zeta \in [0, 1], \tag{2.9}$$

with the same modification for $N_{m,p}$. Alternatively, one can define the B-splines explicitly by applying divided differences to truncated powers [38]. This gives a set of m B-splines that form a basis for the space of *splines*, that is, piecewise polynomials of degree p with $p - 1$ continuous derivatives at the internal knots ξ_ℓ for $\ell = p + 2, \ldots, m$. (Of course, one can also define B-splines on a knot sequence with multiple internal knots which entails that the spline space is not of maximal smoothness.) For $p = 1$, the B-splines are at least $C^0([0, 1])$ which suffices for the discretization of elliptic PDEs of order 2, and for $p = 2$ they are $C^1([0, 1])$ which suffices for $r = 2$. By construction, the B-spline $N_{i,p}$ is supported in the interval $[\xi_i, \xi_{i+p+1}]$.

These definitions are valid for an arbitrary spacing of knots in Ξ (2.7). Recall from standard error estimates in the context of finite elements, see, e.g., [7], that *smooth* solutions of elliptic PDEs can be approximated best with discretizations on a uniform grid. Therefore, in this section, we assume from now on that the grid is *uniform*, i.e., $\xi_{\ell+1} - \xi_\ell = h$ for all $\ell = p + 1, \ldots, m$.

For n space dimensions, we employ tensor products of the one-dimensional B-splines. We take in each space dimension a p-open knot vector Ξ and define on the closure of the parametric domain $\hat{\Omega} = (0, 1)^n$ (which we also denote by $\hat{\Omega}$ for simplicity of presentation) the spline space

$$S_h(\hat{\Omega}) := \mathrm{span}\left\{ B_i(\mathbf{x}) := \prod_{\ell=1}^n N_{i_\ell, p}(x_\ell),\ i = 1, \ldots, N := mn,\ \mathbf{x} \in \hat{\Omega} \right\}$$

$$=: \mathrm{span}\left\{ B_i(\mathbf{x}), i \in \mathscr{I},\ \mathbf{x} \in \hat{\Omega} \right\}. \tag{2.10}$$

In the spirit of isogeometric analysis, we suppose that the computational domain Ω can also described in terms of B-splines. We assume that the computational domain Ω is the image of a mapping $\mathbf{F} : \hat{\Omega} \to \Omega$ with $\mathbf{F} := (F_1, \ldots, F_n)^T$ where each component F_i of $\mathbf{F}$ belongs to $S_{\bar{h}}(\hat{\Omega})$ for some given $\bar{h}$. In many applications, the geometry can be described in terms of a very coarse mesh, namely, $\bar{h} \gg h$. Moreover, we suppose that $\mathbf{F}$ is invertible and satisfies

$$\|D^\alpha \mathbf{F}\|_{L_\infty(\hat{\Omega})} \sim 1 \text{ for } |\alpha| \leq r. \tag{2.11}$$

This assumption on the geometry can be weakened in the sense that the mapping $\mathbf{F}$ can be a piecewise C^∞ function on the mesh with respect to $\bar{h}$, independent of h, or the domain Ω may have a multi-patch representation. This means that one can allow Ω also to be the union of domains Ω_k where each one parametrized by a spline mapping of the parametric domain $\hat{\Omega}$.

We now define the approximation space for (2.6) as

$$V_h^r := \{v_h \in H_0^r(\Omega) : v_h \circ \mathbf{F} \in S_h(\hat{\Omega})\}. \tag{2.12}$$

We will formulate three important properties of this approximation space which will play a crucial role later for the construction of the BPX-type preconditioners. The first one is that we suppose from now on that the B-spline basis is *normalized* with respect to L_2, i.e.,

$$\|B_i\|_{L_2(\hat{\Omega})} \sim 1, \text{ and, thus, also } \|B_i \circ \mathbf{F}^{-1}\|_{L_2(\Omega)} \sim 1 \text{ for all } i \in \mathscr{I}. \tag{2.13}$$

Then one can derive the following facts [10].

Theorem 1 *Let* $\{B_i\}_{i\in\mathscr{I}}$ *be the B-spline basis defined in (2.10) and normalized as in (2.13),* $N = \#\mathscr{I}$ *and* V_h^r *as in (2.12). Then we have*

(S) Uniform stability with respect to $L_2(\Omega)$
  For any $\mathbf{c} \in \ell_2(\mathscr{I})$,

$$\left\| \sum_{i=1}^{N} c_i\, B_i \circ \mathbf{F}^{-1} \right\|_{L_2(\Omega)}^2 \sim \sum_{i=1}^{N} |c_i|^2 =: \|\mathbf{c}\|_{\ell_2}^2, \qquad \mathbf{c} := (c_i)_{i=1,\dots,N};$$

$$(2.14)$$

(J) Direct or Jackson estimates

$$\inf_{v_h \in V_h^r} \|v - v_h\|_{L_2(\Omega)} \lesssim h^s\, |v|_{H^s(\Omega)} \ \textit{for any}\ v \in H^s(\Omega),\ 0 \le s \le r+1,$$

$$(2.15)$$

where $|\cdot|_{H^s(\Omega)}$ *denotes the Sobolev seminorm of highest weak derivatives s;*
(B) Inverse or Bernstein estimates

$$\|v_h\|_{H^s(\Omega)} \lesssim h^{-s}\|v_h\|_{L_2(\Omega)} \ \textit{for any}\ v_h \in V_h^r \ \textit{and}\ 0 \le s \le r. \qquad (2.16)$$

In all these estimates, the constants are independent of h but may depend on $\mathbf{F}$, *i.e.,* Ω, *on the polynomial degree p and on the spatial dimension n.*

In the next section, we construct BPX-type preconditioners for (2.6) in terms of approximations with (2.12) and show their optimality.

2.2.2 Additive Multilevel Preconditioners

The construction of optimal preconditioners are based on a *multiresolution analysis* of the underlying energy function space $H_0^r(\Omega)$. As before, $2r \in \{2, 4\}$ stands for the order of the PDEs we are solving and is always kept fixed.

We first describe the necessary ingredients within an abstract basis-free framework, see, e.g., [24]. Afterwards, we specify the realization for the parametrized tensor product spaces in (2.12).

Let $\mathscr{V}$ be a sequence of strictly nested spaces V_j, starting with some fixed coarsest index $j_0 > 0$, determined by the polynomial degree p which determines the support of the basis functions (which also depends on Ω), and terminating with a highest resolution level J,

$$V_{j_0} \subset V_{j_0+1} \subset \cdots \subset V_j \subset \cdots \subset V_J \subset H_0^r(\Omega). \qquad (2.17)$$

The index j denotes the level of resolution defining approximations on a grid with dyadic grid spacing $h = 2^{-j}$, i.e., we use from now on the notation V_j instead of V_h to indicate different grid spacings. Then, V_J will be the space relative to the finest grid 2^{-J}. We associate with $\mathcal{V}$ a sequence of linear projectors $\mathcal{P} := \{P_j\}_{j \geq j_0}$ with the following properties.

Properties 1 We assume that

(P1) P_j maps $H_0^r(\Omega)$ onto V_j,
(P2) $P_j P_\ell = P_j$ for $j \leq \ell$,
(P3) $\mathcal{P}$ is uniformly bounded on $L_2(\Omega)$, i.e., $\|P_j\|_{L_2(\Omega)} \lesssim 1$ for any $j \geq j_0$ with a constant independent of j.

These conditions are satisfied, for example, for $L_2(\Omega)$-orthogonal projectors, or, in the case of splines, for the quasi-interpolant proposed and analyzed in [65, Chapter 4]. The second condition (P2) ensures that the differences $P_j - P_{j-1}$ are also projectors for any $j > j_0$. We define next a sequence $\mathcal{W} := \{W_j\}_{j \geq j_0}$ of complement spaces

$$W_j := (P_{j+1} - P_j)V_{j+1} \tag{2.18}$$

which then yields the direct (but not necessarily orthogonal) decomposition

$$V_{j+1} = V_j \oplus W_j. \tag{2.19}$$

Thus, for the finest level J, we can express V_J in its *multilevel decomposition*

$$V_J = \bigoplus_{j=j_0-1}^{J-1} W_j \tag{2.20}$$

upon setting $W_{j_0-1} := V_{j_0}$. Setting also $P_{j_0-1} := 0$, the corresponding multilevel representation of any $v \in V_J$ is then

$$v = \sum_{j=j_0}^{J} (P_j - P_{j-1})v. \tag{2.21}$$

We now have the following result which will be used later for the proof of the optimality of the multilevel preconditioners.

Theorem 2 *Let $\mathcal{P}, \mathcal{V}$ be as above where, in addition, we require that for each V_j, $j_0 \leq j \leq J$, a Jackson and Bernstein estimate as in Theorem 1 (J) and (B) hold with $h = 2^{-j}$. Then one has the function space characterization*

$$\|v\|_{H^r(\Omega)} \sim \left(\sum_{j=j_0}^{J} 2^{2rj} \|(P_j - P_{j-1})v\|_{L_2(\Omega)}^2 \right)^{1/2} \qquad \text{for any } v \in V_J. \tag{2.22}$$

Such a result holds for much larger classes of function spaces, Sobolev or even Besov spaces which are subsets of $L_q(\Omega)$ for general q, possibly different from 2 and for any function $v \in H^r(\Omega)$, then with an infinite sum on the right hand side, see, e.g., [24]. The proof of Theorem 2 for such cases heavily relies on tools from approximation theory and can be found in [27, 61].

Next we demonstrate how to exploit the norm equivalence (2.22) in the construction of an optimal multilevel preconditioner. Define for any $v, w \in V_J$ the linear self-adjoint positive-definite operator $C_J : V_J \to V_J$ given by

$$(C_J^{-1} v, w)_{L_2(\Omega)} := \sum_{j=j_0}^{J} 2^{2rj} \left((P_j - P_{j-1})v, (P_j - P_{j-1})w \right)_{L_2(\Omega)}, \qquad (2.23)$$

which we call a multilevel *BPX-type preconditioner*. Let $A_J : V_J \to V_J$ be the finite-dimensional operator defined by $(A_J v, w)_{L_2(\Omega)} := a(v, w)$ for all $v, w \in V_J$, the approximation of A in (2.6) with respect to V_J.

Theorem 3 *With the same prerequisites as in Theorem 2, C_J is an asymptotically optimal symmetric preconditioner for A_J, i.e., $\kappa_2(C_J^{1/2} A_J C_J^{1/2}) \sim 1$ with constants independent of J.*

Proof For the parametric domain $\hat{\Omega}$, the result was proved independently in [27, 61] and is based on the combination of (2.22) together with the well-posedness of the continuous problem (2.6). The result on the physical domain follows then together with (2.11). □

Realizations of the preconditioner defined in (2.23) based on B-splines lead to representations of the complement spaces W_j whose bases are called *wavelets*. For these, efficient implementations of optimal linear complexity involving the Fast Wavelet Transform can be derived explicitly, see Sect. 2.4.

However, since the order of the PDE operator r is positive, one can use here the argumentation from [8] which will allow to work with the same basis functions as for the spaces V_j. The first part of the argument relies on the assumption that the P_j are L_2-*orthogonal* projectors. For a clear distinction, we shall use the notation O_j for L_2-orthogonal projectors and reserve the notation P_j for the linear projectors with Properties 1. Then, the BPX-type preconditioner (2.23) (using the same symbol C_J for simplicity) reads as

$$C_J^{-1} := \sum_{j=j_0}^{J} 2^{2jr} (O_j - O_{j-1}), \qquad (2.24)$$

which is by Theorem 3 a BPX-type preconditioner for the self-adjoint positive definite operator A_J. By the orthogonality of the projectors O_j, we can immediately derive from (2.24) that

$$C_J = \sum_{j=j_0}^{J} 2^{-2jr} (O_j - O_{j-1}). \qquad (2.25)$$

Since $r > 0$, by rearranging the sum, the exponentially decaying scaling factors allow one to replace C_J by the spectrally equivalent operator

$$C_J = \sum_{j=j_0}^{J} 2^{-2jr} O_j \qquad (2.26)$$

(for which we use the same notation C_J). Recall that two linear operators $\mathscr{A} : V_J \to V_J$ and $\mathscr{B} : V_J \to V_J$ are *spectrally equivalent* if they satisfy

$$(\mathscr{A}v, v)_{L_2(\Omega)} \sim (\mathscr{B}v, v)_{L_2(\Omega)}, \quad v \in V_J, \qquad (2.27)$$

with constants independent of J. Thus, the realization of the preconditioner is reduced to a computation in terms of the bases of the spaces V_j instead of W_j. The orthogonal projector O_j can, in turn, be replaced by a simpler local operator which is spectrally equivalent to O_j, see [50] and the derivation below.

Up to this point, the introduction to multilevel preconditioners has been basis-free. We now show how this framework can be used to construct a BPX-preconditioner for the linear system (2.6). Based on the definition (2.12), we construct a sequence of spaces satisfying (2.17) such that $V_J = V_h^r$. In fact, we suppose that for each space dimension we have a sequence of p-open knot vectors $\Xi_{j_0,\ell}, \ldots, \Xi_{J,\ell}$, $\ell = 1, \ldots, n$, which provide a uniform partition of the interval $[0, 1]$ such that $\Xi_{j,\ell} \subset \Xi_{j+1,\ell}$ for $j = j_0, j_0 + 1, \ldots, J$. In particular, we assume that $\Xi_{j+1,\ell}$ is obtained from $\Xi_{j,\ell}$ by dyadic refinement, i.e., the grid spacing for $\Xi_{j,\ell}$ is proportional to 2^{-j} for each $\ell = 1, \ldots, n$. In view of the assumptions on the parametric mapping $\mathbf{F}$, we assume that $\bar{h} = 2^{-j_0}$, i.e., $\mathbf{F}$ can be represented in terms of B-splines on the coarsest level j_0. By construction, we have now achieved that

$$S_{j_0}(\hat{\Omega}) \subset S_{j_0+1}(\hat{\Omega}) \subset \ldots \subset S_J(\hat{\Omega}).$$

Setting $V_j^r := \{v \in H_0^r(\Omega) : v \circ \mathbf{F} \in S_j(\hat{\Omega})\}$, we arrive at a sequence of nested spaces

$$V_{j_0}^r \subset V_{j_0+1}^r \subset \ldots \subset V_J^r.$$

Setting $\mathscr{I}_j := \{1, \ldots, \dim S_j(\hat{\Omega})\}$, we denote by B_i^j, $i \in \mathscr{I}_j$, the set of L_2-normalized B-spline basis functions for the space $S_j(\hat{\Omega})$. Define now the positive definite operator $P_j : L_2(\Omega) \to V_j^r$ as

$$P_j := \sum_{i \in \mathscr{I}_j} (\cdot, B_i^j \circ \mathbf{F}^{-1})_{L_2(\Omega)} B_i^j \circ \mathbf{F}^{-1}. \qquad (2.28)$$

Corollary 1 *For the basis* $\{B_i^j \circ \mathbf{F}^{-1}, i \in \mathbb{I}_j\}$, *the operators* P_j *and the* L_2-*projectors* O_j *are spectrally equivalent for any* j.

Proof The assertion follows by combining (2.11), (2.14), with Remark 3.7.1 from [50], see [8] for the main ingredients. □

Finally, we obtain an explicit representation of the preconditioner C_J in terms of the mapped spline bases for V_j^r, $j = j_0, \ldots, J$,

$$C_J = \sum_{j=j_0}^{J} 2^{-2jr} \sum_{i \in \mathscr{I}_j} (\cdot, B_i^j \circ \mathbf{F}^{-1})_{L_2(\Omega)} B_i^j \circ \mathbf{F}^{-1} \tag{2.29}$$

(denoted again by C_J). Note that this preconditioner involves *all* B-splines from all levels j with an appropriate scaling, i.e., a properly scaled *generating system* for V_J^r.

Remark 1 The *hierarchical basis (HB) preconditioner* introduced for $n = 2$ in [71] for piecewise linear B-splines fits into this framework by choosing *Lagrangian interpolants* in place of the projectors P_j in (2.23). However, since these operators do not satisfy (P3) in Properties 1, they do not yield an asymptotically optimal preconditioner for $n \geq 2$. For $n = 3$, this preconditioner does not have an effect at all.

So far we have not explicitly addressed the dependence of the preconditioned system on p. Since all estimates in Theorem 1 which enter the proof of optimality depend on p, it is to be expected that the absolute values of the condition numbers, i.e., the values of the constants, depend on and increase with p. Indeed, in the next section, we show some numerical results which also aim at studying this dependence.

2.2.3 Realization of the BPX Preconditioner

Now we are in the position to describe the concrete implementation of the BPX preconditioner. Its main ingredient are linear *intergrid operators* which map vectors and matrices between different grids. Specifically, we need to define prolongation and restriction operators.

Since $V_j^r \subset V_{j+1}^r$, each B-spline B_i^j on level j can be represented by a linear combination of B-splines B_k^{j+1} on level $j + 1$. Arranging the B-splines in the set $\{B_i^j, i \in \mathscr{I}_j\}$ into a vector $\mathbf{B}^j$ in a fixed order, this relation denoted as *refinement relation* can be written as

$$\mathbf{B}^j = \mathbf{I}_j^{j+1} \mathbf{B}^{j+1} \tag{2.30}$$

with *prolongation operator* $\mathbf{I}_j^{j+1}$ from the trial space V_j^r to the trial space V_{j+1}^r.
The restriction $\mathbf{I}_{j+1}^j$ is then simply defined as the transposed operator, i.e., $\mathbf{I}_{j+1}^j = (\mathbf{I}_j^{j+1})^T$. In case of piecewise linear B-splines, this definition coincides with the well
known prolongation and restriction operators from finite element textbooks obtained
by interpolation, see, e.g., [7].

We will exemplify the construction in case of quadratic and cubic B-splines on
the interval, see, e.g., [38], as follows. We equidistantly subdivide the interval [0, 1]
into 2^j subintervals and obtain 2^j and $2^j + 1$, respectively, B-splines for $p = 2, 3$
and the corresponding quadratic and cubic spline space V_j^r which is given on this
partition, respectively, see Fig. 2.1 for an illustration. Note that the two boundary
functions which do not vanish at the boundary were removed in order to guarantee
that $V_j^r \subset H_0^r(\Omega)$. Moreover, recall that the B-splines are L_2 normalized according
to (2.13) which means that B_i^j is of the form $B_i^j(\zeta) = 2^{j/2} B(2^j \zeta - i)$ if B_i^j is an
interior function, and correspondingly for the boundary functions.
In case of quadratic B-splines ($p = 2$), the restriction operator $\mathbf{I}_{j+1}^j$ reads

$$
\mathbf{I}_{j+1}^j = 2^{-1/2}
\begin{bmatrix}
\frac{1}{2} & \frac{9}{8} & \frac{3}{8} & & & & & \\
& \frac{1}{4} & \frac{3}{4} & \frac{3}{4} & \frac{1}{4} & & & \\
& & \frac{1}{4} & \frac{3}{4} & \frac{3}{4} & \frac{1}{4} & & \\
& & & \ddots & & \ddots & & \\
& & & & \frac{1}{4} & \frac{3}{4} & \frac{3}{4} & \frac{1}{4} \\
& & & & & & \frac{3}{8} & \frac{9}{8} & \frac{1}{2}
\end{bmatrix}
\in \mathbb{R}^{2^j \times 2^{j+1}}.
$$

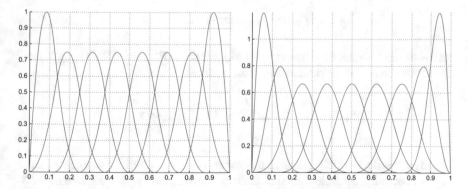

Fig. 2.1   Quadratic ($p = 2$) (left) and cubic ($p = 3$) (right) L_2-normalized B-splines (see (2.13))
on level $j = 3$ on the interval [0, 1], yielding basis functions for $V_j^r \subset H_0^r(\Omega)$

For cubic B-splines ($p = 3$), it has the form

$$
\mathbf{I}_{j+1}^{j} = 2^{-1/2}
\begin{bmatrix}
\frac{1}{2} & \frac{9}{8} & \frac{3}{8} & & & & & & \\
\frac{1}{4} & \frac{11}{12} & \frac{2}{3} & \frac{1}{6} & & & & & \\
 & \frac{1}{8} & \frac{1}{2} & \frac{3}{4} & \frac{1}{2} & \frac{1}{8} & & & \\
 & & & \ddots & \ddots & \ddots & & & \\
 & & & & \frac{1}{8} & \frac{1}{2} & \frac{3}{4} & \frac{1}{2} & \frac{1}{8} \\
 & & & & & & \frac{1}{6} & \frac{2}{3} & \frac{11}{12} & \frac{1}{4} \\
 & & & & & & & \frac{3}{8} & \frac{9}{8} & \frac{1}{2}
\end{bmatrix}
\in \mathbb{R}^{(2^{j}+1) \times (2^{j+1}+1)}.
$$

The normalization factor $2^{-1/2}$ stems from the L_2-normalization (2.13). The matrix entries are scaled in the usual fashion such that their rows sum to two. From these restriction operators for one dimensions, one obtains the related restriction operators on arbitrary unit cubes $[0, 1]^n$ via tensor products. Finally, we set $\mathbf{I}_j^J := \mathbf{I}_{J-1}^J \mathbf{I}_{J-2}^{J-1} \cdots \mathbf{I}_j^{j+1}$ and $\mathbf{I}_j^j := \mathbf{I}_j^{j+1} \mathbf{I}_{j+1}^{j+2} \cdots \mathbf{I}_{J-1}^J$ to define prolongations and restrictions between arbitrary levels j and J.

In order to derive the explicit form of the discretized BPX-preconditioner, for given functions $u_J, v_J \in V_J$ with expansion coefficients $u_{J,k}$ and $v_{J,\ell}$, respectively, we conclude from (2.29) that

$$
(C_J u_J, v_J)_{L_2(\Omega)} = \sum_{k,\ell \in \mathscr{I}_J} u_{J,k} v_{J,\ell} (C_J (B_k^J \circ \mathbf{F}^{-1}), B_\ell^J \circ \mathbf{F}^{-1})_{L_2(\Omega)}
$$

$$
= \sum_{k,\ell \in \mathscr{I}_J} u_{J,k} v_{J,\ell} \sum_{j=j_0}^{J} 2^{-2jr} \sum_{i \in \mathscr{I}_j} (B_k^J \circ \mathbf{F}^{-1}, B_i^j \circ \mathbf{F}^{-1})_{L_2(\Omega)}
$$

$$
\times (B_i^j \circ \mathbf{F}^{-1}, B_\ell^J \circ \mathbf{F}^{-1})_{L_2(\Omega)}.
$$

Next, one can introduce the mass matrix $\mathbf{M}_J = [(B_k^J \circ \mathbf{F}^{-1}, B_\ell^J \circ \mathbf{F}^{-1})_{L_2(\Omega)}]_{k,\ell}$ and obtains by the use of restrictions and prolongations

$$
(C_J u_J, v_J)_{L_2(\Omega)} = \sum_{j=j_0}^{J} 2^{-2jr} \mathbf{u}_J^T \mathbf{M}_J \mathbf{I}_j^J \mathbf{I}_J^j \mathbf{M}_J \mathbf{v}_J.
$$

The mass matrices which appear in this expression can be further suppressed since $\mathbf{M}_J$ is spectrally equivalent to the identity matrix. Finally, the *discretized BPX-preconditioner* to be implemented is of the simple form

$$
\mathbf{C}_J = \sum_{j=j_0}^{J} 2^{-2jr} \mathbf{I}_j^J \mathbf{I}_J^j, \tag{2.31}
$$

involving only restrictions and prolongations. A further simple improvement can be obtained by replacing the scaling factor 2^{-2jr} by $\mathrm{diag}(\mathbf{A}_j)^{-1}$, where $\mathrm{diag}(\mathbf{A}_j)$ denotes the diagonal matrix built from the diagonal entries of the stiffness matrix $\mathbf{A}_j$. This diagonal scaling has the same effect as the levelwise scaling by 2^{-2jr} but improves the condition numbers considerably, particularly if parametric mappings are involved. Thus, the discretized BPX-preconditioner takes on the form

$$\mathbf{C}_J = \sum_{j=j_0}^{J} \mathbf{I}_j^J \, \mathrm{diag}(\mathbf{A}_j)^{-1} \mathbf{I}_J^j \qquad (2.32)$$

which we will use in the subsequent computations presented in Tables 2.1 and 2.2. If the condition number $\kappa(\mathbf{A}_{j_0})$ is already high in absolute numbers on the coarsest level j_0, it is worth to use its exact inverse on the coarse grid, i.e., to apply instead of (2.32) the operator

$$\mathbf{C}_J = \mathbf{I}_{j_0}^J \mathbf{A}_{j_0}^{-1} \mathbf{I}_J^{j_0} + \sum_{j=j_0+1}^{J} \mathbf{I}_j^J \, \mathrm{diag}(\mathbf{A}_j)^{-1} \mathbf{I}_J^j,$$

see [11, 62]. Another substantial improvement of the BPX-preconditioner can be achieved by replacing the diagonal scaling on each level by, e.g., a SSOR preconditioning as follows. We decompose the system matrix as $\mathbf{A}_j = \mathbf{L}_j + \mathbf{D}_j + \mathbf{L}_j^T$ with the diagonal matrix $\mathbf{D}_j$, the lower triangular part $\mathbf{L}_j$, and the upper triangular part $\mathbf{L}_j^T$. Then we replace the diagonal scaling on each level of the BPX-preconditioner (2.32) by the SSOR preconditioner, i.e., instead of (2.32) we apply the preconditioner

$$\mathbf{C}_J = \sum_{j=j_0}^{J} \mathbf{I}_j^J (\mathbf{D}_j + \mathbf{L}_j)^{-T} \mathbf{D}_j (\mathbf{D}_j + \mathbf{L}_j)^{-1} \mathbf{I}_J^j. \qquad (2.33)$$

Table 2.1   Condition numbers of the BPX-preconditioned Laplacian on $\hat{\Omega} = (0,1)^n$ for $n = 1, 2, 3$

Level	Interval ($n = 1$)				Square ($n = 2$)				Cube ($n = 3$)			
	$p=1$	$p=2$	$p=3$	$p=4$	$p=1$	$p=2$	$p=3$	$p=4$	$p=1$	$p=2$	$p=3$	$p=4$
3	7.43	3.81	7.03	5.93	5.93	7.31	22.8	133	3.49	39.5	356	5957
4	8.87	4.40	9.47	7.81	5.00	9.03	40.2	225	4.85	50.8	624	9478
5	10.2	4.67	11.0	9.36	5.70	9.72	51.8	293	5.75	56.6	795	11,887
6	11.3	4.87	12.1	10.7	6.27	10.1	58.7	340	6.40	59.7	895	13,185
7	12.2	5.00	12.7	11.5	6.74	10.4	63.1	371	6.91	61.3	961	13,211
8	13.0	5.10	13.0	11.9	7.14	10.5	66.0	391	7.34	62.2	990	13,234
9	13.7	5.17	13.2	12.1	7.48	10.6	68.0	403	7.70	62.6	1016	13,255
10	14.2	5.22	13.4	12.2	7.77	10.6	69.3	411	7.99	62.9	1040	–

Table 2.2 Condition numbers of the BPX-preconditioned Laplacian on the analytic arc seen on the right hand side

Level	p=1	p=2	p=3	p=4
3	5.04	12.4	31.8	184
	(21.8)	(8.64)	(31.8)	(184)
4	11.1	16.3	54.7	291
	(90.2)	(34.3)	(32.9)	(173)
5	25.3	19.0	70.1	376
	(368)	(139)	(98.9)	(171)
6	31.9	21.4	79.2	436
	(1492)	(560)	(401)	(322)
7	37.4	23.1	84.4	471
	(6015)	(2255)	(1620)	(1297)
8	42.1	24.3	87.3	490
	(241,721)	(9062)	(6506)	(5217)
9	45.7	25.2	89.0	500
	(969,301)	(36,353)	(26,121)	(20,945)
10	48.8	25.9	90.1	505
	(388,690)	(145,774)	(104,745)	(83,975)

The bracketed numbers are the related condition numbers without preconditioning

Table 2.3 Condition numbers of the BPX-preconditioned Laplacian for cubic B-splines on different geometries in case of using a BPX-SSOR preconditioning on each level

Level	Square	Analytic arc	$\mathscr{C}^0$-map of the L-shape	Singular $\mathscr{C}^1$-map of the L-shape
3	3.61	3.65	3.67	3.80
4	6.58	6.97	7.01	7.05
5	8.47	10.2	10.2	14.8
6	9.73	13.1	13.2	32.2
7	10.5	14.9	15.2	77.7
8	11.0	15.9	16.3	180
9	11.2	16.5	17.0	411
10	11.4	16.9	17.7	933

In doing so, the condition numbers can be improved impressively. In Table 2.3, we list the ℓ_2-condition numbers for the BPX-preconditioned Laplacian in case of cubic B-splines in two spatial dimensions. By comparing the numbers with those found in Tables 2.1 and 2.2 one can infer that the related condition numbers are all reduced by a factor about five. Note that the setup, storage and application of the operator defined in (2.33) is still of optimal linear complexity.

Finally, we provide numerical results in order to demonstrate the preconditioning and to specify the dependence on the spatial dimension n and the spline degree p. We consider an approximation of the homogeneous Dirichlet problem for the Poisson equation on the n-dimensional unit cube $\hat{\Omega} = (0, 1)^n$ for $n = 1, 2, 3$. The mesh on level j is obtained by subdividing the cube j-times dyadically into 2^n

subcubes of mesh size $h_j = 2^{-j}$. On this subdivision, we consider the B-splines of degree $p = 1, 2, 3, 4$ as defined in Sect. 2.2.1. The ℓ_2-condition numbers of the related stiffness matrices, preconditioned by the BPX-preconditioner (2.32), are shown in Table 2.1. The condition numbers seem to be independent of the level j, but they depend on the spline degree p and the space dimension n for $n > 1$. For fourth order problems on the sphere, corresponding results for the bi-Laplacian with and without BPX preconditioning were presented in [60].

We study next the dependence of the condition numbers on the parametric mapping **F**. We consider the case $n = 2$ in case of a smooth mapping (see the plot on the right hand side of Table 2.2 for an illustration of the mapping). As one can see from Table 2.2, the condition numbers are at most about a factor of five higher than the related values in Table 2.1. Nearly the same observation holds if we replace the parametric mapping by a $\mathscr{C}^0$-parametrization which maps the unit square onto an L-shaped domain, see [10].

If we consider a singular map **F**, that is, a mapping that does not satisfy (2.11), the condition numbers grow considerably as expected, see [10]. But even in this case, the BPX-preconditioner with SSOR acceleration (2.33) is able to drastically reduce the condition numbers of the system matrix in all examples, see Table 2.3.

For further remarks concerning multiplicative multilevel preconditioners as the so-called multigrid methods in the context of isogeometric analysis together with references, one may consult [55].

2.3   Problem Classes

The variational problems to be investigated further will first be formulated in the following abstract form.

2.3.1   An Abstract Operator Equation

Let $\mathscr{H}$ be a Hilbert space with norm $\| \cdot \|_{\mathscr{H}}$ and let $\mathscr{H}'$ be the normed dual of $\mathscr{H}$ endowed with the norm

$$\|w\|_{\mathscr{H}'} := \sup_{v \in \mathscr{H}} \frac{\langle v, w \rangle}{\|v\|_{\mathscr{H}}} \tag{2.34}$$

where $\langle \cdot, \cdot \rangle$ denotes the dual pairing between $\mathscr{H}$ and $\mathscr{H}'$.

Given $F \in \mathcal{H}'$, we seek a solution to the operator equation

$$\mathcal{L} U = F \qquad\qquad (2.35)$$

where $\mathcal{L} : \mathcal{H} \to \mathcal{H}'$ is a linear operator which is assumed to be a bounded bijection, that is,

$$\|\mathcal{L} V\|_{\mathcal{H}'} \sim \|V\|_{\mathcal{H}}, \qquad V \in \mathcal{H}. \qquad\qquad (2.36)$$

We call the operator equation *well-posed* since (2.35) implies for any given data $F \in \mathcal{H}'$ the existence and uniqueness of the solution $U \in \mathcal{H}$ which depends continuously on the data.

In the following subsections, we describe some problem classes which can be placed into this framework. In particular, these examples will have the format that $\mathcal{H}$ is a product space

$$\mathcal{H} := H_{1,0} \times \cdots \times H_{M,0} \qquad\qquad (2.37)$$

where each of the $H_{i,0} \subseteq H_i$ is a Hilbert space (or a closed subspace of a Hilbert space H_i determined, e.g., by homogeneous boundary conditions). The spaces H_i will be Sobolev spaces living on a domain $\Omega \subset \mathbb{R}^n$ or on (part of) its boundary. According to the definition of $\mathcal{H}$, the elements $V \in \mathcal{H}$ will consist of M components $V = (v_1, \ldots, v_M)^T$, and we define $\|V\|_{\mathcal{H}}^2 := \sum_{i=1}^M \|v_i\|_{H_i}^2$. The dual space $\mathcal{H}'$ is then endowed with the norm

$$\|W\|_{\mathcal{H}'} := \sup_{V \in \mathcal{H}} \frac{\langle V, W \rangle}{\|V\|_{\mathcal{H}}} \qquad\qquad (2.38)$$

where $\langle V, W \rangle := \sum_{i=1}^M \langle v_i, w_i \rangle_i$ in terms of the dual pairing $\langle \cdot, \cdot \rangle_i$ between H_i and H_i'.

We next formulate four classes which fit into this format. The first two concern are elliptic boundary value problems with included essential boundary conditions, and elliptic boundary value problems formulated as saddle point problem with boundary conditions treated by means of Lagrange Multipliers. For an introduction into elliptic boundary value problems and saddle point problems together with the functional analytic background one can, e.g., resort to [7]. Based on these formulations, we afterwards introduce certain control problems. A recurring theme in the derivation of the system of operator equation is the minimization of a quadratic functional subject to linear constraints.

2.3.2   Elliptic Boundary Value Problems

Let $\Omega \subset \mathbb{R}^n$ be a bounded domain with piecewise smooth boundary $\partial\Omega := \Gamma \cup \Gamma_N$. We consider the scalar second order boundary value problem

$$
\begin{aligned}
- \nabla \cdot (\mathbf{a}\nabla y) + cy &= f \quad \text{in } \Omega, \\
y &= g \quad \text{on } \Gamma, \\
(\mathbf{a}\nabla y) \cdot \mathbf{n} &= 0 \quad \text{on } \Gamma_N,
\end{aligned}
\tag{2.39}
$$

where $\mathbf{n} = \mathbf{n}(\mathbf{x})$ is the outward normal at $\mathbf{x} \in \Gamma$, $\mathbf{a} = \mathbf{a}(\mathbf{x}) \in \mathbb{R}^{n\times n}$ is uniformly positive definite and bounded on Ω and $c \in L_\infty(\Omega)$. Moreover, f and g are some given right hand side and boundary data. With the usual definition of the bilinear form

$$
a(v, w) := \int_\Omega (\mathbf{a}\nabla v \cdot \nabla w + cvw)\, d\mathbf{x},
\tag{2.40}
$$

the weak formulation of (2.39) requires in the case $g \equiv 0$ to find $y \in \mathscr{H}$ where

$$
\mathscr{H} := H^1_{0,\Gamma}(\Omega) := \{v \in H^1(\Omega) : v|_\Gamma = 0\},
\tag{2.41}
$$

or

$$
\mathscr{H} := \{v \in H^1(\Omega) : \int_\Omega v(\mathbf{x})\, d\mathbf{x} = 0\} \quad \text{when } \Gamma = \emptyset,
\tag{2.42}
$$

such that

$$
a(y, v) = \langle v, f \rangle, \quad v \in \mathscr{H}.
\tag{2.43}
$$

The Neumann-type boundary conditions on Γ_N are implicitly satisfied in the weak formulation (2.43), therefore called *natural boundary conditions*. In contrast, the Dirichlet boundary conditions on Γ have to be posed explicitly, for this reason called *essential boundary conditions*. The easiest way to achieve this for homogeneous Dirichlet boundary conditions when $g \equiv 0$ is to include them into the solution space as above in (2.41). In the nonhomogeneous case $g \not\equiv 0$ on Γ in (2.39) and $\Gamma \neq \emptyset$, one can reduce the problem to a problem with homogeneous boundary conditions by *homogenization* as follows. Let $w \in H^1(\Omega)$ be such that $w = g$ on Γ. Then $\tilde{y} := y - w$ satisfies $a(\tilde{y}, v) = a(y, v) - a(w, v) = \langle v, f \rangle - a(w, v) =: \langle v, \tilde{f} \rangle$ for all $v \in \mathscr{H}$ defined in (2.41), and on Γ one has $\tilde{y} = g - w \equiv 0$, that is, $\tilde{y} \in \mathscr{H}$. Thus, it suffices to consider the weak form (2.43) with eventually modified right hand side. (A second possibility which allows to treat inhomogeneous boundary conditions explicitly in the context of saddle point problems will be discussed below in Sect. 2.3.3.)

The crucial property is that the bilinear form defined in (2.40) is continuous and elliptic on $\mathcal{H}$,

$$a(v, v) \sim \|v\|_{\mathcal{H}}^2 \qquad \text{for any } v \in \mathcal{H}, \tag{2.44}$$

see, e.g., [7].

By Riesz' representation theorem, the bilinear form defines a linear operator $A : \mathcal{H} \to \mathcal{H}'$ by

$$\langle w, Av \rangle := a(v, w), \qquad v, w \in \mathcal{H}, \tag{2.45}$$

which is under the above assumptions a bounded linear bijection, that is,

$$c_A \|v\|_{\mathcal{H}} \leq \|Av\|_{\mathcal{H}'} \leq C_A \|v\|_{\mathcal{H}} \qquad \text{for any } v \in \mathcal{H}. \tag{2.46}$$

Here we only consider the case where A is symmetric. With corresponding alterations, the material in the subsequent sections can also be derived for the nonsymmetric case with corresponding changes with respect to the employed algorithms.

The relation (2.46) entails that given any $f \in \mathcal{H}'$, there exists a unique $y \in \mathcal{H}$ which solves the linear system

$$Ay = f \qquad \text{in } \mathcal{H}' \tag{2.47}$$

derived from (2.43). This linear operator equation where the operator defines a bounded bijection in the sense of (2.46) is the simplest case of a well-posed variational problem (2.35). Adhering to the notation in Sect. 2.3.1, we have here $M = 1$ and $\mathcal{L} = A$.

2.3.3   Saddle Point Problems Involving Boundary Conditions

A collection of saddle point problems or, more general, multiple field formulations including first order system formulations of the elliptic boundary value problem (2.39) and the three field formulation of the Stokes problem with inhomogeneous boundary conditions have been rephrased as well-posed variational problems in the above sense in [35], see also further references cited therein.

Here a particular saddle point problem derived from (2.39) shall be considered which will be recycled later in the context of control problems. In fact, this formulation is particularly appropriate to handle essential Dirichlet boundary conditions.

Recall from, e.g., [7], that the solution $y \in \mathcal{H}$ of (2.43) is also the unique minimizer of the minimization problem

$$\inf_{v \in \mathcal{H}} \mathcal{J}(v), \qquad \mathcal{J}(v) := \frac{1}{2} a(v, v) - \langle v, f \rangle. \tag{2.48}$$

This means that y is a zero for its first order variational derivative of $\mathcal{J}$, that is, $\delta \mathcal{J}(y; v) = 0$. We denote here and in the following by $\delta^M \mathcal{J}(v; w_1, \ldots, w_M)$ the M-th variation of $\mathcal{J}$ at v in directions $w_1, \ldots, w_M$, see e.g., [72]. In particular, for $M = 1$

$$\delta \mathcal{J}(v; w) := \lim_{\varepsilon \to 0} \frac{\mathcal{J}(v + \varepsilon w) - \mathcal{J}(v)}{\varepsilon} \tag{2.49}$$

is the (Gateaux) derivative of $\mathcal{J}$ at v in direction w.

In order to generalize (2.48) to the case of nonhomogeneous Dirichlet boundary conditions g, we formulate this as minimizing J over $v \in H^1(\Omega)$ subject to constraints in form of the essential boundary conditions $v = g$ on Γ. Using techniques from nonlinear optimization theory, one can employ a *Lagrange multiplier* p to append the constraints to the optimization functional J defined in (2.48). Satisfying the constraint is guaranteed by taking the supremum over all such Lagrange multipliers before taking the infimum. Thus, minimization subject to a constraint leads to the problem of finding a *saddle point* (y, p) of the *saddle point problem*

$$\inf_{v \in H^1(\Omega)} \sup_{q \in (H^{1/2}(\Gamma))'} \mathcal{J}(v) + \langle v - g, q \rangle_\Gamma. \tag{2.50}$$

Some comments on the choice of the Lagrange multiplier space and the dual form $\langle \cdot, \cdot \rangle_\Gamma$ in (2.50) are in order. The boundary expression $v = g$ actually means taking the *trace* of $v \in H^1(\Omega)$ to $\Gamma \subseteq \partial\Omega$ which we explicitly write from now on $\gamma v := v|_\Gamma$. Classical trace theorems which may be found in [43] state that for any $v \in H^1(\Omega)$ one looses '$\frac{1}{2}$ order of smoothness' when taking traces so that one ends up with $\gamma v \in H^{1/2}(\Gamma)$. Thus, when the data g is also such that $g \in H^{1/2}(\Gamma)$, the expression in (2.50) involving the dual form $\langle \cdot, \cdot \rangle_\Gamma := \langle \cdot, \cdot \rangle_{H^{1/2}(\Gamma) \times (H^{1/2}(\Gamma))'}$ is well-defined, and so is the selection of the multiplier space $(H^{1/2}(\Gamma))'$. In case of Dirichlet boundary conditions on the whole boundary of Ω, i.e., the case $\Gamma \equiv \partial\Omega$, one can identify $(H^{1/2}(\Gamma))' = H^{-1/2}(\Gamma)$.

The above formulation (2.50) was first investigated in [2]. Another standard technique from optimization to handle minimization problems under constraints is to append the constraints to $J(v)$ by means of a *penalty parameter* ε as follows, cf. [3]. For the case of homogeneous Dirichlet boundary conditions, one could introduce the functional $J(v) + (2\varepsilon)^{-1} \|\gamma v\|^2_{H^{1/2}(\Gamma)}$. (The original formulation in [3] uses the term $\|\gamma v\|^2_{L_2(\Gamma)}$.) Although the linear system derived from this formulation is still elliptic—the bilinear form is of the type $a(v, v) + \varepsilon^{-1}(\gamma v, \gamma v)_{H^{1/2}(\Gamma)}$—the spectral condition number of the corresponding operator A_ε depends on ε. The choice of ε is typically attached to the discretization of an underlying grid with grid spacing h for Ω of the form $\varepsilon \sim h^\alpha$ when $h \to 0$ for some exponent $\alpha > 0$ chosen such that one retains the optimal approximation order of the underlying scheme. Thus, the spectral condition number of the operators in such systems depends polynomially on (at least) $h^{-\alpha}$. Consequently, iterative solution schemes such as the conjugate gradient method converge as slow as without preconditioning for A, and so far no optimal preconditioners for this situation are known.

It should also be mentioned that the way of treating essential boundary conditions by Lagrange multipliers can be extended to *fictitious domain methods* which may be used for problems with changing boundaries such as shape optimization problems [46, 49]. There one embeds the domain Ω into a larger, simple domain $\Box$, and formulates (2.50) with respect to $H^1(\Box)$ and dual form on the changing boundary Γ [52]. One should note, however, that for Γ a proper subset of $\partial\Omega$, there may occur some ambiguity in the relation between the fictitious domain formulation and the corresponding strong form (2.39).

In order to bring out the role of the trace operator, we define in addition to (2.40) a second bilinear form on $H^1(\Omega) \times (H^{1/2}(\Gamma))'$ by

$$b(v, q) := \int_\Gamma (\gamma v)(s) \, q(s) \, ds \tag{2.51}$$

so that the saddle point problem (2.50) may be rewritten as

$$\inf_{v \in H^1(\Omega)} \sup_{q \in (H^{1/2}(\Gamma))'} \mathscr{J}(v, q), \qquad \mathscr{J}(v, q) := J(v) + b(v, q) - \langle g, q \rangle_\Gamma. \tag{2.52}$$

Computing zeroes of the first order variations of $\mathscr{J}$, now with respect to both v and q, yields the system of equations that a saddle point (y, p) has to satisfy

$$\begin{aligned} a(y, v) + b(v, p) &= \langle v, f \rangle, & v &\in H^1(\Omega), \\ b(y, q) &= \langle g, q \rangle_\Gamma, & q &\in (H^{1/2}(\Gamma))'. \end{aligned} \tag{2.53}$$

Defining the linear operator $B : H^1(\Omega) \to H^{1/2}(\Gamma)$ and its adjoint $B' : (H^{1/2}(\Gamma))' \to (H^1(\Omega))'$ by $\langle Bv, q \rangle_\Gamma = \langle v, B'q \rangle_\Gamma := b(v, q)$, this can be rewritten as the linear operator equation from $\mathscr{H} := H^1(\Omega) \times (H^{1/2}(\Gamma))'$ to $\mathscr{H}'$ as follows: Given $(f, g) \in \mathscr{H}'$, find $(y, p) \in \mathscr{H}$ that solves

$$\begin{pmatrix} A & B' \\ B & 0 \end{pmatrix} \begin{pmatrix} y \\ p \end{pmatrix} = \begin{pmatrix} f \\ g \end{pmatrix}. \tag{2.54}$$

It can be shown that the Lagrange multiplier is given by $p = -\mathbf{n} \cdot \mathbf{a} \nabla y$ and can here be interpreted as a *stress force* on the boundary [2].

Let us briefly investigate the properties of B representing the trace operator. Classical trace theorems from, e.g., [43], state that for any $f \in H^s(\Omega)$, $1/2 < s < 3/2$, one has

$$\|f|_\Gamma\|_{H^{s-1/2}(\Gamma)} \lesssim \|f\|_{H^s(\Omega)}. \tag{2.55}$$

Conversely, for every $g \in H^{s-1/2}(\Gamma)$, there exists some $f \in H^s(\Omega)$ such that $f|_\Gamma = g$ and

$$\|f\|_{H^s(\Omega)} \lesssim \|g\|_{H^{s-1/2}(\Gamma)}. \tag{2.56}$$

Note that the range of s extends accordingly if Γ is more regular. Estimate (2.55) immediately entails for $s = 1$ that $B : H^1(\Omega) \to H^{1/2}(\Gamma)$ is continuous. Moreover, the second property (2.56) means B is surjective, i.e., $\text{range} B = H^{1/2}(\Gamma)$ and $\ker B' = \{0\}$, which yields that the *inf–sup condition*

$$\inf_{q \in (H^{1/2}(\Gamma))'} \sup_{v \in H^1(\Omega)} \frac{\langle Bv, q \rangle_\Gamma}{\|v\|_{H^1(\Omega)} \|q\|_{(H^{1/2}(\Gamma))'}} \gtrsim 1 \tag{2.57}$$

is satisfied.

At this point it will be more convenient to consider (2.54) as a saddle point problem in abstract form on $\mathscr{H} = Y \times Q$. Thus, we identify $Y = H^1(\Omega)$ and $Q = (H^{1/2}(\Gamma))'$ and linear operators $A : Y \to Y'$ and $B : Y \to Q'$.

The abstract theory of saddle point problems states that existence and uniqueness of a solution pair $(y, p) \in \mathscr{H}$ holds if A and B are continuous, A is invertible on $\ker B \subseteq Y$ and the range of B is closed in Q', see, e.g., [7, 9, 42]. The properties for B and the continuity for A have been assured above. In addition, we will always deal here with operators A which are invertible on $\ker B$, which cover the standard cases of the Laplacian ($\mathbf{a} = I$ and $c \equiv 0$) and the Helmholtz operator ($\mathbf{a} = I$ and $c \equiv 1$).

Consequently,

$$\mathscr{L} := \begin{pmatrix} A & B' \\ B & 0 \end{pmatrix} : \mathscr{H} \to \mathscr{H}' \tag{2.58}$$

is linear bijection, and one has the mapping property

$$\left\| \mathscr{L} \begin{pmatrix} v \\ q \end{pmatrix} \right\|_{\mathscr{H}'} \sim \left\| \begin{pmatrix} v \\ q \end{pmatrix} \right\|_{\mathscr{H}} \tag{2.59}$$

for any $(v, q) \in \mathscr{H}$ with constants depending on upper and lower bounds for A, B. Thus, the operator equation (2.54) is established to be a well-posed variational problem in the sense of Sect. 2.3.1: for given $(f, g) \in \mathscr{H}'$, there exists a unique solution $(y, p) \in \mathscr{H} = Y \times Q$ which continuously depends on the data.

2.3.4   Parabolic Boundary Value Problems

More recently, weak full space-time formulation for one linear parabolic equation
became popular which allow us to consider time just as another space variable as
follows.

Let again $\Omega \subset \mathbb{R}^n$ be a bounded Lipschitz domain with boundary $\partial\Omega$, and
denote by $\Omega_T := I \times \Omega$ with time interval $I := (0, T)$ the time-space cylinder for
functions $f = f(t, x)$ depending on time t and space x. The parameter $T < \infty$
will always denote a fixed final time. Let Y be a dense subspace of $H := L_2(\Omega)$
which is continuously embedded in $L_2(\Omega)$ and denote by Y' its topological dual.
The associated dual form is denoted by $\langle \cdot, \cdot \rangle_{Y' \times Y}$ or, shortly $\langle \cdot, \cdot \rangle$. Later we will use
$\langle \cdot, \cdot \rangle$ also for time-space duality with the precise meaning clear from the context.
Norms will be indexed by the corresponding spaces. Following [59], Chapter III,
p. 100, let for a.e. $t \in I$ there be bilinear forms $a(t; \cdot, \cdot) : Y \times Y \to \mathbb{R}$ so that
$t \mapsto a(t; \cdot, \cdot)$ is measurable on I and that $a(t; \cdot, \cdot)$ is continuous and elliptic on Y,
i.e., there exists constants $0 < \alpha_1 \leq \alpha_2 < \infty$ independent of t such that a.e. $t \in I$

$$a(t; v, w) \leq \alpha_2 \|v\|_Y \|w\|_Y, \quad v, w \in Y,$$
$$a(t; v, v) \geq \alpha_1 \|v\|_Y^2, \qquad v \in Y. \tag{2.60}$$

Define accordingly a linear operator $A = A(t) : Y \to Y'$ by

$$\langle A(t)v, w \rangle := a(t; v, w), \qquad v, w \in Y. \tag{2.61}$$

Denoting by $\mathscr{L}(V, W)$ the set of all bounded linear functions from V to W, we
have by (2.60) $A(t) \in \mathscr{L}(Y, Y')$ for a.e. $t \in I$. Typically, $A(t)$ will be a scalar
linear elliptic differential operator of order two on Ω and $Y = H_0^1(\Omega)$. We denote
by $L_2(I; Z)$ the space of all functions $v = v(t, x)$ for which for a.e. $t \in I$ one has
$v(t, \cdot) \in Z$. Instead of $L_2(I; Z)$, we will write this space as the tensor product of
the two separable Hilbert spaces, $L_2(I) \otimes Z$, which, by Theorem 12.6.1 in [1], can
be identified. This fact will be frequently employed also in the sequel.

The standard semi-weak form a linear evolution equation is the following, see
e.g. [40]. Given an initial condition $y_0 \in H$ and right hand side $f \in L_2(I; Y')$, find
y in some function space on Ω_T such that

$$\langle \tfrac{\partial y(t, \cdot)}{\partial t}, v \rangle + \langle A(t) y(t, \cdot), v \rangle = \langle f(t, \cdot), v \rangle \text{ for all } v \in Y \text{ and a.e. } t \in (0, T),$$
$$\langle y(0, \cdot), v \rangle = \langle y_0, v \rangle \qquad \text{for all } v \in H.$$
$$\tag{2.62}$$

For $Y = H_0^1(\Omega)$, the weak formulation of the first equation includes homogeneous
Dirichlet conditions $y(t, \cdot)|_{\partial\Omega} = 0$ for a.e. $t \in I$.

The *space-time variational formulation* for (2.62) will be based on the *solution space*

$$\mathscr{Y} := L_2(I; Y) \cap H^1(I; Y') = (L_2(I) \otimes Y) \cap \left(H^1(I) \otimes Y' \right)$$

$$= \{ w \in L_2(I; Y) : \tfrac{\partial w(t, \cdot)}{\partial t} \in L_2(I; Y') \} \tag{2.63}$$

equipped with the graph norm

$$\| w \|_{\mathscr{Y}}^2 := \| w \|_{L_2(I;Y)}^2 + \| \tfrac{\partial w(t, \cdot)}{\partial t} \|_{L_2(I;Y')}^2 \tag{2.64}$$

and the Cartesian product *space of test functions*

$$\mathscr{V} := L_2(I; Y) \times H = (L_2(I) \otimes Y) \times H \tag{2.65}$$

equipped for $v = (v_1, v_2) \in \mathscr{V}$ with the norm

$$\| v \|_{\mathscr{V}}^2 := \| v_1 \|_{L_2(I;Y)}^2 + \| v_2 \|_H^2 \tag{2.66}$$

Note that $v_1 = v_1(t, x)$ and $v_2 = v_2(x)$.

Integration of (2.62) over $t \in I$ leads to the variational problem to find for given $f \in \mathscr{V}'$ a function $y \in \mathscr{Y}$

$$b(y, v) = \langle f, v \rangle \qquad \text{for all } v = (v_1, v_2) \in \mathscr{V}, \tag{2.67}$$

where the bilinear form $b(\cdot, \cdot) : \mathscr{Y} \times \mathscr{V} \to \mathbb{R}$ is defined by

$$b(w, (v_1, v_2)) := \int_I \left(\langle \tfrac{\partial w(t, \cdot)}{\partial t}, v_1(t, \cdot) \rangle + \langle A(t) \, w(t, \cdot), v_1(t, \cdot) \rangle \right) dt + \langle w(0, \cdot), v_2 \rangle \tag{2.68}$$

and the right hand side $\langle f, \cdot \rangle : \mathscr{V} \to \mathbb{R}$ by

$$\langle f, v \rangle := \int_I \langle f(t, \cdot), v_1(t, \cdot) \rangle \, dt + \langle y_0, v_2 \rangle \tag{2.69}$$

for $v = (v_1, v_2) \in \mathscr{V}$. It was proven in [37, Chapter XVIII, §3] that the operator defined by the bilinear form $b(\cdot, \cdot)$ is an isomorphism with respect to the spaces $\mathscr{Y}$ and $\mathscr{V}$. An alternative, shorter proof given in [66] is based on a characterization of bounded invertibility of linear operators between Hilbert spaces and provides detailed bounds on the norms of the operator and its inverse as follows.

Theorem 4 *The operator* $B \in \mathscr{L}(\mathscr{Y}, \mathscr{V}')$ *defined by* $\langle Bw, v \rangle := b(w, v)$ *for* $w \in \mathscr{Y}$ *and* $v \in \mathscr{V}$ *with* $b(\cdot, \cdot)$ *from (2.68) and spaces* $\mathscr{Y}, \mathscr{V}$ *defined in (2.63), (2.65) is boundedly invertible: There exist constants* $0 < \beta_1 \leq \beta_2 < \infty$ *such that*

$$\|B\|_{\mathscr{Y} \to \mathscr{V}'} \leq \beta_2 \quad and \quad \|B^{-1}\|_{\mathscr{V}' \to \mathscr{Y}} \leq \frac{1}{\beta_1}. \tag{2.70}$$

As proved in [66], the continuity constant β_2 and the inf–sup condition constant β_1 for $b(\cdot, \cdot)$ satisfy

$$\beta_1 \geq \frac{\min(\alpha_1 \alpha_2^{-2}, \alpha_1)}{\sqrt{2 \max(\alpha_1^{-2}, 1) + \varrho^2}}, \qquad \beta_2 \leq \sqrt{2 \max(1, \alpha_2^2) + \varrho^2}, \tag{2.71}$$

where α_1, α_2 are the constants from (2.60) bounding $A(t)$, and ϱ is defined as

$$\varrho := \sup_{0 \neq w \in \mathscr{Y}} \frac{\|w(0, \cdot)\|_H}{\|w\|_{\mathscr{Y}}}.$$

We like to recall from [37, 40] that $\mathscr{Y}$ is continuously embedded in $\mathscr{C}^0(I; H)$ so that the pointwise in time initial condition in (2.62) is well-defined. From this it follows that the constant ρ is bounded uniformly in the choice of $\mathscr{Y} \hookrightarrow H$.

For the sequel, it will be useful to explicitly identify the dual operator $B^* : \mathscr{V} \to \mathscr{Y}'$ of B which is defined by

$$\langle Bw, v \rangle =: \langle w, B^* v \rangle. \tag{2.72}$$

In fact, it follows from the definition of the bilinear form (2.68) on $\mathscr{Y} \times \mathscr{V}$ by integration by parts for the first term with respect to time, and using the dual $A(t)^*$ w.r.t. space that

$$b(w, (v_1, v_2)) = \int_I \left(\langle w(t, \cdot), \tfrac{\partial v_1(t, \cdot)}{\partial t} \rangle + \langle w(t, \cdot), A(t)^* v_1(t, \cdot) \rangle \right) dt$$

$$+ \langle w(0, \cdot), v_2 \rangle + \langle w(t, \cdot), v_2 \rangle |_0^T$$

$$= \int_I \left(\langle w(t, \cdot), \tfrac{\partial v_1(t, \cdot)}{\partial t} \rangle + \langle w(t, \cdot), A(t)^* v_1(t, \cdot) \rangle \right) dt$$

$$+ \langle w(T, \cdot), v_2 \rangle$$

$$=: \langle w, B^* v \rangle. \tag{2.73}$$

Note that the first term of the right hand side defining B^* which involves $\frac{\partial}{\partial t} v_1(t, \cdot)$ is still well-defined with respect to t as an element of $\mathscr{Y}'$ on account of $w \in \mathscr{Y}$.

2.3.5   PDE-Constrained Control Problems: Distributed Control

A class of problems where the numerical solution of systems (2.47) is required repeatedly are certain control problems with PDE-constraints described next. Adhering to the notation from Sect. 2.3.2, consider as a guiding model for the subsequent discussion the objective to minimize a quadratic functional of the form

$$\mathscr{J}(y, u) = \frac{1}{2}\|y - y_*\|^2_{\mathscr{Z}} + \frac{\omega}{2}\|u\|^2_{\mathscr{U}}, \qquad (2.74)$$

subject to linear constraints

$$Ay = f + u \qquad \text{in } H' \qquad (2.75)$$

where $A : H \to H'$ is defined as above in (2.61) satisfying (2.46) and $f \in H$ is given. Reserving the symbol $\mathscr{H}$ for the resulting product space in view of the notation in Sect. 2.3.1, the space H is in this subsection defined as in (2.41) or in (2.42). In order for a solution y of (2.75), the *state* of the system, to be well-defined, the problem formulation has to ensure that the unknown *control u* appearing on the right hand side is at least in H'. This can be achieved by choosing the *control space* $\mathscr{U}$ whose norm appears in (2.74) such that it is as least as smooth as H'. The second ingredient in the functional (2.74) is a data fidelity term which tries to match the system state y to some prescribed target state y_*, measured in some norm which is typically weaker than $\|\cdot\|_H$. Thus, we require that the *observation space* $\mathscr{Z}$ and the control space $\mathscr{U}$ are such that the continuous embeddings

$$\|v\|_{H'} \lesssim \|v\|_{\mathscr{U}}, \quad v \in \mathscr{U}, \qquad \|v\|_{\mathscr{Z}} \lesssim \|v\|_H, \quad v \in H, \qquad (2.76)$$

hold. Mostly one has investigated the simplest cases of norms which occur for $\mathscr{U} = \mathscr{Z} = L_2(\Omega)$ and which are covered by these assumptions [59]. The parameter $\omega \geq 0$ balances the norms in (2.74).

Since the control appears in all of the right hand side of (2.75), such control problems are termed problems with *distributed* control. Although their practical value is of a rather limited nature, distributed control problems help to bring out the basic mechanisms. Note that when the observed data are *compatible* in the sense that $y_* \equiv A^{-1}f$, the control problem has the trivial solution $u \equiv 0$ which yields $\mathscr{J}(y, u) \equiv 0$.

Solution schemes for the control problem (2.74) subject to the constraints (2.75) can be based on the system of operator equations derived next by the same variational principles as employed in the previous section, using a Lagrange multiplier p to enforce the constraints. Defining the Lagrangian functional

$$\text{Lagr}(y, p, u) := \mathscr{J}(y, u) + \langle p, Ay - f - u \rangle \qquad (2.77)$$

on $H \times H \times H'$, the first order necessary conditions or *Karush-Kuhn-Tucker (KKT) conditions* $\delta \operatorname{Lagr}(x) = 0$ for $x = p, y, u$ can be derived as

$$Ay = f + u$$
$$A'p = -S(y - y_*)$$
$$\omega Ru = p.$$

(2.78)

Here the linear operators S and R can be interpreted as Riesz operators defined by the inner products $(\cdot, \cdot)_{\mathscr{Z}}$ and $(\cdot, \cdot)_{\mathscr{U}}$. The system (2.78) may be written in saddle point form as

$$\mathscr{L} V := \begin{pmatrix} \mathscr{A} & \mathscr{B}' \\ \mathscr{B} & 0 \end{pmatrix} V := \begin{pmatrix} S & 0 & A' \\ 0 & \omega R & -I \\ A & -I & 0 \end{pmatrix} \begin{pmatrix} y \\ u \\ p \end{pmatrix} = \begin{pmatrix} Sy_* \\ 0 \\ f \end{pmatrix} =: F$$

(2.79)

on $\mathscr{H} := H \times H \times H'$.

Remark 2 We can also allow for $\mathscr{Z}$ in (2.74) to be a *trace space* on part of the boundary $\partial \Omega$ as long as the corresponding condition (2.76) is satisfied [53].

The class of control problems where the control is exerted through Neumann boundary conditions can also be written in this form since in this case the control still appears on the right hand side of a single operator equation of a form like (2.75), see [29].

Well-posedness of the system (2.79) can now be established by applying the conditions for saddle point problems stated in Sect. 2.3.3. For the control problems here and below we will, however, follow a different route which better supports efficient numerical solution schemes. The idea is as follows. While the PDE constraints (2.75) that govern the system are fixed, there is in many applications some ambiguity with respect to the choice of the spaces $\mathscr{Z}$ and $\mathscr{U}$. L_2 norms are easily realized in finite element discretizations, although in some applications like glass cooling smoother norms for the observation $\| \cdot \|_{\mathscr{Z}}$ are desirable [63]. Once $\mathscr{Z}$ and $\mathscr{U}$ are fixed, there is only a single parameter ω to balance the two norms in (2.74). *Modelling* the objective functional is therefore an issue where more flexibility may be advantageous. Specifically in a multiscale setting, one may want to weight contributions on different scales by multiple parameters.

The wavelet setting which we describe below allows for this flexibility. It is based on formulating the objective functional in terms of weighted wavelet coefficient sequences which are equivalent to $\mathscr{Z}$, $\mathscr{U}$ and which, in addition, support an efficient numerical implementation. Once wavelet discretizations are introduced, we formulate below control problems with such objective functionals.

2.3.6   PDE-Constrained Control Problems: Dirichlet Boundary Control

Even more involved as the control problems with distributed control encountered in the previous section are those problems with Dirichlet boundary control which are, however, practically much more relevant.

An illustrative guiding model for this case is the problem to minimize for some given data y_* the quadratic functional

$$\mathscr{J}(y, u) = \frac{1}{2}\|y - y_*\|_{\mathscr{L}}^2 + \frac{\omega}{2}\|u\|_{\mathscr{U}}^2, \tag{2.80}$$

where, adhering to the notation in Sect. 2.3.2 the state y and the control u are coupled through the linear second order elliptic boundary value problem

$$\begin{aligned} -\nabla \cdot (\mathbf{a}\nabla y) + ky &= f && \text{in } \Omega, \\ y &= u && \text{on } \Gamma, \\ (\mathbf{a}\nabla y)\cdot \mathbf{n} &= 0 && \text{on } \Gamma_N. \end{aligned} \tag{2.81}$$

The appearance of the control u as a Dirichlet boundary condition in (2.81) is referred to as a *Dirichlet boundary control*. In view of the treatment of essential Dirichlet boundary conditions in the context of saddle point problems derived in Sect. 2.3.3, we write the PDE constraints (2.81) in the operator form (2.54) on $Y \times Q$ where $Y = H^1(\Omega)$ and $Q = (H^{1/2}(\Gamma))'$. The model control problem with Dirichlet boundary control then reads as follows: Minimize for given data $y_* \in \mathscr{L}$ and $f \in Y'$ the quadratic functional

$$\mathscr{J}(y, u) = \frac{1}{2}\|y - y_*\|_{\mathscr{L}}^2 + \frac{\omega}{2}\|u\|_{\mathscr{U}}^2 \tag{2.82}$$

subject to

$$\begin{pmatrix} A & B' \\ B & 0 \end{pmatrix} \begin{pmatrix} y \\ p \end{pmatrix} = \begin{pmatrix} f \\ u \end{pmatrix}. \tag{2.83}$$

In view of the problem formulation in Sect. 2.3.5 and the discussion of the choice of the observation space $\mathscr{L}$ and the control space, here we require analogously that $\mathscr{L}$ and $\mathscr{U}$ are such that the continuous embeddings

$$\|v\|_{Q'} \lesssim \|v\|_{\mathscr{U}}, \quad v \in \mathscr{U}, \qquad \|v\|_{\mathscr{L}} \lesssim \|v\|_{Y}, \quad v \in Y, \tag{2.84}$$

hold. In view of Remark 2, also the case of observations on part of the boundary $\partial\Omega$ can be taken into account [54]. Part of the numerical results are for such a situation shown in Fig. 2.4.

Remark 3 It should be mentioned that the simple choice $\mathscr{U} = L_2(\Gamma)$ which is used in many applications of Dirichlet control problems is *not* covered here. There may arise the problem of well-posedness in this case which we briefly discuss. Note that the constraints (2.81) or, in weak form (2.54), guarantee a unique weak solution $y \in Y = H^1(\Omega)$ provided that the boundary term u satisfies $u \in Q' = H^{1/2}(\Gamma)$. In the framework of control problems, this smoothness of u therefore has to be required either by the choice of $\mathscr{U}$ or by the choice of $\mathscr{Z}$ (such as $\mathscr{Z} = H^1(\Omega)$) which would assure $By \in Q'$. In the latter case, we could relax condition (2.84) on $\mathscr{U}$.

In the context of flow control problems, an H^1 norm on the boundary for the control has been used in [45].

Similarly as stated at the end of Sect. 2.3.5, we can derive now by variational principles the first order necessary conditions for a coupled *system* of saddle point problems. Well-posedness of this system can then again be established by applying the conditions for saddle point problems from Sect. 2.3.3 where the inf-sup condition for the saddle point problem (2.54) yields an inf-sup condition for the exterior saddle point problem of interior saddle point problems [51]. However, also in this case, we follow the ideas mentioned at the end of Sect. 2.3.6 and pose a corresponding control problem in terms of wavelet coefficients.

2.3.7 PDE-Constrained Control Problems: Parabolic PDEs

Finally, we consider the following tracking-type control problem constrained by an evolution PDE as formulated in Sect. 2.3.4.

We wish to minimize for some given target state y_* and fixed end time $T > 0$ the quadratic functional

$$J(y, u) := \frac{\omega_1}{2} \|y - y_*\|_{L_2(I;Z)}^2 + \frac{\omega_2}{2} \|y(T, \cdot) - y_*(T, \cdot)\|_Z^2 + \frac{\omega_3}{2} \|u\|_{L_2(I;U)}^2 \qquad (2.85)$$

over the state $y = y(t, x)$ and the control $u = u(t, x)$ subject to

$$By = Eu + f \qquad \text{in } \mathscr{V}' \qquad (2.86)$$

where B is defined by Theorem 4 and $f \in \mathscr{V}'$ is given by (2.69). The real weight parameters $\omega_1, \omega_2 \geq 0$ are such that $\omega_1 + \omega_2 > 0$ and $\omega_3 > 0$. The space Z by which the integral over Ω in the first two terms in (2.85) is indexed is to satisfy $Z \supseteq Y$ with continuous embedding. Although there is in the wavelet framework great flexibility in choosing even fractional Sobolev spaces for Z, for transparency, we pick here $Z = Y$. A more general choice only results in multiplications of vectors in wavelet coordinate with diagonal matrices of the form (2.96) below, see [29]. Moreover, we suppose that the operator E is a linear operator $E : U \rightarrow \mathscr{V}'$ extending $\int_I \langle u(t, \cdot), v_1(t, \cdot) \rangle \, dt$ trivially, that is, $E \equiv (I, 0)^T$. In order to generate a well-posed problem, the space U in (2.85) must be chosen to enforce that Eu is at least in $\mathcal{V}'$. We pick here the natural case $U = Y'$ which is also the weakest possible one. More general cases for both situations which result again in multiplication with diagonal matrices for wavelet coordinate vectors are discussed in [29].

2.4 Wavelets

The numerical solution of the classes of problems introduced above hinges on the availability of appropriate wavelet bases for the function spaces under consideration which are all particular Hilbert spaces. first introduce the three basic properties that we require our wavelet bases to satisfy.

Afterwards, construction principles for wavelets based on multiresolution analysis of function spaces on bounded domains will be given.

2.4.1 Basic Properties

In view of the problem classes considered above, we need to have a wavelet basis for each occurring function space at our disposal. A *wavelet basis* for a Hilbert space H is here understood as a collection of functions

$$\Psi_H := \{\psi_{H,\lambda} : \lambda \in I\!\!I_H\} \subset H \tag{2.87}$$

which are indexed by elements λ from an infinite index set $\in I\!\!I_H$. Each of the λ comprises different information $\lambda = (j, \mathbf{k}, \mathbf{e})$ such as the *refinement scale* or *level of resolution j* and a spatial location $\mathbf{k} = \mathbf{k}(\lambda) \in \mathbb{Z}^n$. In more than one space dimensions, the basis functions are built from taking tensor products of certain univariate functions, and in this case the third index $\mathbf{e}$ contains information on the *type* of wavelet. We will frequently use the symbol $|\lambda| := j$ to have access to the resolution level j. In the univariate case on all of $\mathbb{R}$, $\psi_{H,\lambda}$ is typically generated by means of shifts and dilates of a single function ψ, i.e., $\psi_\lambda = \psi_{j,k} = 2^{j/2}\psi(2^j \cdot -k)$, $j, k \in \mathbb{Z}$, normalized with respect to $\|\cdot\|_{L_2}$. On bounded domains, the structure of the functions is essentially the same up to modifications near the boundary.

The three crucial properties that we will assume the wavelet basis to have for the sequel are the following.

Riesz Basis Property (R) Every $v \in H$ has a unique expansion in terms of Ψ_H,

$$v = \sum_{\lambda \in I\!\!I_H} v_\lambda\, \psi_{H,\lambda} =: \mathbf{v}^T \Psi_H, \quad \mathbf{v} := (v_\lambda)_{\lambda \in I\!\!I_H}, \tag{2.88}$$

and its expansion coefficients satisfy a *norm equivalence*, that is, for any $\mathbf{v} = \{v_\lambda : \lambda \in I\!I_H\}$ one has

$$c_H \, \|\mathbf{v}\|_{\ell_2(I\!I_H)} \leq \|\mathbf{v}^T \Psi_H\|_H \leq C_H \, \|\mathbf{v}\|_{\ell_2(I\!I_H)}, \quad \mathbf{v} \in \ell_2(I\!I_H), \qquad (2.89)$$

where $0 < c_H \leq C_H < \infty$. This means that wavelet expansions induce *isomorphisms* between certain function spaces and sequence spaces. It will be convenient in the following to abbreviate ℓ_2 norms without subscripts as $\| \cdot \| := \| \cdot \|_{\ell_2(I\!I_H)}$ when the index set is clear from the context. If the precise format of the constants does not matter, we write the norm equivalence (2.89) shortly as

$$\|\mathbf{v}\| \sim \|\mathbf{v}^T \Psi_H\|_H, \quad \mathbf{v} \in \ell_2(I\!I_H). \qquad (2.90)$$

Locality (L) The functions $\psi_{H,\lambda}$ are have compact support which decreases with increasing level $j = |\lambda|$, i.e.,

$$diam \, (supp \psi_{H,\lambda}) \sim 2^{-|\lambda|}. \qquad (2.91)$$

Cancellation Property (CP) There exists an integer $\tilde{d} = \tilde{d}_H$ such that

$$\langle v, \psi_{H,\lambda} \rangle \lesssim 2^{-|\lambda|(n/2 - n/p + \tilde{d})} |v|_{W_p^{\tilde{d}}(supp \, \psi_{H,\lambda})}. \qquad (2.92)$$

Thus, integrating against a wavelet has the effect of taking an $\tilde{d}$th order difference which annihilates the smooth part of v. This property is for wavelets defined on Euclidean domains typically realized by constructing Ψ_H in such a way that it possesses a *dual* or *biorthogonal* basis $\tilde{\Psi}_H \subset H'$ such that the multiresolution spaces $\tilde{S}_j := span\{\tilde{\psi}_{H,\lambda} : |\lambda| < j\}$ contain all polynomials of order $\tilde{d}$. Here *dual basis* means that $\langle \psi_{H,\lambda}, \tilde{\psi}_{H,v} \rangle = \delta_{\lambda,v}, \lambda, v \in I\!I_H$.

A few remarks on these properties are in order. In (R), the norm equivalence (2.90) is crucial since it means complete control over a function measured in $\| \cdot \|_H$ from above and below by its expansion coefficients: small changes in the coefficients only causes small changes in the function which, together with the locality (L), also means that local changes stay local. This stability is an important feature which is used for deriving optimal preconditioners and driving adaptive approximations where, again, the locality is crucial. Finally, the cancellation property (CP) entails that smooth functions have small wavelet coefficients which, on account of (2.89) may be neglected in a controllable way. Moreover, (CP) can be used to derive quasi-sparse representations of a wide class of operators.

By duality arguments one can show that (2.89) is equivalent to the existence of a biorthogonal collection which is *dual* or *biorthogonal* to Ψ_H,

$$\tilde{\Psi}_H := \{\tilde{\psi}_{H,\lambda} : \lambda \in I\!I_H\} \subset H', \quad \langle \psi_{H,\lambda}, \tilde{\psi}_{H,\mu} \rangle = \delta_{\lambda,\mu}, \qquad \lambda, \mu \in I\!I_H, \qquad (2.93)$$

which is a Riesz basis for H', that is, for any $\tilde{v} = \tilde{\mathbf{v}}^T \tilde{\Psi}_H \in H'$ one has

$$C_H^{-1} \|\tilde{\mathbf{v}}\| \leq \|\tilde{\mathbf{v}}^T \tilde{\Psi}_H\|_{H'} \leq c_H^{-1} \|\tilde{\mathbf{v}}\|, \qquad (2.94)$$

see [23, 25, 51]. Here and in the sequel the tilde expresses that the collection $\tilde{\Psi}_H$ is a dual basis to a primal one for the space identified by the subscript, so that $\tilde{\Psi}_H = \Psi_{H'}$.

Above in (2.89), we have already introduced the following shorthand notation which simplifies the presentation of many terms. We will view Ψ_H both as in (2.87) as a *collection* of functions as well as a (possibly infinite) column *vector* containing all functions always assembled in some fixed unspecified order. For a countable collection of functions Θ and some single function σ, the term $\langle \Theta, \sigma \rangle$ is to be understood as the column vector with entries $\langle \theta, \sigma \rangle$, $\theta \in \Theta$, and correspondingly $\langle \sigma, \Theta \rangle$ the row vector. For two collections Θ, Σ, the quantity $\langle \Theta, \Sigma \rangle$ is then a (possibly infinite) matrix with entries $(\langle \theta, \sigma \rangle)_{\theta \in \Theta, \sigma \in \Sigma}$ for which $\langle \Theta, \Sigma \rangle = \langle \Sigma, \Theta \rangle^T$. This also implies for a (possibly infinite) matrix $\mathbf{C}$ that $\langle \mathbf{C}\Theta, \Sigma \rangle = \mathbf{C}\langle \Theta, \Sigma \rangle$ and $\langle \Theta, \mathbf{C}\Sigma \rangle = \langle \Theta, \Sigma \rangle \mathbf{C}^T$.

In this notation, the *biorthogonality* or *duality conditions* (2.93) can be reexpressed as

$$\langle \Psi, \tilde{\Psi} \rangle = \mathbf{I} \qquad (2.95)$$

with the infinite identity matrix $\mathbf{I}$.

Wavelets with the above properties can actually obtained in the following way. This concerns, in particular, a scaling depending on the regularity of the space under consideration. In our case, H will always be a Sobolev space $H^s = H^s(\Omega)$ or a closed subspace of $H^s(\Omega)$ determined by homogeneous boundary conditions, or its dual. For $s < 0$, H^s is interpreted as above as the dual of H^{-s}. One typically obtains the wavelet basis Ψ_H for H from an *anchor basis* $\Psi = \{\psi_\lambda : \lambda \in \mathbb{I} = \mathbb{I}_H\}$ which is a Riesz basis for $L_2(\Omega)$, meaning that Ψ is scaled such that $\|\psi_\lambda\|_{L_2(\Omega)} \sim 1$. Moreover, its dual basis $\tilde{\Psi}$ is also a Riesz basis for $L_2(\Omega)$. Ψ and $\tilde{\Psi}$ are constructed in such a way that rescaled versions of *both bases* $\Psi, \tilde{\Psi}$ form Riesz bases for a whole range of (closed subspaces of) Sobolev spaces H^s, for $0 < s < \gamma, \tilde{\gamma}$, respectively. Consequently, one can derive that for each $s \in (-\tilde{\gamma}, \gamma)$ the collection

$$\Psi_s := \{2^{-s|\lambda|}\psi_\lambda : \lambda \in \mathbb{I}\} =: \mathbf{D}^{-s}\Psi \qquad (2.96)$$

is a Riesz basis for H^s [23]. This means that there exist positive finite constants c_s, C_s such that

$$c_s \|\mathbf{v}\| \leq \|\mathbf{v}^T \Psi_s\|_{H^s} \leq C_s \|\mathbf{v}\| \quad \mathbf{v} \in \ell_2(\mathbb{I}), \qquad (2.97)$$

holds for each $s \in (-\tilde{\gamma}, \gamma)$. Such a scaling represented by a diagonal matrix $\mathbf{D}^s$ introduced in (2.96) will play an important role later on. The analogous expression in terms of the dual basis reads

$$\tilde{\Psi}_s := \{2^{s|\lambda|} \tilde{\psi}_\lambda : \lambda \in I\!I\} = \mathbf{D}^s \tilde{\Psi}, \tag{2.98}$$

where $\tilde{\Psi}_s$ forms a Riesz basis of H^s for $s \in (-\gamma, \tilde{\gamma})$. This entails the following fact. For $\tau \in (-\tilde{\gamma}, \gamma)$ the mapping

$$D^\tau : v = \mathbf{v}^T \Psi \mapsto (\mathbf{D}^\tau \mathbf{v})^T \Psi = \mathbf{v}^T \mathbf{D}^\tau \Psi = \sum_{\lambda \in I\!I} v_\lambda \, 2^{\tau|\lambda|} \psi_\lambda \tag{2.99}$$

acts as a shift operator between Sobolev scales which means that

$$\|D^\tau v\|_{H^s} \sim \|v\|_{H^{s+\tau}} \sim \|\mathbf{D}^{s+\tau} \mathbf{v}\|, \quad \text{if } s, \, s+\tau \in (-\tilde{\gamma}, \gamma). \tag{2.100}$$

Concrete constructions of wavelet bases with the above properties for parameters $\gamma, \tilde{\gamma} \le 3/2$ on a bounded Lipschitz domain Ω can be found in [33, 34]. This suffices for the above mentioned examples where the relevant Sobolev regularity indices range between -1 and 1.

2.4.2 Norm Equivalences and Riesz Maps

As we have seen, the scaling provided by $\mathbf{D}^{-s}$ is an important feature to establish norm equivalences (2.97) for the range $s \in (-\tilde{\gamma}, \gamma)$ of Sobolev spaces H^s. However, there are several other norms which are *equivalent* to $\|\cdot\|_{H^s}$ which may later be used in the objective functional (2.74) in the context of control problems. This issue addresses the *mathematical model* which we briefly discuss now.

We first consider norm equivalences for the L_2 norm. Let as before Ψ be the anchor wavelet basis for L_2 for which the *Riesz operator* $\mathbf{R} = \mathbf{R}_{L_2}$ is the (infinite) Gramian matrix with respect to the inner product $(\cdot, \cdot)_{L_2}$ defined as

$$\mathbf{R} := (\Psi, \Psi)_{L_2} = \langle \Psi, \Psi \rangle. \tag{2.101}$$

Expanding Ψ in terms of $\tilde{\Psi}$ and recalling the duality (2.95), this entails

$$\mathbf{I} = \langle \Psi, \tilde{\Psi} \rangle = \left\langle \langle \Psi, \Psi \rangle \tilde{\Psi}, \tilde{\Psi} \right\rangle = \mathbf{R} \langle \tilde{\Psi}, \tilde{\Psi} \rangle \quad \text{or} \quad \mathbf{R}^{-1} = \langle \tilde{\Psi}, \tilde{\Psi} \rangle. \tag{2.102}$$

$\mathbf{R}$ may be interpreted as the transformation matrix for the change of basis from $\tilde{\Psi}$ to Ψ, that is, $\Psi = \mathbf{R}\tilde{\Psi}$.

For any $w = \mathbf{w}^T \Psi \in L_2$, we now obtain the identities

$$\|w\|_{L_2}^2 = (\mathbf{w}^T \Psi, \mathbf{w}^T \Psi)_{L_2} = \mathbf{w}^T \langle \Psi, \Psi \rangle \mathbf{w} = \mathbf{w}^T \mathbf{R} \mathbf{w} = \|\mathbf{R}^{1/2} \mathbf{w}\|^2 =: \|\hat{\mathbf{w}}\|^2. \tag{2.103}$$

Expanding w with respect to the basis $\hat{\psi} := \mathbf{R}^{-1/2}\psi = \mathbf{R}^{1/2}\tilde{\psi}$, that is, $w = \hat{\mathbf{w}}^T \hat{\psi}$, yields $\|w\|_{L_2} = \|\hat{\mathbf{w}}\|$. On the other hand, we get from (2.97) with $s = 0$

$$c_0^2 \|\mathbf{w}\|^2 \leq \|w\|_{L_2}^2 \leq C_0^2 \|\mathbf{w}\|^2. \tag{2.104}$$

From this we can derive the *condition number* $\kappa(\Psi)$ of the wavelet basis in terms of the extreme eigenvalues of $\mathbf{R}$ by defining

$$\kappa(\Psi) := \left(\frac{C_0}{c_0}\right)^2 = \frac{\lambda_{\max}(\mathbf{R})}{\lambda_{\min}(\mathbf{R})} = \kappa(\mathbf{R}) \sim 1, \tag{2.105}$$

where $\kappa(\mathbf{R})$ also denotes the spectral condition number of $\mathbf{R}$ and where the last relation is assured by the asymptotic estimate (2.104). However, the absolute constants will have an impact on numerical results in specific cases.

For a Hilbert space H denote by Ψ_H a wavelet basis for H satisfying (R), (L), (CP) with a corresponding dual basis $\tilde{\Psi}_H$. The (infinite) Gramian matrix with respect to the inner product $(\cdot, \cdot)_H$ inducing $\|\cdot\|_H$ which is defined by

$$\mathbf{R}_H := (\Psi_H, \Psi_H)_H \tag{2.106}$$

will be also called *Riesz operator*. The space L_2 is covered trivially by $\mathbf{R}_0 = \mathbf{R}$. For any function $v := \mathbf{v}^T \Psi_H \in H$ we have then the identity

$$\|v\|_H^2 = (v, v)_H = (\mathbf{v}^T \Psi_H, \mathbf{v}^T \Psi_H)_H = \mathbf{v}^T (\Psi_H, \Psi_H)_H \mathbf{v}$$
$$= \mathbf{v}^T \mathbf{R}_H \mathbf{v} = \|\mathbf{R}_H^{1/2}\mathbf{v}\|^2. \tag{2.107}$$

Note that in general $\mathbf{R}_H$ may not be explicitly computable, in particular, when H is a fractional Sobolev space.

Again referring to (2.97), we obtain as in (2.105) for the more general case

$$\kappa(\Psi_s) := \left(\frac{C_s}{c_s}\right)^2 = \frac{\lambda_{\max}(\mathbf{R}_{H^s})}{\lambda_{\min}(\mathbf{R}_{H^s})} = \kappa(\mathbf{R}_{H^s}) \sim 1 \quad \text{for each } s \in (-\tilde{\gamma}, \gamma). \tag{2.108}$$

Thus, all Riesz operators on the applicable scale of Sobolev spaces are spectrally equivalent. Moreover, comparing (2.108) with (2.105), we get

$$\frac{c_s}{C_0} \|\mathbf{R}^{1/2}\mathbf{v}\| \leq \|\mathbf{R}_{H^s}^{1/2}\mathbf{v}\| \leq \frac{C_s}{c_0} \|\mathbf{R}^{1/2}\mathbf{v}\|. \tag{2.109}$$

Of course, in practice, the constants appearing in this equation may be much sharper, as the bases for Sobolev spaces with different exponents are only obtained by a diagonal scaling which preserves much of the structure of the original basis for L_2.

We summarize these results for further reference.

Proposition 1 *In the above notation, we have for any* $v = \mathbf{v}^T \Psi_s \in H^s$ *the norm equivalences*

$$\|v\|_{H^s} = \|\mathbf{R}_{H^s}^{1/2} \mathbf{v}\| \sim \|\mathbf{R}^{1/2} \mathbf{v}\| \sim \|\mathbf{v}\| \qquad \text{for each } s \in (-\tilde{\gamma}, \gamma). \qquad (2.110)$$

2.4.3 Representation of Operators

A final ingredient concerns the *wavelet representation* of linear operators in terms of wavelets. Let H, V be Hilbert spaces with wavelet bases Ψ_H, Ψ_V and corresponding duals $\tilde{\Psi}_H$, $\tilde{\Psi}_V$, and suppose that $\mathscr{L} : H \to V$ is a linear operator with dual $\mathscr{L}'$: $V' \to H'$ defined by $\langle v, \mathscr{L}' w \rangle := \langle \mathscr{L} v, w \rangle$ for all $v \in H$, $w \in V$.

We shall make frequent use of this representation and its properties.

Remark 4 The wavelet representation of $\mathscr{L} : H \to V$ with respect to the bases Ψ_H, $\tilde{\Psi}_V$ of H, V', respectively, is given by

$$\mathbf{L} := \langle \tilde{\Psi}_V, \mathscr{L} \Psi_H \rangle, \quad \mathscr{L} v = (\mathbf{L} \mathbf{v})^T \Psi_V. \qquad (2.111)$$

Thus, the expansion coefficients of $\mathscr{L} v$ in the basis that spans the range space of $\mathscr{L}$ are obtained by applying the *infinite* matrix $\mathbf{L} = \langle \tilde{\Psi}_V, \mathscr{L} \Psi_H \rangle$ to the coefficient vector of v. Moreover, boundedness of $\mathscr{L}$ implies boundedness of $\mathbf{L}$ in ℓ_2, i.e.,

$$\|\mathscr{L} v\|_V \lesssim \|v\|_H, \quad v \in H, \quad \text{implies} \quad \|\mathbf{L}\| := \sup_{\|\mathbf{v}\|_{\ell_2(\mathbb{I}_H)} \leq 1} \|\mathbf{L} \mathbf{v}\|_{\ell_2(\mathbb{I}_V)} \lesssim 1. \qquad (2.112)$$

Proof Any image $\mathscr{L} v \in V$ can naturally be expanded with respect to Ψ_V as $\mathscr{L} v = \langle \mathscr{L} v, \tilde{\Psi}_V \rangle \Psi_V$. Expanding in addition v in the basis Ψ_H, $v = \mathbf{v}^T \Psi_H$ yields

$$\mathscr{L} v = \mathbf{v}^T \langle \mathscr{L} \Psi_H, \tilde{\Psi}_V \rangle \Psi_V = ((\langle \mathscr{L} \Psi_H, \tilde{\Psi}_V \rangle^T \mathbf{v})^T \Psi_V = ((\langle \tilde{\Psi}_V, \mathscr{L} \Psi_H \rangle \mathbf{v})^T \Psi_V. \qquad (2.113)$$

As for (2.112), we can infer from (2.89) and (2.111) that

$$\|\mathbf{L} \mathbf{v}\|_{\ell_2(\mathbb{I}_V)} \sim \|(\mathbf{L} \mathbf{v})^T \Psi_V\|_V = \|\mathscr{L} v\|_V \lesssim \|v\|_H \sim \|\mathbf{v}\|_{\ell_2(\mathbb{I}_H)},$$

which confirms the claim. $\square$

2.4.4 Multiscale Decomposition of Function Spaces

In this section, the basic construction principles of the biorthogonal wavelets with properties (R), (L) and (CP) are summarized, see, e.g., [24]. Their cornerstones are *multiresolution analyses* of the function spaces under consideration and the concept of *stable completions*. These concepts are free of Fourier techniques and can therefore be applied to derive constructions of wavelets on domains or manifolds which are subsets of $\mathbb{R}^n$.

Multiresolution of L_2 Practical constructions of wavelets typically start out with multiresolution analyses of function spaces. Consider a *multiresolution* $\mathscr{S}$ of L_2 which consists of closed subspaces S_j of L_2, called *trial spaces*, such that they are nested and their union is dense in L_2,

$$S_{j_0} \subset S_{j_0+1} \subset \ldots \subset S_j \subset S_{j+1} \subset \ldots L_2, \qquad \mathrm{clos}_{L_2}\left(\bigcup_{j=j_0}^{\infty} S_j\right) = L_2. \tag{2.114}$$

The index j is the refinement level which appeared already in the elements of the index set $I\!I$ in (2.87), starting with some coarsest level $j_0 \in \mathbb{N}_0$. We abbreviate for a finite subset $\Theta \subset L_2$ the linear span of Θ as

$$S(\Theta) = span\{\Theta\}.$$

Typically the multiresolution spaces S_j have the form

$$S_j = S(\Phi_j), \qquad \Phi_j = \{\phi_{j,k} : k \in \Delta_j\}, \tag{2.115}$$

for some finite index set Δ_j, where the set $\{\Phi_j\}_{j=j_0}^{\infty}$ is *uniformly stable* in the sense that

$$\|\mathbf{c}\|_{\ell_2(\Delta_j)} \sim \|\mathbf{c}^T \Phi_j\|_{L_2}, \qquad \mathbf{c} = \{c_k\}_{k\in\Delta_j} \in \ell_2(\Delta_j), \tag{2.116}$$

holds uniformly in j. Here we have used again the shorthand notation

$$\mathbf{c}^T \Phi_j = \sum_{k\in\Delta_j} c_k\phi_{j,k}$$

and Φ_j denotes both the (column) vector containing the functions $\phi_{j,k}$ as well as the set of functions (2.115).

The collection Φ_j is called *single scale basis* since all its elements live only on one scale j. In the present context of multiresolution analysis, Φ_j is also called *generator basis* or shortly *generators* of the multiresolution. We assume that the $\phi_{j,k}$ are compactly supported with

$$diam(supp\phi_{j,k}) \sim 2^{-j}. \tag{2.117}$$

It follows from (2.116) that they are scaled such that

$$\|\phi_{j,k}\|_{L_2} \sim 1 \tag{2.118}$$

holds. It is known that nestedness (2.114) together with stability (2.116) implies the existence of matrices $\mathbf{M}_{j,0} = (m_{r,k}^j)_{r \in \Delta_{j+1}, k \in \Delta_j}$ such that the two-scale relation

$$\phi_{j,k} = \sum_{r \in \Delta_{j+1}} m_{r,k}^j \phi_{j+1,r}, \quad k \in \Delta_j, \tag{2.119}$$

is satisfied. We can essentially simplify the subsequent presentation of the material by viewing (2.119) as a matrix-vector equation which then attains the compact form

$$\Phi_j = \mathbf{M}_{j,0}^T \Phi_{j+1}. \tag{2.120}$$

Any set of functions satisfying an equation of this form, the *refinement* or *two-scale relation*, will be called *refinable*.

Denoting by $[X, Y]$ the space of bounded linear operators from a normed linear space X into the normed linear space Y, one has that

$$\mathbf{M}_{j,0} \in [\ell_2(\Delta_j), \ell_2(\Delta_{j+1})]$$

is *uniformly sparse* which means that the number of entries in each row or column is uniformly bounded. Furthermore, one infers from (2.116) that

$$\|\mathbf{M}_{j,0}\| = \mathscr{O}(1), \quad j \geq j_0, \tag{2.121}$$

where the corresponding operator norm is defined as

$$\|\mathbf{M}_{j,0}\| := \sup_{\mathbf{c} \in \ell_2(\Delta_j), \ \|\mathbf{c}\|_{\ell_2(\Delta_j)}=1} \|\mathbf{M}_{j,0}\mathbf{c}\|_{\ell_2(\Delta_{j+1})}.$$

Since the union of $\mathscr{S}$ is dense in L_2, a basis for L_2 can be assembled from functions which span any complement between two successive spaces S_j and S_{j+1}, i.e.,

$$S(\Phi_{j+1}) = S(\Phi_j) \oplus S(\Psi_j) \tag{2.122}$$

where

$$\Psi_j = \{\psi_{j,k} : k \in \nabla_j\}, \qquad \nabla_j := \Delta_{j+1} \setminus \Delta_j. \tag{2.123}$$

The functions Ψ_j are called *wavelet functions* or shortly *wavelets* if, among other conditions detailed below, the union $\{\Phi_j \cup \Psi_j\}$ is still uniformly stable in the sense of (2.116). Since (2.122) implies $S(\Psi_j) \subset S(\Phi_{j+1})$, the functions in Ψ_j must also satisfy a matrix-vector relation of the form

$$\Psi_j = \mathbf{M}_{j,1}^T \Phi_{j+1} \tag{2.124}$$

with a matrix $\mathbf{M}_{j,1}$ of size $(\#\Delta_{j+1}) \times (\#\nabla_j)$. Furthermore, (2.122) is equivalent to the fact that the linear operator composed of $\mathbf{M}_{j,0}$ and $\mathbf{M}_{j,1}$,

$$\mathbf{M}_j = (\mathbf{M}_{j,0}, \mathbf{M}_{j,1}), \tag{2.125}$$

is *invertible* as a mapping from $\ell_2(\Delta_j \cup \nabla_j)$ onto $\ell_2(\Delta_{j+1})$. One can also show that the set $\{\Phi_j \cup \Psi_j\}$ is uniformly stable if and only if

$$\|\mathbf{M}_j\|, \|\mathbf{M}_j^{-1}\| = \mathscr{O}(1), \quad j \to \infty. \tag{2.126}$$

The particular cases that will be important for practical purposes are when not only $\mathbf{M}_{j,0}$ and $\mathbf{M}_{j,1}$ are uniformly sparse but also the inverse of $\mathbf{M}_j$. We denote this inverse by $\mathbf{G}_j$ and assume that it is split into

$$\mathbf{G}_j = \mathbf{M}_j^{-1} = \begin{pmatrix} \mathbf{G}_{j,0} \\ \mathbf{G}_{j,1} \end{pmatrix}. \tag{2.127}$$

A special situation occurs when

$$\mathbf{G}_j = \mathbf{M}_j^{-1} = \mathbf{M}_j^T$$

which corresponds to the case of L_2 *orthogonal wavelets* [36]. A systematic construction of more general $\mathbf{M}_j$, $\mathbf{G}_j$ for spline-wavelets can be found in [34], see also [24] for more examples, including the hierarchical basis.

Thus, the identification of the functions Ψ_j which span the complement of $S(\Phi_j)$ in $S(\Phi_{j+1})$ is equivalent to completing a given refinement matrix $\mathbf{M}_{j,0}$ to an invertible matrix $\mathbf{M}_j$ in such a way that (2.126) is satisfied. Any such completion $\mathbf{M}_{j,1}$ is called *stable completion* of $\mathbf{M}_{j,0}$. In other words, the problem of the construction of compactly supported wavelets can equivalently be formulated as an algebraic problem of finding the (uniformly) sparse completion of a (uniformly) sparse matrix $\mathbf{M}_{j,0}$ in such a way that its inverse is also (uniformly) sparse. The fact that inverses of sparse matrices are usually dense elucidates the difficulties in the constructions.

The concept of stable completions has been introduced in [14] for which a special case is known as the *lifting scheme* [69]. Of course, constructions that yield compactly supported wavelets are particularly suited for computations in numerical analysis.

Combining the two-scale relations (2.120) and (2.124), one can see that $\mathbf{M}_j$ performs a change of bases in the space S_{j+1},

$$\begin{pmatrix} \Phi_j \\ \Psi_j \end{pmatrix} = \begin{pmatrix} \mathbf{M}_{j,0}^T \\ \mathbf{M}_{j,1}^T \end{pmatrix} \Phi_{j+1} = \mathbf{M}_j^T \Phi_{j+1}. \tag{2.128}$$

Conversely, applying the inverse of $\mathbf{M}_j$ to both sides of (2.128) results in the *reconstruction identity*

$$\Phi_{j+1} = \mathbf{G}_j^T \begin{pmatrix} \Phi_j \\ \Psi_j \end{pmatrix} = \mathbf{G}_{j,0}^T \Phi_j + \mathbf{G}_{j,1}^T \Psi_j. \tag{2.129}$$

Fixing a *finest resolution level* J, one can repeat the decomposition (2.122) so that $S_J = S(\Phi_J)$ can be written in terms of the functions from the coarsest space supplied with the complement functions from all intermediate levels,

$$S(\Phi_J) = S(\Phi_{j_0}) \oplus \bigoplus_{j=j_0}^{J-1} S(\Psi_j). \tag{2.130}$$

Thus, every function $v \in S(\Phi_J)$ can be written in its *single-scale representation*

$$v = (\mathbf{c}_J)^T \Phi_J = \sum_{k \in \Delta_J} c_{J,k} \phi_{J,k} \tag{2.131}$$

as well as in its *multi-scale form*

$$v = (\mathbf{c}_{j_0})^T \Phi_{j_0} + (\mathbf{d}_{j_0})^T \Psi_{j_0} + \cdots + (\mathbf{d}_{J-1})^T \Psi_{J-1} \tag{2.132}$$

with respect to the *multiscale* or *wavelet basis*

$$\Psi^J := \Phi_{j_0} \cup \bigcup_{j=j_0}^{J-1} \Psi_j =: \bigcup_{j=j_0-1}^{J-1} \Psi_j \tag{2.133}$$

Often the single-scale representation of a function may be easier to compute and evaluate while the multi-scale representation allows one to separate features of the underlying function characterized by different length scales. Since therefore both representations are advantageous, it is useful to determine the transformation between the two representations, commonly referred to as the *Wavelet Transform*,

$$\mathbf{T}_J : \ell_2(\Delta_J) \to \ell_2(\Delta_J), \qquad \mathbf{d}^J \mapsto \mathbf{c}_J, \tag{2.134}$$

where

$$\mathbf{d}^J := (\mathbf{c}_{j_0}, \mathbf{d}_{j_0}, \dots, \mathbf{d}_{J-1})^T.$$

The previous relations (2.128) and (2.129) indicate that this will involve the matrices $\mathbf{M}_j$ and $\mathbf{G}_j$. In fact, $\mathbf{T}_J$ has the representation

$$\mathbf{T}_J = \mathbf{T}_{J,J-1} \cdots \mathbf{T}_{J,j_0}, \tag{2.135}$$

where each factor has the form

$$\mathbf{T}_{J,j} := \begin{pmatrix} \mathbf{M}_j & \mathbf{0} \\ \mathbf{0} & \mathbf{I}^{(\#\Delta_J - \#\Delta_{j+1})} \end{pmatrix} \in \mathbb{R}^{(\#\Delta_J) \times (\#\Delta_J)}. \tag{2.136}$$

Schematically $\mathbf{T}_J$ can be visualized as a pyramid scheme

$$
\begin{array}{ccccccccc}
\mathbf{M}_{j_0,0} & & \mathbf{M}_{j_0+1,0} & & & & & \mathbf{M}_{J-1,0} & \\
\mathbf{c}_{j_0} \longrightarrow & \mathbf{c}_{j_0+1} & \longrightarrow & \mathbf{c}_{j_0+2} \longrightarrow & \cdots & \mathbf{c}_{J-1} & \longrightarrow & & \mathbf{c}_J \\
\mathbf{M}_{j_0,1} & & \mathbf{M}_{j_0+1,1} & & & & & \mathbf{M}_{J-1,1} & \\
\nearrow & & \nearrow & & \nearrow & \cdots & & \nearrow & \\
\mathbf{d}_{j_0} & \mathbf{d}_{j_0+1} & & \mathbf{d}_{j_0+2} & & \mathbf{d}_{J-1} & & &
\end{array} \tag{2.137}
$$

Accordingly, the inverse transform $\mathbf{T}_J^{-1}$ can be written also in product structure (2.135) in reverse order involving the matrices $\mathbf{G}_j$ as follows:

$$\mathbf{T}_J^{-1} = \mathbf{T}_{J,j_0}^{-1} \cdots \mathbf{T}_{J,J-1}^{-1}, \tag{2.138}$$

where each factor has the form

$$\mathbf{T}_{J,j}^{-1} := \begin{pmatrix} \mathbf{G}_j & \mathbf{0} \\ \mathbf{0} & \mathbf{I}^{(\#\Delta_J - \#\Delta_{j+1})} \end{pmatrix} \in \mathbb{R}^{(\#\Delta_J) \times (\#\Delta_J)}. \tag{2.139}$$

The corresponding pyramid scheme is then

$$
\begin{array}{ccccccccc}
\mathbf{G}_{J-1,0} & & \mathbf{G}_{J-2,0} & & & & \mathbf{G}_{j_0,0} & \\
\mathbf{c}_J \longrightarrow & \mathbf{c}_{J-1} & \longrightarrow & \mathbf{c}_{J-2} \longrightarrow & \cdots & \longrightarrow & \mathbf{c}_{j_0} \\
\mathbf{G}_{J-1,1} & & \mathbf{G}_{J-2,1} & & & & \mathbf{G}_{j_0,1} & \\
\searrow & & \searrow & & \searrow & \cdots & \searrow & \\
& \mathbf{d}_{J-1} & & \mathbf{d}_{J-2} & & \mathbf{d}_{J-1} & & \mathbf{d}_{j_0}
\end{array} \tag{2.140}
$$

Remark 5 Property (2.126) and the fact that $\mathbf{M}_j$ and $\mathbf{G}_j$ can be applied in $(\#\Delta_{j+1})$ operations uniformly in j entails that the complexity of applying $\mathbf{T}_J$ or $\mathbf{T}_J^{-1}$ using the pyramid scheme is of order $\mathcal{O}(\#\Delta_J) = \mathcal{O}(\dim S_J)$ uniformly in J. For this reason, $\mathbf{T}_J$ is called the *Fast Wavelet Transform* (FWT). Note that there is no need to explicitly assemble $\mathbf{T}_J$ or $\mathbf{T}_J^{-1}$.

In Table 2.4 spectral condition numbers for the Fast Wavelet Transform (FWT) for different constructions of biorthogonal wavelets on the interval computed in [62] are displayed.

Since $\cup_{j \geq j_0} S_j$ is dense in L_2, a basis for the whole space L_2 is obtained when letting $J \to \infty$ in (2.133),

$$\Psi := \bigcup_{j=j_0-1}^{\infty} \Psi_j = \{\psi_{j,k} : (j,k) \in \mathit{II}\}, \qquad \Psi_{j_0-1} := \Phi_{j_0}$$

$$\mathit{II} := \left\{ \{j_0\} \times \Delta_{j_0} \right\} \cup \bigcup_{j=j_0}^{\infty} \left\{ \{j\} \times \nabla_j \right\}.$$

(2.141)

The next theorem from [23] illustrates the relation between Ψ and $\mathbf{T}_J$.

Theorem 5 *The multiscale transformations $\mathbf{T}_J$ are well-conditioned in the sense*

$$\|\mathbf{T}_J\|, \|\mathbf{T}_J^{-1}\| = \mathcal{O}(1), \quad J \geq j_0, \tag{2.142}$$

if and only if the collection Ψ defined by (2.141) is a Riesz basis for L_2, i.e., every $v \in L_2$ has unique expansions

$$v = \sum_{j=j_0-1}^{\infty} \langle v, \tilde{\Psi}_j \rangle \Psi_j = \sum_{j=j_0-1}^{\infty} \langle v, \Psi_j \rangle \tilde{\Psi}_j, \tag{2.143}$$

where $\tilde{\Psi}$ defined analogously as in (2.141) is also a Riesz basis for L_2 which is biorthogonal or dual to Ψ,

$$\langle \Psi, \tilde{\Psi} \rangle = \mathbf{I} \tag{2.144}$$

such that

$$\|v\|_{L_2} \sim \|\langle \tilde{\Psi}, v \rangle\|_{\ell_2(\mathit{II})} \sim \|\langle \Psi, v \rangle\|_{\ell_2(\mathit{II})}. \tag{2.145}$$

We briefly explain next how the functions in $\tilde{\Psi}$, denoted as *wavelets dual to Ψ*, or *dual wavelets*, can be determined. Assume that there is a second multiresolution $\tilde{\mathscr{S}}$ of L_2 satisfying (2.114) where

$$\tilde{S}_j = S(\tilde{\Phi}_j), \qquad \tilde{\Phi}_j = \{\tilde{\phi}_{j,k} : k \in \Delta_j\} \tag{2.146}$$

and $\{\tilde{\Phi}_j\}_{j=j_0}^{\infty}$ is uniformly stable in j in the sense of (2.116). Let the functions in $\tilde{\Phi}_j$ also have compact support satisfying (2.117). Furthermore, suppose that the biorthogonality conditions

$$\langle \Phi_j, \tilde{\Phi}_j \rangle = \mathbf{I} \qquad (2.147)$$

hold. We will often refer to Φ_j as the *primal* and to $\tilde{\Phi}_j$ as the *dual generators*. The nestedness of the $\tilde{S}_j$ and the stability again implies that $\tilde{\Phi}_j$ is refinable with some matrix $\tilde{\mathbf{M}}_{j,0}$, similar to (2.120),

$$\tilde{\Phi}_j = \tilde{\mathbf{M}}_{j,0}^T \tilde{\Phi}_{j+1}. \qquad (2.148)$$

The problem of determining biorthogonal wavelets now consists in finding bases $\Psi_j, \tilde{\Psi}_j$ for the complements of $S(\Phi_j)$ in $S(\Phi_{j+1})$, and of $S(\tilde{\Phi}_j)$ in $S(\tilde{\Phi}_{j+1})$, such that

$$S(\Phi_j) \perp S(\tilde{\Psi}_j), \qquad S(\tilde{\Phi}_j) \perp S(\Psi_j) \qquad (2.149)$$

and

$$S(\Psi_j) \perp S(\tilde{\Psi}_r), \quad j \neq r, \qquad (2.150)$$

holds. The connection between the concept of stable completions and the dual generators and wavelets is made by the following result which is a special case from [14].

Proposition 2 *Suppose that the biorthogonal collections $\{\Phi_j\}_{j=j_0}^{\infty}$ and $\{\tilde{\Phi}_j\}_{j=j_0}^{\infty}$ are both uniformly stable and refinable with refinement matrices $\mathbf{M}_{j,0}, \tilde{\mathbf{M}}_{j,0}$, i.e.,*

$$\Phi_j = \mathbf{M}_{j,0}^T \Phi_{j+1}, \qquad \tilde{\Phi}_j = \tilde{\mathbf{M}}_{j,0}^T \tilde{\Phi}_{j+1}, \qquad (2.151)$$

and satisfy the duality condition (2.147). Assume that $\check{\mathbf{M}}_{j,1}$ is any stable completion of $\mathbf{M}_{j,0}$ such that

$$\check{\mathbf{M}}_j := (\mathbf{M}_{j,0}, \check{\mathbf{M}}_{j,1}) = \check{\mathbf{G}}_j^{-1} \qquad (2.152)$$

satisfies (2.126).
   Then

$$\mathbf{M}_{j,1} := (\mathbf{I} - \mathbf{M}_{j,0}\tilde{\mathbf{M}}_{j,0}^T)\check{\mathbf{M}}_{j,1} \qquad (2.153)$$

is also a stable completion of $\mathbf{M}_{j,0}$, *and* $\mathbf{G}_j = \mathbf{M}_j^{-1} = (\mathbf{M}_{j,0}, \mathbf{M}_{j,1})^{-1}$ *has the form*

$$\mathbf{G}_j = \begin{pmatrix} \check{\mathbf{M}}_{j,0}^T \\ \check{\mathbf{G}}_{j,1} \end{pmatrix}. \tag{2.154}$$

Moreover, the collections of functions

$$\Psi_j := \mathbf{M}_{j,1}^T \Phi_{j+1}, \qquad \tilde{\Psi}_j := \check{\mathbf{G}}_{j,1} \tilde{\Phi}_{j+1} \tag{2.155}$$

form biorthogonal systems,

$$\langle \Psi_j, \tilde{\Psi}_j \rangle = \mathbf{I}, \qquad \langle \Psi_j, \tilde{\Phi}_j \rangle = \langle \Phi_j, \tilde{\Psi}_j \rangle = \mathbf{0}, \tag{2.156}$$

so that

$$S(\Psi_j) \perp S(\tilde{\Psi}_r), \quad j \neq r, \qquad S(\Phi_j) \perp S(\tilde{\Psi}_j), \quad S(\tilde{\Phi}_j) \perp S(\Psi_j). \tag{2.157}$$

In particular, the relations (2.147), (2.156) imply that the collections

$$\Psi = \bigcup_{j=j_0-1}^{\infty} \Psi_j, \qquad \tilde{\Psi} := \bigcup_{j=j_0-1}^{\infty} \tilde{\Psi}_j := \tilde{\Phi}_{j_0} \cup \bigcup_{j=j_0}^{\infty} \tilde{\Psi}_j \tag{2.158}$$

are biorthogonal,

$$\langle \Psi, \tilde{\Psi} \rangle = \mathbf{I}. \tag{2.159}$$

Remark 6 It is important to note that the properties needed in addition to (2.159) in order to ensure (2.145) are neither properties of the complements nor of their bases Ψ, $\tilde{\Psi}$ but of the multiresolution sequences $\mathscr{S}$ and $\tilde{\mathscr{S}}$. These can be phrased as approximation and regularity properties and appear in Theorem 6.

We briefly recall yet another useful point of view. The operators

$$P_j v := \langle v, \tilde{\Phi}_j \rangle \Phi_j = \langle v, \tilde{\Psi}^j \rangle \Psi^j = \langle v, \tilde{\Phi}_{j_0} \rangle \Phi_{j_0} + \sum_{r=j_0}^{j-1} \langle v, \tilde{\Psi}_r \rangle \Psi_r$$

$$P_j' v := \langle v, \Phi_j \rangle \tilde{\Phi}_j = \langle v, \Psi^j \rangle \tilde{\Psi}^j = \langle v, \Phi_{j_0} \rangle \tilde{\Phi}_{j_0} + \sum_{r=j_0}^{j-1} \langle v, \Psi_r \rangle \tilde{\Psi}_r \tag{2.160}$$

are projectors onto

$$S(\Phi_j) = S(\Psi^j) \qquad \text{and} \qquad S(\tilde{\Phi}_j) = S(\tilde{\Psi}^j) \tag{2.161}$$

respectively, which satisfy

$$P_r P_j = P_r, \quad P'_r P'_j = P'_r, \quad r \le j. \tag{2.162}$$

Remark 7 Let $\{\Phi_j\}_{j=j_0}^{\infty}$ be uniformly stable. The P_j defined by (2.160) are uniformly bounded if and only if $\{\tilde{\Phi}_j\}_{j=j_0}^{\infty}$ is also uniformly stable. Moreover, the P_j satisfy (2.162) if and only if the $\tilde{\Phi}_j$ are refinable as well. Note that then (2.147) implies

$$\mathbf{M}_{j,0}^T \tilde{\mathbf{M}}_{j,0} = \mathbf{I}. \tag{2.163}$$

In terms of the projectors, the uniform stability of the complement bases Ψ_j, $\tilde{\Psi}_j$ means that

$$\|(P_{j+1} - P_j)v\|_{L_2} \sim \|\langle \tilde{\Psi}_j, v \rangle\|_{\ell_2(\nabla_j)}, \quad \|(P'_{j+1} - P'_j)v\|_{L_2} \sim \|\langle \Psi_j, v \rangle\|_{\ell_2(\nabla_j)}, \tag{2.164}$$

so that the L_2 norm equivalence (2.145) is equivalent to

$$\|v\|_{L_2}^2 \sim \sum_{j=j_0}^{\infty} \|(P_j - P_{j-1})v\|_{L_2}^2 \sim \sum_{j=j_0}^{\infty} \|(P'_j - P'_{j-1})v\|_{L_2}^2 \tag{2.165}$$

for any $v \in L_2$, where $P_{j_0-1} = P'_{j_0-1} := 0$.

The whole concept derived so far lives from both Φ_j and $\tilde{\Phi}_j$. It should be pointed out that in the algorithms one actually does not need $\tilde{\Phi}_j$ explicitly for computations.

We recall next results that guarantee norm equivalences of the type (2.89) for Sobolev spaces.

Multiresolution of Sobolev Spaces Let now $\mathscr{S}$ be a multiresolution sequence consisting of closed subspaces of H^s with the property (2.114) whose union is dense in H^s. The following result from [23] ensures under which conditions norm equivalences hold for the H^s-norm.

Theorem 6 *Let* $\{\Phi_j\}_{j=j_0}^{\infty}$ *and* $\{\tilde{\Phi}_j\}_{j=j_0}^{\infty}$ *be uniformly stable, refinable, biorthogonal collections and let the* $P_j : H^s \to S(\Phi_j)$ *be defined by (2.160). If the Jackson-type estimate*

$$\inf_{v_j \in S_j} \|v - v_j\|_{L_2} \lesssim 2^{-sj} \|v\|_{H^s}, \quad v \in H^s, \ 0 < s \le \bar{d}, \tag{2.166}$$

and the Bernstein inequality

$$\|v_j\|_{H^s} \lesssim 2^{sj} \|v_j\|_{L_2}, \quad v_j \in S_j, \ s < \bar{t}, \tag{2.167}$$

hold for

$$S_j = \left\{ \begin{matrix} S(\Phi_j) \\ S(\tilde{\Phi}_j) \end{matrix} \right\} \text{ with order } \bar{d} = \left\{ \begin{matrix} d \\ \tilde{d} \end{matrix} \right\} \text{ and } \bar{t} = \left\{ \begin{matrix} t \\ \tilde{t} \end{matrix} \right\}, \tag{2.168}$$

then for

$$0 < \sigma := \min\{d, t\}, \qquad 0 < \tilde{\sigma} := \min\{\tilde{d}, \tilde{t}\}, \tag{2.169}$$

one has

$$\|v\|_{H^s}^2 \sim \sum_{j=j_0}^{\infty} 2^{2sj} \|(P_j - P_{j-1})v\|_{L_2}^2, \quad s \in (-\tilde{\sigma}, \sigma). \tag{2.170}$$

Recall that we always write $H^s = (H^{-s})'$ for $s < 0$.
The regularity of $\mathscr{S}$ and $\tilde{\mathscr{S}}$ is characterized by

$$t := \sup\{s : S(\Phi_j) \subset H^s, \ j \ge j_0\}, \qquad \tilde{t} := \sup\{s : S(\tilde{\Phi}_j) \subset H^s, \ j \ge j_0\} \tag{2.171}$$

Recalling the representation (2.164), we can immediately derive the following fact.

Corollary 2 *Suppose that the assumptions in Theorem 6 hold. Then we have the norm equivalence*

$$\|v\|_{H^s}^2 \sim \sum_{j=j_0-1}^{\infty} 2^{2sj} \|\langle \tilde{\Psi}_j, v \rangle\|_{\ell_2(\nabla_j)}^2, \quad s \in (-\tilde{\sigma}, \sigma). \tag{2.172}$$

In particular for $s = 0$ the Riesz basis property of the Ψ, $\tilde{\Psi}$ relative to L_2(2.145) is recovered. For many applications it suffices to have (2.170) or (2.172) only for certain $s > 0$ for which one only needs to require (2.166) and (2.167) for $\{\Phi_j\}_{j=j_0}^{\infty}$. The Jackson estimates (2.166) of order $\tilde{d}$ for $S(\tilde{\Phi}_j)$ imply the cancellation properties (CP) (2.92), see, e.g., [26].

Remark 8 When the wavelets live on $\Omega \subset \mathbb{R}^n$, (2.166) means that all polynomials up to order $\tilde{d}$ are contained in $S(\tilde{\Phi}_j)$. One also says that $S(\tilde{\Phi}_j)$ is *exact* of order $\tilde{d}$. On account of (2.144), this implies that the wavelets $\psi_{j,k}$ are orthogonal to polynomials up to order $\tilde{d}$ or have $\tilde{d}$th order *vanishing moments*. By Taylor expansion, this in turn yields (2.92).

We will later use the following generalization of the discrete norms (2.165). Let for $s \in \mathbb{R}$

$$\|\|v\|\|_s := \left(\sum_{j=j_0}^{\infty} 2^{2sj} \|(P_j - P_{j-1})v\|_{L_2}^2 \right)^{1/2} \tag{2.173}$$

which by the relations (2.164) is also equivalent to

$$|v|_s := \left(\sum_{j=j_0-1}^{\infty} 2^{2sj} \|\langle \tilde{\Psi}_j, v \rangle\|_{\ell_2(\nabla_j)}^2 \right)^{1/2} . \tag{2.174}$$

In this notation, (2.170) and (2.172) read

$$\|v\|_{H^s} \sim \|\|v\|\|_s \sim |v|_s. \tag{2.175}$$

In terms of such discrete norms, Jackson and Bernstein estimates hold with constants equal to one [51], which turns out to be useful later in Sect. 2.5.2.

Lemma 1 *Let $\{\Phi_j\}_{j=j_0}^{\infty}$ and $\{\tilde{\Phi}_j\}_{j=j_0}^{\infty}$ be uniformly stable, refinable, biorthogonal collections and let the P_j be defined by (2.160). Then the estimates*

$$|v - P_j v|_{s'} \leq 2^{-(j+1)(s-s')} |v|_s, \qquad v \in H^s, \ s' \leq s \leq d, \tag{2.176}$$

and

$$|v_j|_s \leq 2^{j(s-s')} |v_j|_{s'}, \qquad v_j \in S(\Phi_j), \ s' \leq s \leq d, \tag{2.177}$$

are valid, and correspondingly for the dual side.

The same results hold for the norm $\|\| \cdot \|\|$ defined in (2.173).

Reverse Cauchy–Schwarz Inequalities The biorthogonality condition (2.147) implies together with direct and inverse estimates the following reverse Cauchy–Schwarz inequalities for finite-dimensional spaces [28]. It will be one essential ingredient for the discussion of the LBB condition in Sect. 2.5.2.

Lemma 2 *Let the assumptions in Theorem 6 be valid such that the norm equivalence (2.170) holds for $(-\tilde{\sigma}, \sigma)$ with $\sigma, \tilde{\sigma}$ defined in (2.169). Then for any $v \in S(\Phi_j)$ there exists some $\tilde{v}^* = \tilde{v}^*(v) \in S(\tilde{\Phi}_j)$ such that*

$$\|v\|_{H^s} \|\tilde{v}^*\|_{H^{-s}} \lesssim \langle v, \tilde{v}^* \rangle \tag{2.178}$$

for any $0 \leq s < \min(\sigma, \tilde{\sigma})$.

The proof of this result given in [28] for $s = 1/2$ in terms of the projectors P_j defined in (2.160) and corresponding duals P'_j immediately carries over to more general s. Recalling the representation (2.161) in terms of wavelets, the reverse Cauchy inequality (2.178) attains the following sharp form.

Lemma 3 ([51]) *Let the assumptions of Lemma 1 hold. Then for every $v \in S(\Phi_j)$ there exists some $\tilde{v}^* = \tilde{v}^*(v) \in S(\tilde{\Phi}_j)$ such that*

$$|v|_s \, |\tilde{v}^*|_{-s} = \langle v, \tilde{v}^* \rangle \tag{2.179}$$

for any $0 \le s \le \min(\sigma, \tilde{\sigma})$.

Proof Every $v \in S(\Phi_j)$ can be written as

$$v = \sum_{r=j_0-1}^{j-1} 2^{sr} \sum_{k \in \nabla_r} v_{r,k} \psi_{r,k}.$$

Setting now

$$\tilde{v}^* := \sum_{r=j_0-1}^{j-1} 2^{-sr} \sum_{k \in \nabla_r} v_{r,k} \tilde{\psi}_{r,k}$$

with the same coefficients $v_{j,k}$, the definition of $|\cdot|_s$ yields by biorthogonality (2.159)

$$|v|_s \, |\tilde{v}^*|_{-s} = \sum_{r=j_0-1}^{j-1} \sum_{k \in \nabla_r} |v_{j,k}|^2.$$

Combining this with the observation

$$\langle v, \tilde{v}^* \rangle = \sum_{r=j_0-1}^{j-1} \sum_{k \in \nabla_r} |v_{j,k}|^2$$

confirms (2.179). □

Remark 9 The previous proof reveals that the identity (2.179) is also true for elements from infinite-dimensional spaces H^s and $(H^s)'$ for which Ψ and $\tilde{\Psi}$ are Riesz bases.

Biorthogonal Wavelets on $\mathbb{R}$ The construction of biorthogonal spline-wavelets on $\mathbb{R}$ from [18] for $L_2 = L_2(\mathbb{R})$ employs the multiresolution framework introduced at the beginning of this section. There the $\phi_{j,k}$ are generated through the dilates and translates of a single function $\phi \in L_2$,

$$\phi_{j,k} = 2^{j/2}\phi(2^j \cdot -k).$$ (2.180)

This corresponds to the idea of a *uniform* virtual underlying grid, explaining the terminology *uniform refinements*. B-Splines on uniform grids are known to satisfy refinement relations (2.119) in addition to being compactly supported and having L_2-stable integer translates. For computations, they have the additional advantage that they can be expressed as piecewise polynomials. In the context of variational formulations for second order boundary value problems, a well-used example are the nodal finite elements $\phi_{j,k}$ generated by the cardinal B-Spline of order two, i.e., the piecewise linear continuous function commonly called the 'hat function'. For cardinal B-Splines as generators, a whole class of dual generators $\tilde{\phi}_{j,k}$ (of arbitrary smoothness at the expense of larger supports) can be constructed which are also generated by one single function $\tilde{\phi}$ through translates and dilates. By Fourier techniques, one can construct from $\phi, \tilde{\phi}$ then a pair of biorthogonal wavelets $\psi, \tilde{\psi}$ whose dilates and translates built as in (2.180) constitute Riesz bases for $L_2(\mathbb{R})$.

By taking tensor products of these functions, one can generate biorthogonal wavelet bases for $L_2(\mathbb{R}^n)$.

Biorthogonal Wavelets on Domains Some constructions that exist by now have as a core ingredient tensor products of one-dimensional wavelets on an *interval* derived from the biorthogonal wavelets from [18] on $\mathbb{R}$. On finite intervals in $\mathbb{R}$, the corresponding constructions are usually based on keeping the elements of $\Phi_j, \tilde{\Phi}_j$ supported *inside* the interval while modifying those translates overlapping the end points of the interval so as to preserve a desired degree of polynomial exactness. A general detailed construction satisfying all these requirements has been proposed in [34]. Here just the main ideas for constructing a biorthogonal pair $\Phi_j, \tilde{\Phi}_j$ and corresponding wavelets satisfying the above requirements are sketched, where we apply the techniques derived at the beginning of this section.

We start out with those functions from two collections of biorthogonal generators $\Phi_j^{\mathbb{R}}, \tilde{\Phi}_j^{\mathbb{R}}$ for some fixed $j \geq j_0$ living on the whole real line whose support has nonempty intersection with the interval $(0, 1)$. In order to treat the boundary effects separately, we assumed that the coarsest resolution level j_0 is large enough so that, in view of (2.117), functions overlapping one end of the interval vanish at the other. One then leaves as many functions from the collection $\Phi_j^{\mathbb{R}}, \tilde{\Phi}_j^{\mathbb{R}}$ living in the interior of the interval untouched and modifies only those near the interval ends. Note that keeping just the restrictions to the interval of those translates overlapping the end points would destroy stability (and also the cardinality of the primal and dual basis functions living on $(0, 1)$ since their supports do not have the same size). Therefore, modifications at the end points are necessary; also, just discarding them from the collections (2.115), (2.146) would produce an error near the end points. The basic idea is essentially the same for all constructions of orthogonal and biorthogonal wavelets on $\mathbb{R}$ adapted to an interval. Namely, one takes *fixed* linear combinations of all functions in $\Phi_j^{\mathbb{R}}, \tilde{\Phi}_j^{\mathbb{R}}$ living near the ends of the interval in such a way that monomials up to the exactness order are reproduced there and such that the generator bases have the same cardinality. Because of the boundary modifications, the collections of generators are there no longer biorthogonal. However, one can show in the case of cardinal B-Splines as primal generators (which is a widely used class for numerical analysis) that biorthogonalization is indeed possible. This yields collections denoted by $\Phi_j^{(0,1)}$, $\tilde{\Phi}_j^{(0,1)}$ which then satisfy (2.147) on $(0, 1)$ and all assumptions required in Proposition 2.

For the construction of corresponding wavelets, first an *initial* stable completion $\check{\mathbf{M}}_{j,1}$ is computed by applying Gaussian eliminations to factor $\mathbf{M}_{j,0}$ and then to find a uniformly stable inverse of $\check{\mathbf{M}}_j$. Here we exploit that for cardinal B-Splines as generators the refinement matrices $\mathbf{M}_{j,0}$ are totally positive. Thus, they can be stably decomposed by Gaussian elimination without pivoting. Application of Proposition 2 then gives the corresponding biorthogonal wavelets $\Psi_j^{(0,1)}$, $\tilde{\Psi}_j^{(0,1)}$ on $(0, 1)$ which satisfy the requirements in Corollary 2. It turns out that these wavelets coincide in the interior of the interval again with those on all of $\mathbb{R}$ from [18]. An example of the primal wavelets for $d = 2$ generated by piecewise linear continuous functions is displayed in Fig. 2.2 on the left. After constructing these basic versions, one can then perform local transformations near the ends of the interval in order to improve the condition or L_2 stability constants, see [11, 62] for corresponding results and numerical examples.

We display spectral condition numbers for the FWT for two different constructions of biorthogonal wavelets on the interval computed in [62] in Table 2.4. The first column denotes the finest level on which the spectral condition numbers of the FWT are computed. The next column contains the numbers for the construction of biorthogonal spline-wavelets on the interval from [34] for the case $d = 2$, $\tilde{d} = 4$ while the last column displays the numbers for a scaled version derived in [11]. We will see later in Sect. 2.5.1 how the transformation $\mathbf{T}_J$ is used for preconditioning.

Along these lines, also biorthogonal generators and wavelets with homogeneous (Dirichlet) boundary conditions can be constructed. Since the $\Phi_j^{(0,1)}$ are locally near the boundary monomials which all vanish at 0, 1 except for one, removing the one from $\Phi_j^{(0,1)}$ which corresponds to the constant function produces a collection of generators with homogeneous boundary conditions at 0, 1. In order for the

Table 2.4 Computed spectral condition numbers [62] for the Fast Wavelet Transform for different constructions of biorthogonal wavelets on the interval [11, 34]

j	$\kappa_2(\mathbf{T}_{\mathrm{DKU}})$	$\kappa_2(\mathbf{T}_{\mathrm{B}})$	j	$\kappa_2(\mathbf{T}_{\mathrm{DKU}})$	$\kappa_2(\mathbf{T}_{\mathrm{B}})$
4	4.743e+00	4.640e+00	11	1.097e+01	8.011e+00
5	6.221e+00	6.024e+00	12	1.103e+01	8.034e+00
6	8.154e+00	6.860e+00	13	1.106e+01	8.046e+00
7	9.473e+00	7.396e+00	14	1.107e+01	8.051e+00
8	1.023e+01	7.707e+00	15	1.108e+01	8.054e+00
9	1.064e+01	7.876e+00	16	1.108e+01	8.056e+00
10	1.086e+01	7.965e+00			

moment conditions (2.92) still to hold for the Ψ_j, the dual generators have to have *complementary* boundary conditions. A corresponding construction has been carried out in [30] and implemented in [11]. Homogeneous boundary conditions of higher order can be generated accordingly.

By taking tensor products of the wavelets on $(0, 1)$, in this manner biorthogonal wavelets for Sobolev spaces on $(0, 1)^n$ with or without homogeneous boundary conditions are obtained. This construction can be further extended to any other domain or manifold which is the image of a regular parametric mapping of the unit cube. Some results on the construction of wavelets on manifolds are summarized in [25]. There are essentially two approaches. The first idea is based on domain decomposition and consists in 'gluing' generators across interelement boundaries, see, e.g., [13, 31]. These approaches all have in common that the norm equivalences (2.172) for $H^s = H^s(\Gamma)$ can be shown to hold only for the range $-1/2 < s < 3/2$, due to the fact that duality arguments apply only for this range because of the nature of a modified inner product to which biorthogonality refers. The other approach which overcomes the above limitations on the ranges for which the norm equivalences hold has been developed in [32] based on previous characterizations of function spaces as Cartesian products from [15]. The construction in [32] has been optimized and implemented to construct wavelet bases on the sphere in [56, 64], see Fig. 2.2.

Of course, there are also different attempts to construct wavelet bases with the above properties without using tensor products. A construction of biorthogonal spline-wavelets on triangles introduced by [68] has been implemented in two spatial dimensions with an application to the numerical solution of a linear elliptic boundary value problem in [48].

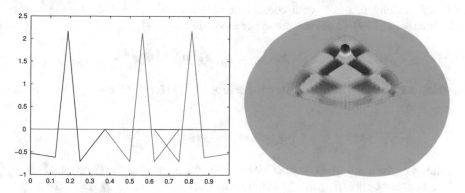

Fig. 2.2 Primal wavelets for $d = 2$ on $[0, 1]$ (left) and on a sphere (right) from [64]

2.5   Problems in Wavelet Coordinates

2.5.1   Elliptic Boundary Value Problems

We now consider the wavelet representation of the elliptic boundary value problem
from Sect. 2.3.2. Let for $\mathscr{H}$ given by (2.41) or (2.42) $\Psi_{\mathscr{H}}$ be a wavelet basis
with corresponding dual $\tilde{\Psi}_{\mathscr{H}}$ which satisfies the properties (R), (L) and (CP)
from Sect. 2.4.1. Following the recipe from Sect. 2.4.3, expanding $y = \mathbf{y}^T \Psi_{\mathscr{H}}$,
$f = \mathbf{f}^T \tilde{\Psi}_{\mathscr{H}}$ and recalling (2.45), the wavelet representation of the elliptic boundary
value problem (2.47) is given by

$$\mathbf{Ay} = \mathbf{f} \tag{2.181}$$

where

$$\mathbf{A} := a(\Psi_{\mathscr{H}}, \Psi_{\mathscr{H}}), \qquad \mathbf{f} := \langle \Psi_{\mathscr{H}}, f \rangle. \tag{2.182}$$

Then the mapping property (2.46) and the Riesz basis property (R) yield the
following fact.

Proposition 3 *The infinite matrix* $\mathbf{A}$ *is a boundedly invertible mapping from* $\ell_2 =$
$\ell_2(\mathbb{I}_{\mathscr{H}})$ *into itself, and there exists finite positive constants* $c_{\mathbf{A}} \leq C_{\mathbf{A}}$ *such that*

$$c_{\mathbf{A}} \|\mathbf{v}\| \leq \|\mathbf{Av}\| \leq C_{\mathbf{A}} \|\mathbf{v}\|, \qquad \mathbf{v} \in \ell_2(\mathbb{I}_{\mathscr{H}}). \tag{2.183}$$

Proof For any $v \in \mathscr{H}$ with coefficient vector $\mathbf{v} \in \ell_2$, we have by the lower
estimates in (2.89), (2.46) and the upper inequality in (2.94), respectively,

$$\|\mathbf{v}\| \leq c_{\mathscr{H}}^{-1} \|v\|_{\mathscr{H}} \leq c_{\mathscr{H}}^{-1} c_A^{-1} \|Av\|_{\mathscr{H}'} = c_{\mathscr{H}}^{-1} c_A^{-1} \|(\mathbf{Av})^T \tilde{\Psi}_{\mathscr{H}}\|_{\mathscr{H}'} \leq c_{\mathscr{H}}^{-2} c_A^{-1} \|\mathbf{Av}\|$$

where we have used the wavelet representation (2.111) for A. Likewise, the converse
estimate

$$\|\mathbf{Av}\| \leq C_{\mathscr{H}} \|Av\|_{\mathscr{H}'} \leq C_{\mathscr{H}} C_A \|v\|_{\mathscr{H}} \leq C_{\mathscr{H}}^2 C_A \|\mathbf{v}\|$$

follows by the lower inequality in (2.94) and the upper estimates in (2.46) and (2.89).
The constants appearing in (2.183) are therefore identified as $c_{\mathbf{A}} := c_{\mathscr{H}}^2 c_A$ and
$C_{\mathbf{A}} := c_{\mathscr{H}}^2 C_A$.   □

In the present situation where $\mathbf{A}$ is defined via the elliptic bilinear form $a(\cdot, \cdot)$,
Proposition 3 entails the following result with respect to *preconditioning*. Let for
$\mathbb{I} = \mathbb{I}_{\mathscr{H}}$ the symbol Λ denote *any* finite subset of the index set $\mathbb{I}$. For the
corresponding set of wavelets $\Psi_\Lambda := \{\psi_\lambda : \lambda \in \Lambda\}$ denote by $S_\Lambda := span\Psi_\Lambda$
the respective finite-dimensional subspace of $\mathscr{H}$. For the wavelet representation of

A in terms of Ψ_Λ,

$$\mathbf{A}_\Lambda := a(\Psi_\Lambda, \Psi_\Lambda), \tag{2.184}$$

we obtain the following result.

Proposition 4 *If $a(\cdot, \cdot)$ is $\mathscr{H}$-elliptic according to (2.44), the finite matrix $\mathbf{A}_\Lambda$ is symmetric positive definite and its spectral condition number is bounded uniformly in Λ, i.e.,*

$$\kappa_2(\mathbf{A}_\Lambda) \le \frac{C_\mathbf{A}}{c_\mathbf{A}}, \tag{2.185}$$

where $c_\mathbf{A}, C_\mathbf{A}$ are the constants from (2.183).

Proof Clearly, since $\mathbf{A}_\Lambda$ is just a finite section of $\mathbf{A}$, we have $\|\mathbf{A}_\Lambda\| \le \|\mathbf{A}\|$. On the other hand, by assumption, $a(\cdot, \cdot)$ is $\mathscr{H}$-elliptic which entails that $a(\cdot, \cdot)$ is also elliptic on every finite subspace $S_\Lambda \subset \mathscr{H}$. Thus, we infer $\|\mathbf{A}_\Lambda^{-1}\| \le \|\mathbf{A}^{-1}\|$, and we have

$$c_\mathbf{A} \|\mathbf{v}_\Lambda\| \le \|\mathbf{A}_\Lambda \mathbf{v}_\Lambda\| \le C_\mathbf{A} \|\mathbf{v}_\Lambda\|, \qquad \mathbf{v}_\Lambda \in S_\Lambda. \tag{2.186}$$

Together with the definition $\kappa_2(\mathbf{A}_\Lambda) := \|\mathbf{A}_\Lambda\| \|\mathbf{A}_\Lambda^{-1}\|$ we obtain the claimed estimate. $\qquad\square$

In other words, representations of A with respect to properly scaled wavelet bases for $\mathscr{H}$ entail well-conditioned system matrices $\mathbf{A}_\Lambda$ independent of Λ. This in turn means that the convergence speed of an iterative solver applied to the corresponding finite system

$$\mathbf{A}_\Lambda \mathbf{y}_\Lambda = \mathbf{f}_\Lambda \tag{2.187}$$

does not deteriorate as $\Lambda \to \infty$.

In summary, ellipticity implies stability of the Galerkin discretizations for any set $\Lambda \subset \mathbb{I}$. This is not the case for finite versions of the saddle point problems discussed in Sect. 2.5.2.

Fast Wavelet Transform Let us briefly summarize how in the situation of uniform refinements, i.e., when $S(\Phi_J) = S(\Psi^J)$, the Fast Wavelet Transformation (FWT) $\mathbf{T}_J$ can be used for preconditioning linear elliptic operators, together with a diagonal scaling induced by the norm equivalence (2.172) [27]. Here we recall the notation from Sect. 2.4.4 where the wavelet basis is in fact the (unscaled) anchor basis from Sect. 2.4.1. Thus, the norm equivalence (2.89) using the scaled wavelet basis Ψ_H is the same as (2.172) in the anchor basis. Recall that the norm equivalence (2.172) implies that every $v \in H^s$ can be expanded uniquely in terms of the Ψ and its expansion coefficients $\mathbf{v}$ satisfy

$$\|v\|_{H^s} \sim \|\mathbf{D}^s \mathbf{v}\|_{\ell_2}$$

where $\mathbf{D}^s$ is a diagonal matrix with entries $\mathbf{D}^s_{(j,k),(j',k')} = 2^{sj}\delta_{j,j'}\delta_{k,k'}$. For $\mathscr{H} \subset H^1(\Omega)$, the case $s = 1$ is relevant.

In a stable Galerkin scheme for (2.43) with respect to $S(\Psi^J) = S(\Psi_\Lambda)$, we have therefore already identified the diagonal (scaling) matrix $\mathbf{D}_J$ consisting of the finite portion of the matrix $\mathbf{D} = \mathbf{D}^1$ for which $j_0 - 1 \le j \le J - 1$. The representation of A with respect to the (unscaled) wavelet basis Ψ^J can be expressed in terms of the Fast Wavelet Transform $\mathbf{T}_J$, that is,

$$\langle \Psi^J, A\Psi^J \rangle = \mathbf{T}_J^T \langle \Phi_J, A\Phi_J \rangle \mathbf{T}_J, \tag{2.188}$$

where Φ_J is the single-scale basis for $S(\Psi^J)$. Thus, we first set up the operator equation as in Finite Element settings in terms of the single-scale basis Φ_J. Applying the Fast Wavelet Transform $\mathbf{T}_J$ together with $\mathbf{D}_J$ yields that the operator

$$\mathbf{A}_J := \mathbf{D}_J^{-1} \mathbf{T}_J^T \langle \Phi_J, A\Phi_J \rangle \mathbf{T}_J \mathbf{D}_J^{-1} \tag{2.189}$$

has uniformly bounded condition numbers independent of J. This can be seen by combining the properties of A according to (2.46) with the norm equivalences (2.89) and (2.94).

It is known that the boundary adaptations of the generators and wavelets aggravate the absolute values of the condition numbers. Nevertheless, these constants can be greatly reduced by sophisticated biorthogonalizations of the boundary adapted functions [11]. Numerical tests confirm that the absolute constants can further be improved by taking instead of $\mathbf{D}_J^{-1}$ the inverse of the diagonal of $\langle \Psi^J, A\Psi^J \rangle$ for the scaling in (2.189) [11, 17, 62]. Table 2.5 displays the condition numbers for discretizations of an operator in two spatial dimensions for boundary adapted biorthogonal spline-wavelets in the case $d = 2, \tilde{d} = 4$ with such a scaling.

2.5.2   Saddle Point Problems Involving Boundary Conditions

As in the previous situation, we first derive an infinite wavelet representation of the saddle point problem introduced in Sect. 2.3.3.

Let for $\mathscr{H} = Y \times Q$ with $Y = H^1(\Omega)$, $Q = (H^{1/2}(\Gamma))'$ two collections of wavelet bases Ψ_Y, Ψ_Q be available, each satisfying (R), (L) and (CP), with respective duals $\tilde{\Psi}_Y$, $\tilde{\Psi}_Q$. Similar to the previous case, we expand $y = \mathbf{y}^T \Psi_Y$ and $p = \mathbf{p}^T \Psi_Q$ and test with the elements from Ψ_Y, Ψ_Q. Then (2.54) attains the form

$$\mathbf{L} \begin{pmatrix} \mathbf{y} \\ \mathbf{p} \end{pmatrix} := \begin{pmatrix} \mathbf{A} & \mathbf{B}^T \\ \mathbf{B} & \mathbf{0} \end{pmatrix} \begin{pmatrix} \mathbf{y} \\ \mathbf{p} \end{pmatrix} = \begin{pmatrix} \mathbf{f} \\ \mathbf{g} \end{pmatrix}, \tag{2.190}$$

where

$$\mathbf{A} := \langle \Psi_Y, A\Psi_Y \rangle \quad \mathbf{f} := \langle \Psi_Y, f \rangle,$$
$$\mathbf{B} := \langle \Psi_Q, B\Psi_Y \rangle, \quad \mathbf{g} := \langle \Psi_Q, g \rangle.$$

(2.191)

In view of the above assertions, the operator $\mathbf{L}$ is an ℓ_2-automorphism, i.e., for every $(\mathbf{v}, \mathbf{q}) \in \ell_2(\mathbb{I}) = \ell_2(\mathbb{I}_Y \times \mathbb{I}_Q)$ we have

$$c_{\mathbf{L}} \left\| \begin{pmatrix} \mathbf{v} \\ \mathbf{q} \end{pmatrix} \right\| \leq \left\| \mathbf{L} \begin{pmatrix} \mathbf{v} \\ \mathbf{q} \end{pmatrix} \right\| \leq C_{\mathbf{L}} \left\| \begin{pmatrix} \mathbf{v} \\ \mathbf{q} \end{pmatrix} \right\|$$

(2.192)

with constants $c_{\mathbf{L}}, C_{\mathbf{L}}$ only depending on $c_{\mathscr{L}}, C_{\mathscr{L}}$ from (2.59) and the constants in the norm equivalences (2.89) and (2.94).

For saddle point problems with an operator $\mathbf{L}$ satisfying (2.192), finite sections are in general not uniformly stable in the sense of (2.186). In fact, for discretizations on uniform grids, the validity of the corresponding mapping property relies on a suitable stability condition, see e.g. [9, 42]. The relevant facts derived in [28] are as follows.

The bilinear form $a(\cdot, \cdot)$ defined in (2.40) is for $c > 0$ elliptic on all of $Y = H^1(\Omega)$ and, hence, also on any finite-dimensional subspace of Y. Let there be two multiresolution analyses $\mathscr{Y}$ of $H^1(\Omega)$ and $\mathscr{Q}$ of Q where the discrete spaces are $Y_j \subset H^1(\Omega)$ and $Q_A =: Q_\ell \subset (H^{1/2}(\Gamma))'$. With the notation from Sect. 2.4.4 and in addition superscripts referring to the domain on which the functions live, these spaces are represented by

$$Y_j = S(\Phi_j^{\Omega}) = S(\Psi^{j,\Omega}), \quad \tilde{Y}_j = S(\tilde{\Phi}_j^{\Omega}) = S(\tilde{\Psi}^{j,\Omega}),$$
$$Q_\ell = S(\Phi_\ell^{\Gamma}) = S(\Psi^{\ell,\Gamma}), \quad \tilde{Q}_\ell = S(\tilde{\Phi}_\ell^{\Gamma}) = S(\tilde{\Psi}^{\ell,\Gamma}).$$

(2.193)

Here the indices j and ℓ refer to mesh sizes on the domain and the boundary,

$$h_\Omega \sim 2^{-j} \qquad \text{and} \qquad h_\Gamma \sim 2^{-\ell}.$$

The discrete inf–sup condition, the *LBB condition*, for the pair Y_j, Q_ℓ requires that there exists a constant $\beta_1 > 0$ *independent* of j and ℓ such that

$$\inf_{q \in Q_\ell} \sup_{v \in Y_j} \frac{b(v, q)}{\|v\|_{H^1(\Omega)} \|q\|_{(H^{1/2}(\Gamma))'}} \geq \beta_1 > 0$$

(2.194)

holds. We have investigated in [28] the general case in arbitrary spatial dimensions where the Q_ℓ are *not* trace spaces of Y_j. Employing the reverse Cauchy-Schwarz inequalities from Sect. 2.4.4, one can show that (2.194) is satisfied provided that $h_\Gamma(h_\Omega)^{-1} = 2^{j-\ell} \geq c_\Omega > 1$, similar to a condition which was known for bivariate polygons and particular finite elements [2, 41].

Table 2.5 Spectral
condition numbers of the
operators **A** and **L** for
different constructions of
biorthogonal wavelets on the
interval [62]

j	$\kappa_2(\mathbf{A}_{\mathrm{DKU}})$	$\kappa_2(\mathbf{A}_{\mathrm{B}})$	$\kappa_2(\mathbf{L}_{\mathrm{DKU}})$	$\kappa_2(\mathbf{L}_{\mathrm{DKU}})$
3	5.195e+02	1.898e+01	1.581e+02	4.147e+01
4	6.271e+02	1.066e+02	1.903e+02	1.050e+02
5	6.522e+02	1.423e+02	1.997e+02	1.399e+02
6	6.830e+02	1.820e+02	2.112e+02	1.806e+02
7	7.037e+02	2.162e+02	2.318e+02	2.145e+02
8	7.205e+02	2.457e+02	2.530e+02	2.431e+02
9	7.336e+02	2.679e+02	2.706e+02	2.652e+02

It should be mentioned that the obstructions caused by the LBB condition can be avoided by means of stabilization techniques proposed, e.g., in [67] where, however, the location of the boundary of Ω relative to the mesh is somewhat constrained. Another stabilization strategy based on wavelets has been investigated in [6]. A related approach which systematically avoids restrictions of the LBB type is based on least squares techniques [35].

It is particularly interesting that adaptive schemes based on wavelets like the one in Sect. 2.6.2 can be designed in such a way that the LBB condition is *automatically* enforced which was first observed in [22]. More on this subject can be found in [26].

In order to get an impression of the value of the constants for the condition numbers for $\mathbf{A}_\Lambda$ in (2.185) and the corresponding ones for the saddle point operator on uniform grids (2.192), we mention an example investigated and implemented in [62]. In this example, $\Omega = (0, 1)^2$ and Γ is one face of its boundary. In Table 2.5 from [62], the spectral condition numbers of **A** and **L** with respect to two different constructions of wavelets for the case $d = 2$ and $\tilde{d} = 4$ are displayed. We see next to the first column in which the refinement level j is listed the spectral condition numbers of **A** with the wavelet construction from [34] denoted by $\mathbf{A}_{\mathrm{DKU}}$ and with the modification introduced in [11] and a further transformation [62] denoted by $\mathbf{A}_{\mathrm{B}}$. The last columns contain the respective numbers for the saddle point matrix **L** where $\kappa_2(\mathbf{L}) := \sqrt{\kappa(\mathbf{L}^T\mathbf{L})}$.

2.5.3 Control Problems: Distributed Control

We now discuss appropriate wavelet formulations for PDE-constrained control problems with distributed control as introduced in Sect. 2.3.5. Let for any space $\mathcal{V} \in \{H, \mathcal{Z}, \mathcal{U}\}$ $\Psi_{\mathcal{V}}$ denote a wavelet basis with the properties (R), (L), (CP) for $\mathcal{V}$ with dual basis $\tilde{\Psi}_{\mathcal{V}}$.

Let $\mathcal{Z}, \mathcal{U}$ satisfy the embedding (2.76). In terms of wavelet bases, the corresponding canonical injections correspond in view of (2.96) to a multiplication by a diagonal matrix. That is, let $\mathbf{D}_{\mathcal{Z}}, \mathbf{D}_H$ be such that

$$\Psi_{\mathcal{Z}} = \mathbf{D}_{\mathcal{Z}}\Psi_H, \quad \tilde{\Psi}_H = \mathbf{D}_H\Psi_{\mathcal{U}}. \tag{2.195}$$

Since $\mathscr{L}$ possibly induces a weaker and $\mathscr{U}$ a stronger topology, the diagonal matrices $\mathbf{D}_{\mathscr{L}}$, $\mathbf{D}_H$ are such that their entries are nondecreasing in scale, and there is a finite constant C such that

$$\|\mathbf{D}_{\mathscr{L}}^{-1}\|, \|\mathbf{D}_H^{-1}\| \leq C. \tag{2.196}$$

For instance, for $H = H^\alpha$, $\mathscr{L} = H^\beta$, or for $H' = H^{-\alpha}$, $\mathscr{U} = H^{-\beta}$, $0 \leq \beta \leq \alpha$, $\mathbf{D}_{\mathscr{L}}$, $\mathbf{D}_H$ have entries $(\mathbf{D}_{\mathscr{L}})_{\lambda,\lambda} = (\mathbf{D}_H)_{\lambda,\lambda} = (\mathbf{D}^{\alpha-\beta})_{\lambda,\lambda} = 2^{(\alpha-\beta)|\lambda|}$.

We expand y in Ψ_H and u in a wavelet basis $\Psi_{\mathscr{U}}$ for $\mathscr{U} \subset H'$,

$$u = \mathbf{u}^T \Psi_U = (\mathbf{D}_H^{-1}\mathbf{u})^T \Psi_{H'}. \tag{2.197}$$

Following the derivation in Sect. 2.5.1, the linear constraints (2.75) attain the form

$$\mathbf{A}\mathbf{y} = \mathbf{f} + \mathbf{D}_H^{-1}\mathbf{u} \tag{2.198}$$

where

$$\mathbf{A} := a(\Psi_H, \Psi_H), \qquad \mathbf{f} := \langle \Psi_H, f \rangle. \tag{2.199}$$

Recall that $\mathbf{A}$ has been assumed to be symmetric. The objective functional (2.80) is stated in terms of the norms $\| \cdot \|_{\mathscr{L}}$ and $\| \cdot \|_{\mathscr{U}}$. For an exact representation of these norms, corresponding Riesz operators $\mathbf{R}_{\mathscr{L}}$ and $\mathbf{R}_{\mathscr{U}}$ defined analogously to (2.106) would come into play which may not be explicitly computable if $\mathscr{L}$, $\mathscr{U}$ are fractional Sobolev spaces. On the other hand, as mentioned before, such a cost functional in many cases serves the purpose of yielding unique solutions while there is some ambiguity in its exact formulation. Hence, in search for a formulation which best supports numerical realizations, it is often sufficient to employ norms which are *equivalent* to $\| \cdot \|_{\mathscr{L}}$ and $\| \cdot \|_{\mathscr{U}}$. In view of the discussion in Sect. 2.4.2, we can work for the norms $\| \cdot \|_{\mathscr{L}}$, $\| \cdot \|_{\mathscr{U}}$ only with the diagonal scaling matrices $\mathbf{D}^s$ induced by the regularity of $\mathscr{L}$, $\mathscr{U}$, or we can in addition include the Riesz map $\mathbf{R}$ defined in (2.101). In the numerical studies in [11], a somewhat better quality of the solution is observed when $\mathbf{R}$ is included. In order to keep track of the appearance of the Riesz maps in the linear systems derived below, we choose here the latter variant.

Moreover, we expand the given observation function $y_* \in \mathscr{L}$ as

$$y_* = \langle y_*, \tilde{\Psi}_{\mathscr{L}} \rangle \Psi_{\mathscr{L}} =: (\mathbf{D}_{\mathscr{L}}^{-1}\mathbf{y}_*)^T \Psi_{\mathscr{L}} = \mathbf{y}_*^T \Psi_H. \tag{2.200}$$

The way the vector $\mathbf{y}_*$ is defined here for notational convenience may by itself actually have infinite norm in ℓ_2. However, its occurrence will always include premultiplication by $\mathbf{D}_{\mathscr{L}}^{-1}$ which is therefore always well-defined. In view of (2.110), we obtain the relations

$$\|y - y_*\|_{\mathscr{L}} \sim \|\mathbf{R}^{1/2}\mathbf{D}_{\mathscr{L}}^{-1}(\mathbf{y} - \mathbf{y}_*)\| \sim \|\mathbf{D}_{\mathscr{L}}^{-1}(\mathbf{y} - \mathbf{y}_*)\|. \tag{2.201}$$

Note that here $\mathbf{R} = \langle \Psi, \Psi \rangle$ (and not $\mathbf{R}^{-1}$) comes into play since y, y_* have been expanded in a scaled version of the primal wavelet basis Ψ. Hence, equivalent norms for $\| \cdot \|_{\mathscr{Y}}$ may involve $\mathbf{R}$. As for describing equivalent norms for $\| \cdot \|_{\mathscr{U}}$, recall that u is expanded in the basis Ψ_U for $U \subset H'$. Consequently, $\mathbf{R}^{-1}$ is the natural matrix to take into account when considering equivalent norms, i.e., we choose here

$$\|u\|_{\mathscr{U}} \sim \|\mathbf{R}^{-1/2}\mathbf{u}\|. \tag{2.202}$$

Finally, we formulate the following control problem in (infinite) wavelet coordinates.

(DCP) *For given data* $\mathbf{D}_{\mathscr{Y}}^{-1}\mathbf{y}_* \in \ell_2(\mathbb{I}_{\mathscr{Y}})$, $\mathbf{f} \in \ell_2(\mathbb{I}_H)$, *and weight parameter* $\omega > 0$, *minimize the quadratic functional*

$$\check{\mathbf{J}}(\mathbf{y}, \mathbf{u}) := \tfrac{1}{2} \|\mathbf{R}^{1/2}\mathbf{D}_{\mathscr{Y}}^{-1}(\mathbf{y} - \mathbf{y}_*)\|^2 + \tfrac{\omega}{2} \|\mathbf{R}^{-1/2}\mathbf{u}\|^2 \tag{2.203}$$

over $(\mathbf{y}, \mathbf{u}) \in \ell_2(\mathbb{I}_H) \times \ell_2(\mathbb{I}_H)$ *subject to the linear constraints*

$$\mathbf{A}\mathbf{y} = \mathbf{f} + \mathbf{D}_H^{-1}\mathbf{u}. \tag{2.204}$$

Remark 10 Problem (DCP) can be viewed as (discretized yet still infinite-dimensional) *representation* of the linear-quadratic control problem (2.74) together with (2.75) in wavelet coordinates in the following sense. The functional $\check{\mathbf{J}}(\mathbf{y}, \mathbf{u})$ defined in (2.203) is equivalent to the functional $J(y, u)$ from (2.74) in the sense that there exist constants $0 < c_J \leq C_J < \infty$ such that

$$c_J \check{\mathbf{J}}(\mathbf{y}, \mathbf{u}) \leq J(y, u) \leq C_J \check{\mathbf{J}}(\mathbf{y}, \mathbf{u}) \tag{2.205}$$

holds for any $y = \mathbf{y}^T \Psi_H \in H$, given $y_* = (\mathbf{D}_{\mathscr{Y}}^{-1}\mathbf{y}_*)^T \Psi_{\mathscr{Y}} \in \mathscr{Y}$ and any $u = \mathbf{u}^T \Psi_{\mathscr{U}} \in \mathscr{U}$. Moreover, in the case of compatible data $y_* = A^{-1}f$ yielding $J(y, u) \equiv 0$, the respective minimizers coincide, and $\mathbf{y}_* = \mathbf{A}^{-1}\mathbf{f}$ yields $\check{\mathbf{J}}(\mathbf{y}, \mathbf{u}) \equiv \mathbf{0}$. In this sense the new functional (2.203) captures the essential features of the model minimization functional.

Once problem (DCP) is posed, we can apply variational principles to derive necessary and sufficient conditions for a unique solution. All control problems considered here are in fact simple in this regard, as we have to minimize a quadratic functional subject to linear constraints, for which the necessary conditions are also sufficient. In principle, there are two ways to derive the optimality conditions for (DCP). We have encountered in Sect. 2.3.5 already the technique via the Lagrangian.

We define for (DCP) the *Lagrangian* introducing the *Lagrange multiplier, adjoint variable* or *adjoint state* $\mathbf{p}$ as

$$\mathbf{Lagr}(\mathbf{y}, \mathbf{p}, \mathbf{u}) := \check{\mathbf{J}}(\mathbf{y}, \mathbf{u}) + \langle \mathbf{p}, \mathbf{A}\mathbf{y} - \mathbf{f} - \mathbf{D}_H^{-1}\mathbf{u} \rangle. \tag{2.206}$$

Then the KKT conditions $\delta \mathbf{Lagr}(\mathbf{w}) = \mathbf{0}$ for $\mathbf{w} = \mathbf{p}, \mathbf{y}, \mathbf{u}$ are, respectively,

$$\mathbf{Ay} = \mathbf{f} + \mathbf{D}_H^{-1}\mathbf{u}, \qquad (2.207\text{a})$$

$$\mathbf{A}^T\mathbf{p} = -\mathbf{D}_{\mathscr{L}}^{-1}\mathbf{RD}_{\mathscr{L}}^{-1}(\mathbf{y} - \mathbf{y}_*) \qquad (2.207\text{b})$$

$$\omega\mathbf{R}^{-1}\mathbf{u} = \mathbf{D}_H^{-1}\mathbf{p}. \qquad (2.207\text{c})$$

The first system resulting from the variation with respect to the Lagrange multiplier always recovers the original constraints (2.204) and will be referred to as the *primal system* or the *state equation*. Accordingly, we call (2.207b) the *adjoint* or *dual system*, or the *costate equation*. The third Eq. (2.207c) is sometimes denoted as the *design equation*. Although $\mathbf{A}$ is symmetric, we continue to write $\mathbf{A}^T$ for the operator of the adjoint system to distinguish it from the primal system.

The coupled system (2.207) later is to be solved. However, in order to derive convergent iterations and deduce complexity estimates, a different formulation will be advantageous. It is based on the fact that $\mathbf{A}$ is according to Proposition 3 a boundedly invertible mapping on ℓ_2. Thus, we can formally invert (2.198) to obtain $\mathbf{y} = \mathbf{A}^{-1}\mathbf{f} + \mathbf{A}^{-1}\mathbf{D}_H^{-1}\mathbf{u}$. Substitution into (2.203) yields a functional depending only on $\mathbf{u}$,

$$\mathbf{J}(\mathbf{u}) := \tfrac{1}{2}\,\|\mathbf{R}^{1/2}\mathbf{D}_{\mathscr{L}}^{-1}\left(\mathbf{A}^{-1}\mathbf{D}_H^{-1}\mathbf{u} - (\mathbf{y}_* - \mathbf{A}^{-1}\mathbf{f})\right)\|^2 + \tfrac{\omega}{2}\,\|\mathbf{R}^{-1/2}\mathbf{u}\|^2. \qquad (2.208)$$

Employing the abbreviations

$$\mathbf{Z} := \mathbf{R}^{1/2}\mathbf{D}_{\mathscr{L}}^{-1}\mathbf{A}^{-1}\mathbf{D}_H^{-1}, \qquad (2.209\text{a})$$

$$\mathbf{G} := -\mathbf{R}^{1/2}\mathbf{D}_{\mathscr{L}}^{-1}(\mathbf{A}^{-1}\mathbf{f} - \mathbf{y}_*), \qquad (2.209\text{b})$$

the functional simplifies to

$$\mathbf{J}(\mathbf{u}) = \tfrac{1}{2}\|\mathbf{Zu} - \mathbf{G}\|^2 + \tfrac{\omega}{2}\,\|\mathbf{R}^{-1/2}\mathbf{u}\|^2. \qquad (2.210)$$

Proposition 5 ([53]) *The functional* $\mathbf{J}$ *is twice differentiable with first and second variation*

$$\delta\mathbf{J}(\mathbf{u}) = (\mathbf{Z}^T\mathbf{Z} + \omega\mathbf{R}^{-1})\mathbf{u} - \mathbf{Z}^T\mathbf{G}, \qquad \delta^2\mathbf{J}(\mathbf{u}) = \mathbf{Z}^T\mathbf{Z} + \omega\mathbf{R}^{-1}. \qquad (2.211)$$

In particular, $\mathbf{J}$ *is convex so that a unique minimizer exists.*

Setting

$$\mathbf{Q} := \mathbf{Z}^T\mathbf{Z} + \omega\mathbf{R}^{-1}, \qquad \mathbf{g} := \mathbf{Z}^T\mathbf{G}, \qquad (2.212)$$

the unique minimizer $\mathbf{u}$ of (2.210) is given by solving

$$\delta \mathbf{J}(\mathbf{u}) = \mathbf{0} \tag{2.213}$$

or, equivalently, the system

$$\mathbf{Q}\mathbf{u} = \mathbf{g}. \tag{2.214}$$

By definition (2.212), $\mathbf{Q}$ is a symmetric positive definite (infinite) matrix. Hence, finite versions of (2.214) could be solved by gradient or conjugate gradient iterative schemes. As the convergence speed of any such iteration depends on the spectral condition number of $\mathbf{Q}$, it is important to note that the following result.

Proposition 6 *The (infinite) matrix* $\mathbf{Q}$ *is uniformly bounded on* ℓ_2, *i.e., there exist constants* $0 < c_Q \le C_Q < \infty$ *such that*

$$c_Q \|\mathbf{v}\| \le \|\mathbf{Q}\mathbf{v}\| \le C_Q \|\mathbf{v}\|, \qquad \mathbf{v} \in \ell_2. \tag{2.215}$$

The proof follows from (2.46) and (2.196) [29]. Of course, in order to make such iterative schemes for (2.214) practically feasible, the explicit inversion of $\mathbf{A}$ in the definition of $\mathbf{Q}$ has to be avoided and replaced by an iterative solver in turn. This is where the system (2.207) will come into play. In particular, the third equation (2.207c) has the following interpretation which will turn out to be very useful later.

Proposition 7 *If we solve for a given control vector* $\mathbf{u}$ *successively (2.204) for* $\mathbf{y}$ *and (2.207b) for* $\mathbf{p}$, *then the residual for (2.214) attains the form*

$$\mathbf{Q}\mathbf{u} - \mathbf{g} = \omega \mathbf{R}^{-1}\mathbf{u} - \mathbf{D}_U^{-1}\mathbf{p}. \tag{2.216}$$

Proof Solving consecutively (2.204) and (2.207b) and recalling the definitions of $\mathbf{Z}$, $\mathbf{g}$ (2.209a), (2.212) we obtain

$$
\begin{aligned}
\mathbf{D}_H^{-1}\mathbf{p} &= -\mathbf{D}_H^{-1}(\mathbf{A}^{-T}\mathbf{D}_{\mathscr{Y}}^{-1}\mathbf{R}\mathbf{D}_{\mathscr{Y}}^{-1}(\mathbf{y} - \mathbf{y}_*)) \\
&= -\mathbf{Z}^T\mathbf{R}^{1/2}\mathbf{D}_{\mathscr{Y}}^{-1}(\mathbf{A}^{-1}\mathbf{f} + \mathbf{A}^{-1}\mathbf{D}_H^{-1}\mathbf{u} - \mathbf{y}_*) \\
&= \mathbf{Z}^T\mathbf{G} - \mathbf{Z}^T\mathbf{R}^{1/2}\mathbf{D}_{\mathscr{Y}}^{-1}\mathbf{A}^{-1}\mathbf{D}_H^{-1}\mathbf{u} \\
&= \mathbf{g} - \mathbf{Z}^T\mathbf{Z}\mathbf{u}.
\end{aligned}
$$

Hence, the residual $\mathbf{Q}\mathbf{u} - \mathbf{g}$ attains the form

$$\mathbf{Q}\mathbf{u} - \mathbf{g} = (\mathbf{Z}^T\mathbf{Z} + \omega \mathbf{R}^{-1})\mathbf{u} - \mathbf{g} = \omega \mathbf{R}^{-1}\mathbf{u} - \mathbf{D}_H^{-1}\mathbf{p},$$

where we have used the definition of $\mathbf{Q}$ from (2.212). $\qquad\square$

Having derived the optimality conditions (2.207), the next issue is their efficient numerical solution. In view of the fact that the system (2.207) still involves infinite matrices and vectors, this also raises the question how to derive computable finite versions. By now we have investigated two scenarios.

The first version with respect to *uniform discretizations* is based on choosing finite-dimensional subspaces of the function spaces under consideration. The second version which deals with *adaptive discretizations* is actually based on the infinite system (2.207). In both scenarios, a fully iterative numerical scheme for the solution of (2.207) is designed along the following lines. The basic iteration scheme is a *gradient* or *conjugate gradient iteration* for (2.214) as an *outer iteration* where each application of $\mathbf{Q}$ is in turn realized by solving the primal and the dual system (2.204) and (2.207b) also by a gradient or conjugate gradient method as *inner iterations*.

For *uniform* discretizations for which we wanted to test numerically the role of equivalent norms and the influence of Riesz maps in the cost functional (2.203), we have used in [12] as central iterative scheme the conjugate gradient (CG) method. Since the interior systems are only solved up to discretization error accuracy, the whole procedure may therefore be viewed as an *inexact conjugate gradient (CG) method*. We stress already at this point that the iteration numbers of such a method do *not* depend on the discretization level as finite versions of all involved operators are also uniformly well-conditioned in the sense of (2.215). In each step of the outer iteration, the error will be reduced by a fixed factor ρ. Combined with a *nested iteration strategy*, it will be shown that this yields an asymptotically optimal method in the amount of arithmetic operations.

Starting from the infinite coupled system (2.207), we have investigated in [29] *adaptive schemes* which, given any prescribed accuracy $\varepsilon > 0$, solve (2.207) such that the error for $\mathbf{y}, \mathbf{u}, \mathbf{p}$ is controlled by ε. Here we have used a *gradient scheme* as basic iterative scheme since it somehow simplifies the analysis, see Sect. 2.6.2.

2.5.4   Control Problems: Dirichlet Boundary Control

Having derived a representation in wavelet coordinates for both the saddle point problem from Sect. 2.3.3 and the PDE-constrained control problem in the previous section, it is straightforward to find also an appropriate representation of the control problem with Dirichlet boundary control introduced in Sect. 2.3.6. In order not to be overburdened with notation, we specifically choose the control space on the boundary as $\mathscr{U} := Q(= (H^{1/2}(\Gamma))')$. For the more general situation covered by (2.84), a diagonal matrix with nondecreasing entries like in (2.195) would come into play to switch between $\mathscr{U}$ and Q. Thus, the exact wavelet representation of the constraints (2.83) is given by the system (2.190), where we exchange the given Dirichlet boundary term $\mathbf{g}$ by $\mathbf{u}$ in the present situation to express the dependence on the control in the right hand side, i.e.,

$$L \begin{pmatrix} \mathbf{y} \\ \mathbf{p} \end{pmatrix} := \begin{pmatrix} \mathbf{A} & \mathbf{B}^T \\ \mathbf{B} & \mathbf{0} \end{pmatrix} \begin{pmatrix} \mathbf{y} \\ \mathbf{p} \end{pmatrix} = \begin{pmatrix} \mathbf{f} \\ \mathbf{u} \end{pmatrix}. \tag{2.217}$$

The derivation of a representer of the initial objective functional (2.82) is under the embedding condition (2.84) $\|v\|_{\mathscr{X}} \lesssim \|v\|_Y$ for $v \in Y$ now the same as in the previous section, where all reference to the space H is to be exchanged by reference to Y. We end up with the following minimization problem in wavelet coordinates for the case of Dirichlet boundary control. **(DCP)** *For given data* $\mathbf{D}_{\mathscr{X}}^{-1}\mathbf{y}_* \in \ell_2(I\!\!I_{\mathscr{X}})$, $\mathbf{f} \in \ell_2(I\!\!I_Y)$, *and weight parameter* $\omega > 0$, *minimize the quadratic functional*

$$\check{\mathbf{J}}(\mathbf{y}, \mathbf{u}) := \tfrac{1}{2} \|\mathbf{R}^{1/2}\mathbf{D}_{\mathscr{X}}^{-1}(\mathbf{y} - \mathbf{y}_*)\|^2 + \tfrac{\omega}{2} \|\mathbf{R}^{-1/2}\mathbf{u}\|^2 \tag{2.218}$$

over $(\mathbf{y}, \mathbf{u}) \in \ell_2(I\!\!I_Y) \times \ell_2(I\!\!I_Y)$ *subject to the linear constraints (2.217),*

$$L \begin{pmatrix} \mathbf{y} \\ \mathbf{p} \end{pmatrix} = \begin{pmatrix} \mathbf{f} \\ \mathbf{u} \end{pmatrix}.$$

The corresponding Karush-Kuhn-Tucker conditions can be derived by the same variational principles as in the previous section by defining a Lagrangian in terms of the functional $\check{\mathbf{J}}(\mathbf{y}, \mathbf{u})$ and appending the constraints (2.198) with the help of additional Lagrange multipliers $(\mathbf{z}, \boldsymbol{\mu})^T$, see [53]. We obtain in this case a system of coupled saddle point problems

$$L \begin{pmatrix} \mathbf{y} \\ \mathbf{p} \end{pmatrix} = \begin{pmatrix} \mathbf{f} \\ \mathbf{u} \end{pmatrix} \tag{2.219a}$$

$$L^T \begin{pmatrix} \mathbf{z} \\ \boldsymbol{\mu} \end{pmatrix} = \begin{pmatrix} -\omega \mathbf{D}_{\mathscr{X}}^{-1} \mathbf{R} \mathbf{D}_{\mathscr{X}}^{-1}(\mathbf{y} - \mathbf{y}_*) \\ 0 \end{pmatrix} \tag{2.219b}$$

$$\mathbf{u} = \boldsymbol{\mu}. \tag{2.219c}$$

Again, the first system appearing here, the *primal system*, are just the constraints (2.198) while (2.95) will be referred to as the *dual* or *adjoint system*. The specific form of the right hand side of the dual system emerges from the particular formulation of the minimization functional (2.218). The (here trivial) equation (2.219c) stems from measuring $\mathbf{u}$ just in ℓ_2, representing measuring the control in its natural trace norm. Instead of replacing $\boldsymbol{\mu}$ by $\mathbf{u}$ in (2.95) and trying to solve the resulting equations, (2.219c) will be essential to devise an inexact gradient scheme. In fact, since L in (2.198) is an invertible operator, we can rewrite $\check{\mathbf{J}}(\mathbf{y}, \mathbf{u})$ by formally inverting (2.198) as a functional of $\mathbf{u}$, that is, $\mathbf{J}(\mathbf{u}) := \check{\mathbf{J}}(\mathbf{y}(\mathbf{u}), \mathbf{u})$ as above. The following result will be very useful for the design of the outer–inner iterative solvers

Proposition 8 *The first variation of* **J** *satisfies*

$$\delta J(u) = u - \mu, \tag{2.220}$$

where (u, μ) *are part of the solution of (2.219). Moreover,* **J** *is convex so that a unique minimizer exists.*

Hence, Eq. (2.219c) is just $\delta J(u) = 0$. For a unified treatment below of both control problems considered in these notes, it will be useful to rewrite (2.219c) like in (2.214) as a condensed equation for the control u alone. We formally invert (2.217) and (2.219b) and obtain

$$Qu = g \tag{2.221}$$

with the abbreviations

$$Q := Z^T Z + \omega I, \quad g := Z^T (y_* - T_\square L^{-1} I_\square f) \tag{2.222}$$

and

$$Z := T_\square L^{-1} I_\square, \qquad I_\square := \begin{pmatrix} 0 \\ I \end{pmatrix}, \qquad T_\square := (T \ 0). \tag{2.223}$$

Proposition 9 *The vector* u *as part of the solution vector* (y, p, z, μ, u) *of (2.219) coincides with the unique solution* u *of the condensed equations (2.221).*

2.6   Iterative Solution

Each of the four problem classes discussed above lead to the problem to finally solve a system

$$\delta J(q) = 0 \tag{2.224}$$

or, equivalently, a linear system

$$Mq = b, \tag{2.225}$$

where $M : \ell_2 \to \ell_2$ is a (possibly infinite) symmetric positive definite matrix satisfying

$$c_M \|v\| \le \|Mv\| \le C_M \|v\|, \quad v \in \ell_2, \tag{2.226}$$

for some constants $0 < c_\mathbf{M} \leq C_\mathbf{M} < \infty$ and where $\mathbf{b} \in \ell_2$ is some given right hand side.

A simple *gradient method* for solving (2.224) is

$$\mathbf{q}_{k+1} := \mathbf{q}_k - \alpha\, \delta \mathbf{J}(\mathbf{q}_k), \qquad k = 0, 1, 2, \ldots \tag{2.227}$$

with some initial guess $\mathbf{q}_0$. In all of the previously considered situations, it has been asserted that there exists a fixed parameter α, depending on bounds for the second variation of $\mathbf{J}$, such that (2.227) converges and reduces the error in each step by at least a fixed factor $\rho < 1$, i.e.,

$$\|\mathbf{q} - \mathbf{q}_{k+1}\| \leq \rho \|\mathbf{q} - \mathbf{q}_k\|, \quad k = 0, 1, 2, \ldots, \tag{2.228}$$

where ρ is determined by

$$\rho := \|\mathbf{I} - \alpha \mathbf{M}\| < 1.$$

Hence, the scheme (2.227) is a convergent iteration for the possibly infinite system (2.225). Next we will need to discuss how to reduce the infinite systems to computable finite versions.

2.6.1 Finite Systems on Uniform Grids

Let us first consider finite-dimensional trial spaces with respect to uniform discretizations. For each of the Hilbert spaces H, this means in the wavelet setting to pick the index set of all indices up to some *highest refinement level* J, i.e.,

$$I\!I_{J,H} := \{\lambda \in I\!I_H : |\lambda| \leq J\} \subset I\!I_H$$

satisfying $N_{J,H} := \#I\!I_{J,H} < \infty$. The representation of operators is then built as in Sect. 2.4.3 with respect to this truncated index set which corresponds to deleting all rows and columns that refer to indices λ such that $|\lambda| > J$, and correspondingly for functions. There is by construction also a *coarsest level* of resolution denoted by j_0.

Computationally the representation of operators according to (2.111) is in the case of uniform grids always realized as follows. First, the operator is set up in terms of the *generator basis* on the finest level J. This generator basis simply consists of tensor products of B-Splines, or linear combinations of these near the boundaries. The representation of an operator in the *wavelet basis* is then achieved by applying the Fast Wavelet Transform (FWT) which needs $\mathcal{O}(N_{J,H})$ arithmetic operations and is therefore asymptotically optimal, see, e.g., [24, 34, 51] and Sect. 2.4.4.

In order not to overburden the notation, let in this subsection the resulting system for $N = N_{J,H}$ unknowns again be denoted by

$$\mathbf{Mq} = \mathbf{b}, \tag{2.229}$$

where now $\mathbf{M} : \mathbb{R}^N \to \mathbb{R}^N$ is a symmetric positive definite matrix satisfying (2.226) on $\mathbb{R}^N$. It will be convenient to abbreviate the residual using an approximation $\tilde{\mathbf{q}}$ to $\mathbf{q}$ for (2.229) as

$$\text{RESD}(\tilde{\mathbf{q}}) := \mathbf{M}\tilde{\mathbf{q}} - \mathbf{b}. \tag{2.230}$$

We will employ a basic conjugate gradient method that iteratively computes an approximate solution $\mathbf{q}_K$ to (2.229) with given initial vector $\mathbf{q}_0$ and given tolerance $\varepsilon > 0$ such that

$$\|\mathbf{Mq}_K - \mathbf{b}\| = \|\text{RESD}(\mathbf{q}_K)\| \le \varepsilon, \tag{2.231}$$

where K denotes the number of iterations used. Later we specify ε depending on the discretization for which (2.229) is set up. The following scheme CG contains a routine $\text{APP}(\eta_k, \mathbf{M}, \mathbf{d}_k)$ which in view of the problem classes discussed above is to have the property that it approximately computes the product $\mathbf{Md}_k$ up to a tolerance $\eta_k = \eta_k(\varepsilon)$ depending on ε, i.e., the output $\mathbf{m}_k$ of $\text{APP}(\eta_k, \mathbf{M}, \mathbf{d}_k)$ satisfies

$$\|\mathbf{m}_k - \mathbf{Md}_k\| \le \eta_k. \tag{2.232}$$

For the cases where $\mathbf{M} = \mathbf{A}$, this is simply the matrix-vector multiplication $\mathbf{Md}_k$. For the situations where $\mathbf{M}$ may involve the solution of an additional system, this multiplication will be only approximative. The routine is as follows.
CG $[\varepsilon, \mathbf{q}_0, \mathbf{M}, \mathbf{b}] \to \mathbf{q}_K$

(I) SET $\mathbf{d}_0 := \mathbf{b} - \mathbf{Mq}_0$ AND $\mathbf{r}_0 := -\mathbf{d}_0$. LET $k = 0$.
(II) WHILE $\|\mathbf{r}_k\| > \varepsilon$

$$\mathbf{m}_k := \text{APP}(\eta_k(\varepsilon), \mathbf{M}, \mathbf{d}_k)$$

$$\alpha_k := \frac{(\mathbf{r}_k)^T \mathbf{r}_k}{(\mathbf{d}_k)^T \mathbf{m}_k} \qquad \mathbf{q}_{k+1} := \mathbf{q}_k + \alpha_k \mathbf{d}_k$$

$$\mathbf{r}_{k+1} := \mathbf{r}_k + \alpha_k \mathbf{m}_k \qquad \beta_k := \frac{(\mathbf{r}_{k+1})^T \mathbf{r}_{k+1}}{(\mathbf{r}_k)^T \mathbf{r}_k} \qquad (2.233)$$

$$\mathbf{d}_{k+1} := -\mathbf{r}_{k+1} + \beta_k \mathbf{d}_k$$

$$k := k + 1$$

(III) SET $K := k - 1$.

Let us briefly discuss in the case $\mathbf{M} = \mathbf{A}$ that the final iterate $\mathbf{q}_K$ indeed satisfies (2.231). From the newly computed iterate $\mathbf{q}_{k+1} = \mathbf{q}_k + \alpha_k \mathbf{d}_k$ it follows by applying $\mathbf{M}$ on both sides that $\mathbf{M}\mathbf{q}_{k+1} - \mathbf{b} = \mathbf{M}\mathbf{q}_k - \mathbf{b} + \alpha_k \mathbf{M}\mathbf{d}_k$ which is the same as $\text{RESD}(\mathbf{q}_{k+1}) = \text{RESD}(\mathbf{q}_k) + \alpha_k \mathbf{M}\mathbf{d}_k$. By the initialization for $\mathbf{r}_k$ used above, this in turn is the updating term for $\mathbf{r}_k$, hence, $\mathbf{r}_k = \text{RESD}(\mathbf{q}_k)$. After the stopping criterion based on $\mathbf{r}_k$ is met, the final iterate $\mathbf{q}_K$ observes (2.231).

The routine CG computes the *residual* up to the stopping criterion ε. From the residual, we can in view of (2.226) estimate the *error* in the solution as

$$\|\mathbf{q} - \mathbf{q}_K\| = \|\mathbf{M}^{-1}(\mathbf{b} - \mathbf{M}\mathbf{q}_K)\| \leq \|\mathbf{M}^{-1}\| \, \|\text{RESD}(\mathbf{q}_K)\| \leq \frac{\varepsilon}{c_{\mathbf{M}}}, \qquad (2.234)$$

that is, it may deviate from the norm of the residual from a factor proportional to the smallest eigenvalue of $\mathbf{M}$.

Distributed Control Let us now apply the solution scheme to the situation from Sect. 2.5.3 where $\mathbf{Q}$ now involves the inversion of finite-dimensional systems (2.207a) and (2.207b). The material in the remainder of this subsection is essentially contained in [12].

We begin with a specification of the approximate computation of the right hand side $\mathbf{b}$ which also contains applications of $\mathbf{A}^{-1}$.

RHS $[\zeta, \mathbf{A}, \mathbf{f}, \mathbf{y}_*] \to \mathbf{b}_\zeta$

(I) CG $[\frac{c_A}{2C} \frac{c_A}{C^2 C_0^2} \zeta, \mathbf{0}, \mathbf{A}, \mathbf{f}] \to \mathbf{b}_1$

(II) CG $[\frac{c_A}{2C} \zeta, \mathbf{0}, \mathbf{A}^T, -\mathbf{D}_{\mathscr{L}}^{-1} \mathbf{R} \mathbf{D}_{\mathscr{L}}^{-1} (\mathbf{b}_1 - \mathbf{y}_*)] \to \mathbf{b}_2$

(III) $\mathbf{b}_\zeta := \mathbf{D}_H^{-1} \mathbf{b}_2$.

The tolerances used within the two conjugate gradient methods depend on the constants c_A, C, C_0 from (2.46), (2.196) and (2.104), respectively. Since the additional factor $c_A (C C_0)^{-2}$ in the stopping criterion in step (I) in comparison to step (II) is in general smaller than one, this means that the primal system needs to be solved more accurately than the adjoint system in step (II).

Proposition 10 *The result $\mathbf{b}_\zeta$ of RHS $[\zeta, \mathbf{A}, \mathbf{f}, \mathbf{y}_*]$ satisfies*

$$\|\mathbf{b}_\zeta - \mathbf{b}\| \leq \zeta. \qquad (2.235)$$

Proof Recalling the definition (2.212) of $\mathbf{b}$, step (III) and step (II) yield

$$\begin{aligned}
\|\mathbf{b}_\zeta - \mathbf{b}\| &\leq \|\mathbf{D}_H^{-1}\| \, \|\mathbf{b}_2 - \mathbf{D}_H \mathbf{b}\| \\
&\leq C \|\mathbf{A}^{-T}\| \, \|\mathbf{A}^T \mathbf{b}_2 - \mathbf{D}_{\mathscr{L}}^{-1} \mathbf{R} \mathbf{D}_{\mathscr{L}}^{-1} (\mathbf{A}^{-1} \mathbf{f} - \mathbf{b}_1 + \mathbf{b}_1 - \mathbf{y}_*)\| \qquad (2.236) \\
&\leq \frac{C}{c_A} \left(\frac{c_A}{2C} \zeta + \|\mathbf{D}_{\mathscr{L}}^{-1} \mathbf{R} \mathbf{D}_{\mathscr{L}}^{-1} (\mathbf{A}^{-1} \mathbf{f} - \mathbf{b}_1)\| \right).
\end{aligned}$$

Employing the upper bounds for $\mathbf{D}_{\mathscr{L}}^{-1}$ and $\mathbf{R}$, we arrive at

$$\|\mathbf{b}_{\zeta} - \mathbf{b}\| \leq \frac{C}{c_{\mathbf{A}}} \left(\frac{c_{\mathbf{A}}}{2C} \zeta + C^2 C_0^2 \|\mathbf{A}^{-1}\| \|\mathbf{f} - \mathbf{Ab}_1\| \right)$$

$$\leq \frac{C}{c_{\mathbf{A}}} \left(\frac{c_{\mathbf{A}}}{2C} \zeta + \frac{C^2 C_0^2}{c_{\mathbf{A}}} \frac{c_{\mathbf{A}}}{2C} \frac{c_{\mathbf{A}}}{C^2 C_0^2} \zeta \right) = \zeta.$$

(2.237)

$\square$

Accordingly, an approximation $\mathbf{m}_{\eta}$ to the matrix-vector product $\mathbf{Qd}$ is the output of the following routine APP.

APP $[\eta, \mathbf{Q}, \mathbf{d}] \rightarrow \mathbf{m}_{\eta}$

(I)   CG $[\frac{c_{\mathbf{A}}}{3C} \frac{c_{\mathbf{A}}}{C^2 C_0^2} \eta, \mathbf{0}, \mathbf{A}, \mathbf{f} + \mathbf{D}_H^{-1} \mathbf{d}] \rightarrow \mathbf{y}_{\eta}$

(II)   CG $[\frac{c_{\mathbf{A}}}{3C} \eta, \mathbf{0}, \mathbf{A}^T, -\mathbf{D}_{\mathscr{L}}^{-1} \mathbf{RD}_Z^{-1}(\mathbf{y}_{\eta} - \mathbf{y}_*)] \rightarrow \mathbf{p}_{\eta}$

(III) $\mathbf{m}_{\eta} := \mathbf{g}_{\eta/3} + \omega \mathbf{R}^{-1} \mathbf{d} - \mathbf{D}_H^{-1} \mathbf{p}_{\eta}.$

The choice of the tolerances for the interior application of CG in steps (i) and (ii) will become clear from the following result.

Proposition 11 *The result* $\mathbf{m}_{\eta}$ *of* APP$[\eta, \mathbf{Q}, \mathbf{d}]$ *satisfies*

$$\|\mathbf{m}_{\eta} - \mathbf{Qd}\| \leq \eta.$$

(2.238)

Proof Denote by $\mathbf{y_d}$ the exact solution of (2.207a) with $\mathbf{d}$ in place of $\mathbf{u}$ on the right hand side, and by $\mathbf{p_d}$ the exact solution of (2.207b) with $\mathbf{y_d}$ on the right hand side. Then we deduce from step (iii) and (2.216) combined with (2.104) and (2.196)

$$\|\mathbf{m}_{\eta} - \mathbf{Qd}\| = \|\mathbf{g}_{\eta/3} - \mathbf{g} + \omega \mathbf{R}^{-1} \mathbf{d} - \mathbf{D}_U^{-1} \mathbf{p}_{\eta} - (\mathbf{Qd} - \mathbf{g})\|$$

$$\leq \frac{1}{3}\eta + \|\omega \mathbf{R}^{-1} \mathbf{d} - \mathbf{D}_U^{-1} \mathbf{p}_{\eta} - (\omega \mathbf{R}^{-1} \mathbf{d} - \mathbf{D}_U^{-1} \mathbf{p_d})\|$$

(2.239)

$$\leq \frac{1}{3}\eta + C\|\mathbf{p_d} - \mathbf{p}_{\eta}\|.$$

Denote by $\hat{\mathbf{p}}$ the exact solution of (2.207b) with $\mathbf{y}_{\eta}$ on the right hand side. Then we have $\mathbf{p_d} - \hat{\mathbf{p}} = -\mathbf{A}^{-T} \mathbf{D}_Z^{-1} \mathbf{RD}_Z^{-1}(\mathbf{y_d} - \mathbf{y}_{\eta})$. It follows by (2.46), (2.104) and (2.196) that

$$\|\mathbf{p_d} - \hat{\mathbf{p}}\| \leq \frac{C^2 C_0^2}{c_{\mathbf{A}}} \|\mathbf{y_d} - \mathbf{y}_{\eta}\| \leq \frac{1}{3C} \eta,$$

(2.240)

where the last estimate follows by the choice of the threshold in step (i). Finally, the combination (2.239) and (2.240) together with (2.235) and the stopping criterion in step (ii) readily confirms that

$$\|\mathbf{m}_\eta - \mathbf{Qd}\| \leq \frac{1}{3}\eta + C\left(\|\mathbf{p_d} - \hat{\mathbf{p}}\| + \|\hat{\mathbf{p}} - \mathbf{p}_\eta\|\right)$$

$$\leq \frac{1}{3}\eta + C\left(\frac{1}{3C}\eta + \frac{1}{3C}\eta\right) = \eta. \qquad \square$$

The effect of perturbed applications of $\mathbf{M}$ in CG and more general Krylov subspace schemes with respect to convergence has been investigated in a numerical linear algebra context for a given linear system (2.229) in several papers, see, e.g., [70]. Here we have chosen the η_i to be proportional to the outer accuracy ε incorporating a safety factor accounting for the values of β_i and $\|\mathbf{r}_i\|$.

Finally, we can formulate a full nested iteration strategy for finite systems (2.207) on uniform grids which employs outer and inner CG routines as follows. The scheme starts at the coarsest level of resolution j_0 with some initial guess $\mathbf{u}_0^{j_0}$ and successively solves (2.214) with respect to each level j until the norm of the current residual is below the discretization error on that level.

In wavelet coordinates, $\|\cdot\|$ corresponds to the energy norm. If we employ as in [12] on the primal side for approximation linear combinations of B-splines of order d (degree $d - 1$, see Sect. 2.2.1), the discretization error is for smooth solutions expected to be proportional to $2^{-(d-1)j}$ (compare (2.15)). Then the refinement level is successively increased until on the finest level J a prescribed tolerance proportional to the discretization error $2^{-(d-1)J}$ is met. In the following, superscripts on vectors denote the refinement level on which this term is computed. The given data $\mathbf{y}_*^j$, $\mathbf{f}^j$ are supposed to be accessible on all levels. On the coarsest level, the solution of (2.214) is computed exactly up to double precision by QR decomposition. Subsequently, the results from level j are prolongated onto the next higher level $j + 1$. Using wavelets, this is accomplished by simply adding zeros: wavelet coordinates have the character of differences, this prolongation corresponds to the exact representation in higher resolution wavelet coordinates. The resulting *Nested-Iteration-Incomplete-Conjugate-Gradient* Algorithm is the following.
NEICG $[J] \rightarrow \mathbf{u}^J$

(I) INITIALIZATION FOR COARSEST LEVEL $j := j_0$

    (1) COMPUTE RIGHT HAND SIDE $\mathbf{g}^{j_0} = (\mathbf{Z}^T\mathbf{G})^{j_0}$ BY QR DECOMPOSITION USING (2.209).

    (2) COMPUTE SOLUTION $\mathbf{u}^{j_0}$ OF (2.214) BY QR DECOMPOSITION.

(II) WHILE $j < J$

    (1) PROLONGATE $\mathbf{u}^j \rightarrow \mathbf{u}_0^{j+1}$ BY ADDING ZEROS, SET $j := j + 1$.

    (2) COMPUTE RIGHT HAND SIDE USING RHS $[2^{-(d-1)j}, \mathbf{A}, \mathbf{f}^j, \mathbf{y}_*^j] \rightarrow \mathbf{g}^j$.

    (3) COMPUTE SOLUTION OF (2.214) USING CG $[2^{-(d-1)j}, \mathbf{u}_0^j, \mathbf{Q}, \mathbf{g}^j] \rightarrow \mathbf{u}^j$.

Recall that step (II.3) requires multiple calls of APP[η, $\mathbf{Q}$, $\mathbf{d}$], which in turn invokes both CG[..., $\mathbf{A}$, ...] as well as CG[..., $\mathbf{A}^T$, ...] in each application.

On account of (2.46) and (2.215), finite versions of the system matrices $\mathbf{A}$ and $\mathbf{Q}$ have uniformly bounded condition numbers, entailing that each CG routine employed in the process reduces the error by a fixed rate $\rho < 1$ in each iteration step. Let $N_J \sim 2^{nJ}$ be the total number of unknowns (for $\mathbf{y}^J$, $\mathbf{u}^J$ and $\mathbf{p}^J$) on the highest level J. Employing the CG method only on the highest level, one needs $\mathcal{O}(J) = \mathcal{O}(\log \varepsilon)$ iterations to achieve the prescribed discretization error accuracy $\varepsilon_J = 2^{-(d-1)J}$. As each application of $\mathbf{A}$ and $\mathbf{Q}$ requires $\mathcal{O}(N_J)$ operations, the solution of (2.214) by CG only on the finest level requires $\mathcal{O}(J\,N_J)$ arithmetic operations.

Theorem 7 ([12]) *If the residual (2.216) is computed up to discretization error proportional to $2^{-(d-1)j}$ on each level j and the corresponding solutions are taken as initial guesses for the next higher level, NEICG is an asymptotically optimal method in the sense that it provides the solution $\mathbf{u}^J$ up to discretization error on level J in an overall amount of $\mathcal{O}(N_J)$ arithmetic operations.*

Proof In the above notation, nested iteration allows one to get rid of the factor J in the total amount of operations. Starting with the exact solution on the coarsest level j_0, in view of the uniformly bounded condition numbers of $\mathbf{A}$ and $\mathbf{Q}$, one needs only a fixed amount of iterations to reduce the error up to discretization error accuracy $\varepsilon_j = 2^{-(d-1)j}$ on each subsequent level j, taking the solution from the previous level as initial guess. Thus, on each level, one needs $\mathcal{O}(N_j)$ operations to realize discretization error accuracy. Since the spaces are nested and the number of unknowns on each level grows like $N_j \sim 2^{nj}$, by a geometric series argument the total number of arithmetic operations stays proportional to $\mathcal{O}(N_J)$.   □

Numerical Examples As an illustration of the ingredients for a distributed control problem, we consider the following example taken from [12] with the Helmholtz operator in (2.39) ($\mathbf{a} = I$, $c = 1$) and homogeneous Dirichlet boundary condition. A non-constant right hand side $f(x) := 1 + 2.3 \exp(-15|x - \frac{1}{2}|)$ is chosen, and the target state is set to a constant $y_* \equiv 1$. We first investigate the role the different norms $\| \cdot \|_{\mathscr{L}}$ and $\| \cdot \|_{\mathscr{U}}$ in (2.74), encoded in the diagonal matrices $\mathbf{D}_{\mathscr{L}}, \mathbf{D}_H$ from (2.195), have on the solution. We see in Fig. 2.3 for the choice $\mathscr{U} = L_2$ and $\mathscr{L} = H^s(0, 1)$ for different values of s varying between 0 and 1 the solution y (left) and the corresponding control u (right) for fixed weight $\omega = 1$. As s is increased, a stronger tendency of y towards the prescribed state $y_* \equiv 1$ can be observed which is, however, deterred from reaching this state by the homogeneous boundary conditions. Extensive studies of this type can be found in [11, 12].

As an example displaying the performance of the proposed fully iterative scheme NEICG in two spatial dimensions, Table 2.6 from [12] is included. This is an example of a control problem for the Helmholtz operator with Neumann boundary conditions. The stopping criterion for the outer iteration (relative to $\| \cdot \|$ which corresponds to the energy norm) on level j is chosen to be proportional to 2^{-j}. The second column displays the final value of the residual of the outer CG scheme

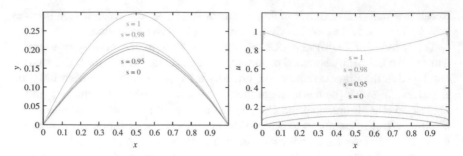

Fig. 2.3 Distributed control problem for elliptic problem with Dirichlet boundary conditions, a peak as right hand side f, $y_* \equiv 1$, $\omega = 0$, $\mathcal{U} = L_2$ and varying $\mathcal{Z} = H^s(0, 1)$

Table 2.6 Iteration history for a two-dimensional distributed control problem with Neumann boundary conditions, $\omega = 1$, $\mathcal{Z} = H^1(\Omega)$, $\mathcal{U} = (H^{0.5}(\Omega))'$

j	$\|\mathbf{r}_K^j\|$	#O	#E	#A	#R	$\|R(\mathbf{y}^J) - \mathbf{y}^j\|$	$\|\mathbf{y}^J - P(\mathbf{y}^j)\|$	$\|R(\mathbf{u}^J) - \mathbf{u}^j\|$	$\|\mathbf{u}^J - P(\mathbf{u}^j)\|$
3						6.86e−03	1.48e−02	1.27e−04	4.38e−04
4	1.79e−05	5	12	5	8	2.29e−03	7.84e−03	4.77e−05	3.55e−04
5	1.98e−05	5	14	6	9	6.59e−04	3.94e−03	1.03e−05	2.68e−04
6	4.92e−06	7	13	5	9	1.74e−04	1.96e−03	2.86e−06	1.94e−04
7	3.35e−06	7	12	5	9	4.55e−05	9.73e−04	9.65e−07	1.35e−04
8	2.42e−06	7	11	5	10	1.25e−05	4.74e−04	7.59e−07	8.88e−05
9	1.20e−06	8	11	5	10	4.55e−06	2.12e−04	4.33e−07	5.14e−05
10	4.68e−07	9	10	5	9	3.02e−06	3.02e−06	2.91e−07	2.91e−07

on this level, i.e., $\|\mathbf{r}_K^j\| = \|\mathrm{RESD}(\mathbf{u}_K^j)\|$. The next three columns show the number of outer CG iterations (#O) for $\mathbf{Q}$ according to the APP scheme followed by the maximum number of inner iterations for the primal system (#E), the adjoint system (#A) and the design equation (#R). We see very well the effect of the uniformly bounded condition numbers of the involved operators. The last columns display different versions of the actual error in the state $\mathbf{y}$ and the control $\mathbf{u}$ when compared to the fine grid solution (R denotes restriction of the fine grid solution to the actual grid, and P prolongation). Here we can see the effect of the constants appearing in (2.234), that is, the error is very well controlled via the residual. More results for up to three spatial dimensions can be found in [11, 12].

Dirichlet Boundary Control For the system of saddle point problems (2.219) arising from the control problem with Dirichlet boundary control in Sect. 2.3.6, also a fully iterative algorithm NEICG can be designed along the above lines. Again the design equation (2.219c) for $\mathbf{u}$ serves as the equation for which a basic iterative scheme (2.227) can be posed. Of course, the CG method for $\mathbf{A}$ then has to be replaced by a convergent iterative scheme for saddle point operators $\mathbf{L}$ like Uzawa's algorithm. Also the discretization has to be chosen such that the LBB condition is satisfied, see Sect. 2.5.2. Details can be found in [53]. Alternatively, since $\mathbf{L}$ has a uniformly bounded condition number, the CG scheme can, in principle, also be

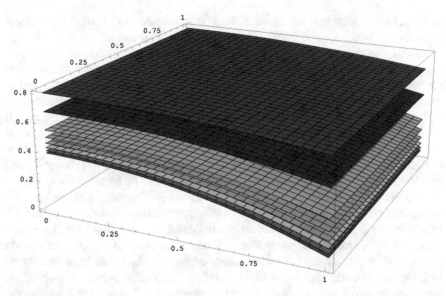

Fig. 2.4 State y of the Dirichlet boundary control problem using the objective functional $J(y, u) = \frac{1}{2}\|y - y_*\|^2_{H^s(\Gamma_y)} + \frac{1}{2}\|u\|^2_{H^{1/2}(\Gamma)}$ for $s = 0.1, 0.2, 0.3, 0.4, 0.5, 0.7, 0.9$ (from bottom to top) on resolution level $J = 5$

applied to $\mathbf{L}^T\mathbf{L}$. The performance of wavelet schemes on uniform grids for such systems of saddle point problems arising from optimal control is currently under investigation [62].

Numerical Example For illustration of the choice of different norms for the Dirichlet boundary control problem, consider the following example taken from [62]. Here we actually have the situation of controlling the system through the control boundary Γ on the right hand side of Fig. 2.4 while a prescribed state $y_* \equiv 1$ on the observation boundary Γ_y opposite the control boundary is to be achieved. The right hand side is chosen as constant $f \equiv 1$, and $\omega = 1$. Each layer in Fig. 2.4 corresponds to the state y for different values of s when the observation term is measured in $H^s(\Gamma_y)$, that is, the objective functional (2.82) contains a term $\|y - y_*\|^2_{H^s(\Gamma_y)}$ for $s = 1/10, 2/10, 3/10, 4/10, 5/10, 7/10, 9/10$ from bottom to top. We see that as the smoothness index s for the observation increases, the state moves towards the target state at the observation boundary.

2.6.2 Adaptive Schemes

In case of the appearance of singularities caused by the data or the domain, a prescribed accuracy may require discretizations with respect to uniform grids to spend a large amount of degrees of freedom in areas where the solution is actually smooth. Hence, although the above numerical scheme NEICG is of optimal linear complexity, the degrees of freedom are not implanted in an optimal way. In these situations, one expects adaptive schemes to work favourably which judiciously place degrees of freedom where singularities occur. Thus, the guiding line for adaptive schemes is to reduce the total amount of degrees of freedom when compared to discretizations on a uniform grid. This does not mean that the previous investigations with respect to uniform discretizations are dispensable. In fact, the above results on conditioning carry over to the adaptive case, the solvers are still linear in the amount of arithmetic operations and, in particular, one expects to recover the uniform situation when the solutions are smooth. Much on adaptivity for variational problems and the relation to nonlinear approximation can be found in [26].

The starting point for adaptive wavelet schemes systematically derived for variational problems in [19–21] is the infinite formulation in wavelet coordinates as derived for the different problem classes in Sect. 2.5. These algorithms have been proven to be optimal in the sense that they match the optimal work/ accuracy rate of the wavelet-best N-term approximation, a concept which has been introduced in [19]. The schemes start out with formulating algorithmic ingredients which are then step by step reduced to computable quantities. We follow in this section the material for the distributed control problem from [29]. An extension to Dirichlet control problem involving saddle point problems can be found in [54]. It should be pointed out that the theory is neither confined to symmetric $\mathbf{A}$ nor to the positive definite case.

Algorithmic Ingredients We start out again with a very simple iterative scheme for the design equation. In view of (2.215) and the fact that $\mathbf{Q}$ is positive definite, there exists a fixed positive parameter α such that in the *Richardson iteration* (which is a special case of a gradient method)

$$\mathbf{u}^{k+1} = \mathbf{u}^k + \alpha(\mathbf{g} - \mathbf{Q}\mathbf{u}^k) \tag{2.241}$$

the error is reduced in each step by at least a factor

$$\rho := \|\mathbf{I} - \alpha\mathbf{Q}\| < 1, \tag{2.242}$$

$$\|\mathbf{u} - \mathbf{u}^{k+1}\| \le \rho \|\mathbf{u} - \mathbf{u}^k\|, \quad k = 0, 1, 2, \ldots, \tag{2.243}$$

where $\mathbf{u}$ is the exact solution of (2.214). As the involved system is still infinite, we aim at carrying out this iteration approximately with dynamically updated accuracy tolerances.

The central idea of the wavelet-based adaptive schemes is to start from the infinite system in wavelet coordinates (2.207) and step by step reduce the routines to computable versions of applying the infinite matrix $\mathbf{Q}$ and the evaluation of the right hand side $\mathbf{g}$ of (2.214) involving the inversion of $\mathbf{A}$. The main conceptual tools from [19–21] are the following.

We first assume that we have a routine at our disposal with the following property. Later it will be shown how to realize this routine in the concrete case.

RES $[\eta, \mathbf{Q}, \mathbf{g}, \mathbf{v}] \to \mathbf{r}_\eta$ DETERMINES FOR A GIVEN TOLERANCE $\eta > 0$ A FINITELY SUPPORTED SEQUENCE $\mathbf{r}_\eta$ SATISFYING

$$\|\mathbf{g} - \mathbf{Q}\mathbf{v} - \mathbf{r}_\eta\| \le \eta. \tag{2.244}$$

The schemes considered below will also contain the following routine.

COARSE $[\eta, \mathbf{w}] \to \mathbf{w}_\eta$ DETERMINES FOR ANY FINITELY SUPPORTED INPUT VECTOR $\mathbf{w}$ A VECTOR $\mathbf{w}_\eta$ WITH SMALLEST POSSIBLE SUPPORT SUCH THAT

$$\|\mathbf{w} - \mathbf{w}_\eta\| \le \eta. \tag{2.245}$$

This ingredient will eventually play a crucial role in controlling the complexity of the scheme although its role is not yet apparent at this stage. A detailed description of COARSE can be found in [19]. The basic idea is to first sort the entries of $\mathbf{w}$ by size. Then one subtracts squares of their moduli until the sum reaches η^2, starting from the smallest entry. A quasi-sorting based on binary binning can be shown to avoid the logarithmic term in the sorting procedure at the expense of the resulting support size being at most a fixed constant of the minimal size, see [4].

Next a *perturbed iteration* is designed which converges in the following sense: for every target accuracy ε, the scheme produces after finitely many steps a finitely supported approximate solution with accuracy ε. To obtain a correctly balanced interplay between the routines RES and COARSE, we need the following control parameter. Given (an estimate of) the reduction rate ρ and the step size parameter α from (2.242), let K denote the minimal integer ℓ for which $\rho^{\ell-1}(\alpha\ell + \rho) \le \frac{1}{10}$.

Denoting in the following always by $\mathbf{u}$ the exact solution of (2.214), a perturbed version of (2.241) for a fixed target accuracy $\varepsilon > 0$ is the following.

SOLVE $[\varepsilon, \mathbf{Q}, \mathbf{g}, \overline{\mathbf{q}}^0, \varepsilon_0] \to \overline{\mathbf{q}}_\varepsilon$

(I) GIVEN AN INITIAL GUESS $\overline{\mathbf{q}}^0$ AND AN ERROR BOUND $\|\mathbf{q} - \overline{\mathbf{q}}^0\| \le \varepsilon_0$; SET $j = 0$.

(II) IF $\varepsilon_j \le \varepsilon$, STOP AND SET $\overline{\mathbf{q}}_\varepsilon := \overline{\mathbf{q}}^j$. OTHERWISE SET $\mathbf{v}^0 := \overline{\mathbf{q}}^j$.

(II.1) FOR $k = 0, \dots, K - 1$ COMPUTE RES $[\rho^k \varepsilon_j, \mathbf{Q}, \mathbf{g}, \mathbf{v}^k] \to \mathbf{r}^k$ AND

$$\mathbf{v}^{k+1} := \mathbf{v}^k + \alpha \mathbf{r}^k. \tag{2.246}$$

(II.2) APPLY COARSE $[\frac{2}{5}\varepsilon_j, \mathbf{v}^K] \to \overline{\mathbf{q}}^{j+1}$; SET $\varepsilon_{j+1} := \frac{1}{2}\varepsilon_j$, $j + 1 \to j$ AND GO TO (II).

In the case that no particular initial guess is known, we initialize $\overline{\mathbf{q}}^0 = \mathbf{0}$, set $\varepsilon_0 := c_{\mathbf{Q}}^{-1}\|\mathbf{g}\|$ and briefly write then SOLVE $[\varepsilon, \mathbf{Q}, \mathbf{g}] \to \overline{\mathbf{q}}_\varepsilon$.

In a straightforward manner, perturbation arguments yield the convergence of this algorithm [20, 21].

Proposition 12 *The iterates* $\overline{\mathbf{q}}^j$ *generated by* SOLVE $[\varepsilon, \mathbf{Q}, \mathbf{g}]$ *satisfy*

$$\|\mathbf{q} - \overline{\mathbf{q}}^j\| \le \varepsilon_j \qquad \text{for any} \quad j \ge 0, \tag{2.247}$$

where $\varepsilon_j = 2^{-j}\varepsilon_0$.

In order to derive appropriate numerical realizations of SOLVE, recall that (2.214) is equivalent to the KKT conditions (2.207). Although the matrix $\mathbf{A}$ is always assumed to be symmetric here, the distinction between the system matrices for the primal and the dual system, $\mathbf{A}$ and $\mathbf{A}^T$, may be helpful.

The strategy for approximating in each step the residual $\mathbf{g} - \mathbf{Q}\mathbf{u}^k$, that is, realization of the routine RES for the problem (2.214), is based upon the result stated in Proposition 7. In turn, this requires solving the two auxiliary systems in (2.207). Since the residual only has to be approximated, these systems will have to be solved only approximately. These approximate solutions, in turn, will be provided again by employing SOLVE but this time with respect to suitable residual schemes tailored to the systems in (2.207). In our special case, the matrix $\mathbf{A}$ is symmetric positive definite, and the choice of wavelet bases ensures the validity of (2.46). Thus, (2.242) holds for $\mathbf{A}$ and $\mathbf{A}^T$ so that the scheme SOLVE can indeed be invoked. Although we conceptually use the fact that a gradient iteration for the reduced problem (2.214) reduces the error for $\mathbf{u}$ in each step by a fixed amount, employing (2.207) for the evaluation of the residuals will generate as byproducts approximate solutions to the exact solution triple $(\mathbf{y}, \mathbf{p}, \mathbf{u})$ of (2.207). Under this hypothesis, we formulate next the ingredients for suitable versions SOLVE$_{\text{PRM}}$ and SOLVE$_{\text{ADJ}}$ of SOLVE for the systems in (2.207). Specifically, this requires identifying residual routines RES$_{\text{PRM}}$ and RES$_{\text{ADJ}}$ for the systems SOLVE$_{\text{PRM}}$ and SOLVE$_{\text{ADJ}}$. The main task in both cases is to apply the operators $\mathbf{A}, \mathbf{A}^T, \mathbf{D}_H^{-1}$ and $\mathbf{R}^{1/2}\mathbf{D}_{\mathscr{L}}^{-1}$. Again we assume for the moment that routines for the application of these operators are available, i.e., that for any $\mathbf{L} \in \{\mathbf{A}, \mathbf{A}^T, \mathbf{D}_H^{-1}, \mathbf{R}^{1/2}\mathbf{D}_{\mathscr{L}}^{-1}\}$ we have a scheme at our disposal with the following property.

APPLY $[\eta, \mathbf{L}, \mathbf{v}] \rightarrow \mathbf{w}_\eta$ DETERMINES FOR ANY FINITELY SUPPORTED INPUT VECTOR $\mathbf{v}$ AND ANY TOLERANCE $\eta > 0$ A FINITELY SUPPORTED OUTPUT $\mathbf{w}_\eta$ WHICH SATISFIES

$$\|\mathbf{L}\mathbf{v} - \mathbf{w}_\eta\| \le \eta. \tag{2.248}$$

The scheme SOLVE$_{\text{PRM}}$ for the first system in (2.207) is then defined by

$$\text{SOLVE}_{\text{PRM}} [\eta, \mathbf{A}, \mathbf{D}_H^{-1}, \mathbf{f}, \mathbf{v}, \overline{\mathbf{y}}^0, \varepsilon_0] := \text{SOLVE} [\eta, \mathbf{A}, \mathbf{f} + \mathbf{D}_H^{-1}\mathbf{v}, \overline{\mathbf{y}}^0, \varepsilon_0],$$

where $\overline{\mathbf{y}}^0$ is an initial guess for the solution $\mathbf{y}$ of $\mathbf{A}\mathbf{y} = \mathbf{f} + \mathbf{D}_H^{-1}\mathbf{v}$ with accuracy ε_0. The scheme RES for Step (II) in SOLVE is in this case realized by a new routine RES$_{\text{PRM}}$ defined as follows.

RES$_{\text{PRM}}$ $[\eta, \mathbf{A}, \mathbf{D}_H^{-1}, \mathbf{f}, \mathbf{v}, \overline{\mathbf{y}}] \rightarrow \mathbf{r}_\eta$ DETERMINES FOR ANY POSITIVE TOLERANCE η, A GIVEN FINITELY SUPPORTED $\mathbf{v}$ AND ANY FINITELY SUPPORTED INPUT $\overline{\mathbf{y}}$ A

FINITELY SUPPORTED APPROXIMATE RESIDUAL $\mathbf{r}_\eta$ SATISFYING (2.244),

$$\|\mathbf{f} + \mathbf{D}_H^{-1}\mathbf{v} - \mathbf{A}\bar{\mathbf{y}} - \mathbf{r}_\eta\| \leq \eta, \tag{2.249}$$

AS FOLLOWS:

(I) APPLY $[\frac{1}{3}\eta, \mathbf{A}, \bar{\mathbf{y}}] \to \mathbf{w}_\eta$;
(II) COARSE $[\frac{1}{3}\eta, \mathbf{f}] \to \mathbf{f}_\eta$;
(III) APPLY $[\frac{1}{3}\eta, \mathbf{D}_H^{-1}, \mathbf{v}] \to \mathbf{z}_\eta$;
(IV) set $\mathbf{r}_\eta := \mathbf{f}_\eta + \mathbf{z}_\eta - \mathbf{w}_\eta$.

By triangle inequality, one can for RES$_{\text{PRM}}$ and the subsequent variants of RES show that indeed (2.249) or (2.244) holds.

Similarly, one needs a version of SOLVE for the approximate solution of the second system (2.207b), $\mathbf{A}^T\mathbf{p} = -\mathbf{D}_{\mathscr{L}}^{-1}\mathbf{RD}_{\mathscr{L}}^{-1}(\mathbf{y} - \mathbf{y}_*)$, which depends on an approximate solution $\bar{\mathbf{y}}$ of the primal system and possibly on some initial guess $\bar{\mathbf{p}}^0$ with accuracy ε_0. Here we set

$$\text{SOLVE}_{\text{ADJ}}\,[\eta, \mathbf{A}, \mathbf{D}_{\mathscr{L}}^{-1}, \mathbf{y}_*, \bar{\mathbf{y}}, \bar{\mathbf{p}}^0, \varepsilon_0] := \text{SOLVE}\,[\eta, \mathbf{A}^T, \mathbf{D}_{\mathscr{L}}^{-1}\mathbf{RD}_{\mathscr{L}}^{-1}(\mathbf{y} - \bar{\mathbf{y}}), \bar{\mathbf{p}}^0, \varepsilon_0].$$

As usual we assume that the data $\mathbf{f}, \mathbf{y}_*$ are approximated in a preprocessing step with sufficient accuracy. A suitable residual approximation scheme RES$_{\text{ADJ}}$ for Step (II) of this version of SOLVE is the following where the main issue is the approximate evaluation of the right hand side.

RES$_{\text{ADJ}}$ $[\eta, \mathbf{A}, \mathbf{D}_{\mathscr{L}}^{-1}, \mathbf{y}_*, \bar{\mathbf{y}}, \bar{\mathbf{p}}] \to \mathbf{r}_\eta$ DETERMINES FOR ANY POSITIVE TOLERANCE η, GIVEN FINITELY SUPPORTED DATA $\bar{\mathbf{y}}, \mathbf{y}_*$ AND ANY FINITELY SUPPORTED INPUT $\bar{\mathbf{p}}$ AN APPROXIMATE RESIDUAL $\mathbf{r}_\eta$ SATISFYING (2.244), I.E.,

$$\| -\mathbf{D}_{\mathscr{L}}^{-1}\mathbf{RD}_{\mathscr{L}}^{-1}(\bar{\mathbf{y}} - \mathbf{y}_*) - \mathbf{A}^T\bar{\mathbf{p}} - \mathbf{r}_\eta\| \leq \eta, \tag{2.250}$$

AS FOLLOWS:

(i) APPLY $[\frac{1}{3}\eta, \mathbf{A}^T, \bar{\mathbf{p}}] \to \mathbf{w}_\eta$;
(ii) APPLY $[\frac{1}{6}\eta, \mathbf{D}_{\mathscr{L}}^{-1}, \bar{\mathbf{y}}] \to \mathbf{z}_\eta$; COARSE $[\frac{1}{6}\eta, \mathbf{y}_*] \to (\mathbf{y}_*)_\eta$;
    SET $\mathbf{d}_\eta := (\mathbf{y}_Z)_\eta - \mathbf{z}_\eta$;
    APPLY $[\frac{1}{6}\eta, \mathbf{D}_{\mathscr{L}}^{-1}, \mathbf{d}_\eta] \to \hat{\mathbf{v}}_\eta$; APPLY $[\frac{1}{6}\eta, \mathbf{R}, \hat{\mathbf{v}}_\eta] \to \mathbf{v}_\eta$;
(iii) SET $\mathbf{r}_\eta := \mathbf{v}_\eta - \mathbf{w}_\eta$.

Finally, we can define the residual scheme for the version of SOLVE applied to (2.214). We shall refer to this specification as SOLVE$_{\text{DCP}}$ with corresponding residual scheme is RES$_{\text{DCP}}$. Since the scheme is based on Proposition 7, it will involve several parameters stemming from the auxiliary systems (2.207).

RES$_{\text{DCP}}$ $[\eta, \mathbf{Q}, \mathbf{g}, \tilde{\mathbf{y}}, \delta_y, \tilde{\mathbf{p}}, \delta_p, \mathbf{v}, \delta_v] \to (\mathbf{r}_\eta, \tilde{\mathbf{y}}, \delta_y, \tilde{\mathbf{p}}, \delta_p)$ DETERMINES FOR ANY APPROXIMATE SOLUTION TRIPLE $(\tilde{\mathbf{y}}, \tilde{\mathbf{p}}, \mathbf{v})$ OF THE SYSTEM (2.207) SATISFYING

$$\|\mathbf{y} - \tilde{\mathbf{y}}\| \leq \delta_y, \quad \|\mathbf{p} - \tilde{\mathbf{p}}\| \leq \delta_p, \quad \|\mathbf{u} - \mathbf{v}\| \leq \delta_v, \tag{2.251}$$

AN APPROXIMATE RESIDUAL $\mathbf{r}_\eta$ SUCH THAT

$$\|\mathbf{g} - \mathbf{Q}\mathbf{v} - \mathbf{r}_\eta\| \leq \eta. \qquad (2.252)$$

MOREOVER, THE INITIAL APPROXIMATIONS $\tilde{\mathbf{y}}, \tilde{\mathbf{p}}$ ARE OVERWRITTEN BY NEW APPROXIMATIONS $\tilde{\mathbf{y}}, \tilde{\mathbf{p}}$ SATISFYING (2.251) WITH NEW BOUNDS δ_y AND δ_p DEFINED IN (2.253) BELOW, AS FOLLOWS:

(I) SOLVE$_{\text{PRM}}$ $[\frac{1}{3}c_A \, \eta, \mathbf{A}, \mathbf{D}_H^{-1}, \mathbf{f}, \mathbf{v}, \tilde{\mathbf{y}}, \delta_y] \to \mathbf{y}_\eta$;

(II) SOLVE$_{\text{ADJ}}$ $[\frac{1}{3}\eta, \mathbf{A}, \mathbf{D}_{\mathscr{L}}^{-1}, \mathbf{y}_*, \mathbf{y}_\eta, \tilde{\mathbf{p}}, \delta_p] \to \mathbf{p}_\eta$;

(III) APPLY $[\frac{1}{3}\eta, \mathbf{D}_H^{-1}, \mathbf{p}_\eta] \to \mathbf{q}_\eta$; SET $\mathbf{r}_\eta := \mathbf{q}_\eta - \omega\mathbf{v}$;

(IV) SET $\xi_y := c_A^{-1} \delta_v + \frac{1}{3}c_A\eta$, $\xi_p := c_A^{-2} \delta_v + \frac{2}{3}\eta$; REPLACE $\tilde{\mathbf{y}}, \delta_y$ AND $\tilde{\mathbf{p}}, \delta_p$ BY

$$\begin{aligned} \tilde{\mathbf{y}} &:= \text{COARSE}[4\xi_y, \mathbf{y}_\eta], & \delta_y &:= 5\xi_y, \\ \tilde{\mathbf{p}} &:= \text{COARSE}[4\xi_p, \mathbf{p}_\eta], & \delta_p &:= 5\xi_p. \end{aligned} \qquad (2.253)$$

Step (IV) already indicates the conditions on the tolerance η and the accuracy bound δ_v under which the new error bounds in (2.253) are actually tighter. The precise relation between η and δ_v in the context of SOLVE$_{\text{DCP}}$ is not apparent yet and emerges as well as the claimed estimates (2.252) and (2.253) from the complexity analysis in [29].

Finally, the scheme SOLVE$_{\text{DCP}}$ attains the following form with the error reduction factor ρ from (2.242) and α from (2.241).

SOLVE$_{\text{DCP}}$ $[\varepsilon, \mathbf{Q}, \mathbf{g}] \to \bar{\mathbf{u}}_\varepsilon$

(I) LET $\bar{\mathbf{q}}^0 := \mathbf{0}$ AND $\varepsilon_0 := c_A^{-1}(\|\mathbf{y}_Z\| + c_A^{-1}\|\mathbf{f}\|)$.
Let $\tilde{\mathbf{y}} := \mathbf{0}$, $\tilde{\mathbf{p}} := \mathbf{0}$ AND SET $j = 0$.
DEFINE $\delta_y := \delta_{y,0} := c_A^{-1}(\|\mathbf{f}\| + \varepsilon_0)$ AND $\delta_p := \delta_{p,0} := c_A^{-1}(\delta_{y,0} + \|\mathbf{y}_Z\|)$.

(II) IF $\varepsilon_j \leq \varepsilon$, STOP AND SET $\bar{\mathbf{u}}_\varepsilon := \bar{\mathbf{u}}^j$, $\bar{\mathbf{y}}_\varepsilon = \tilde{\mathbf{y}}$, $\bar{\mathbf{p}}_\varepsilon = \tilde{\mathbf{p}}$.
OTHERWISE SET $\mathbf{v}^0 := \bar{\mathbf{u}}^j$.

   (II.1) FOR $k = 0, \ldots, K - 1$, COMPUTE
   RES$_{\text{DCP}}$ $[\rho^k \varepsilon_j, \mathbf{Q}, \mathbf{g}, \tilde{\mathbf{y}}, \delta_y, \tilde{\mathbf{p}}, \delta_p, \mathbf{v}^k, \delta_k] \to (\mathbf{r}^k, \tilde{\mathbf{y}}, \delta_y, \tilde{\mathbf{p}}, \delta_p)$,
   WHERE $\delta_0 := \varepsilon_j$ AND $\delta_k := \rho^{k-1}(\alpha k + \rho)\varepsilon_j$;
   SET

$$\mathbf{v}^{k+1} := \mathbf{v}^k + \alpha\mathbf{r}^k. \qquad (2.254)$$

   (II.2) COARSE $[\frac{2}{5}\varepsilon_j, \mathbf{v}^K] \to \bar{\mathbf{u}}^{j+1}$; set $\varepsilon_{j+1} := \frac{1}{2}\varepsilon_j$, $j + 1 \to j$ and go to (ii).

By overwriting $\tilde{\mathbf{y}}, \tilde{\mathbf{p}}$ at the last stage prior to the termination of SOLVE$_{\text{DCP}}$ one has $\delta_v \leq \varepsilon, \eta \leq \varepsilon$, so that the following fact is an immediate consequence of (2.253).

Proposition 13 *The outputs $\bar{\mathbf{y}}_\varepsilon$ and $\bar{\mathbf{p}}_\varepsilon$ produced by SOLVE$_{\text{DCP}}$ in addition to $\mathbf{u}_\varepsilon$ are approximations to the exact solutions $\mathbf{y}, \mathbf{p}$ of (2.207) satisfying*

$$\|\mathbf{y} - \bar{\mathbf{y}}_\varepsilon\| \leq 5\varepsilon \, (c_A^{-1} + \tfrac{1}{3}c_A), \qquad \|\mathbf{p} - \bar{\mathbf{p}}_\varepsilon\| \leq 5\varepsilon \, (c_A^{-2} + \tfrac{2}{3}).$$

Complexity Analysis Proposition 12 states that the routine SOLVE converges for an arbitrary given accuracy provided that there is a routine RES satisfying the property (2.244). Then we have broken down step by step the necessary ingredients to derive computable versions which satisfy these requirements. What we finally want to show is that the routines are *optimal* in the sense that they provide the optimal work/accuracy rate in terms of best N-term approximation. The complexity analysis given next also reveals the role of the routine COARSE within the algorithms and the particular choices of the thresholds in Step (IV) of RES_{DCP}.

In order to be able to assess the quality of the adaptive algorithm, the notion of *optimality* has to be clarified first in the present context.

Definition 1 The scheme SOLVE has an *optimal work/accuracy rate s* if the following holds: Whenever the error of *best N-term approximation* satisfies

$$\|\mathbf{q} - \mathbf{q}_N\| := \min_{\#supp\mathbf{v} \leq N} \|\mathbf{q} - \mathbf{v}\| \lesssim N^{-s},$$

then the solution $\overline{\mathbf{q}}_\varepsilon$ is generated by SOLVE at an expense that also stays proportional to $\varepsilon^{-1/s}$ and in that sense matches the best N-term approximation rate.

Note that this implies that $\#supp \overline{\mathbf{q}}_\varepsilon$ also stays proportional to $\varepsilon^{-1/s}$. Thus, our benchmark is that whenever the solution of (2.214) can be approximated by N terms at rate s, SOLVE recovers that rate asymptotically. If $\mathbf{q}$ is known, the wavelet-best N-term approximation $\mathbf{q}_N$ of $\mathbf{q}$ is given by picking the N largest terms in modulus from $\mathbf{q}$, of course. However, when $\mathbf{q}$ is the (unknown) solution of (2.214) this information is certainly not available.

Since we are here in the framework of sequence spaces ℓ_2, the formulation of appropriate criteria for complexity will be based on a characterization of sequences which are *sparse* in the following sense. We consider sequences $\mathbf{v}$ for which the best N-term approximation error decays at a particular rate (*Lorentz spaces*). That is, for any given threshold $0 < \eta \leq 1$, the number of terms exceeding that threshold is controlled by some function of this threshold. In particular, set for some $0 < \tau < 2$

$$\ell_\tau^w := \{\mathbf{v} \in \ell_2 : \#\{\lambda \in \mathbb{I} : |v_\lambda| > \eta\} \leq C_\mathbf{v}\,\eta^{-\tau}, \text{ for all } 0 < \eta \leq 1\}. \tag{2.255}$$

This determines a strict subspace of ℓ_2 only when $\tau < 2$. Smaller τ's indicate sparser sequences. Let $C_\mathbf{v}$ for a given $\mathbf{v} \in \ell_\tau^w$ be the smallest constant for which (2.255) holds. Then one has $|\mathbf{v}|_{\ell_\tau^w} := \sup_{n \in \mathbb{N}} n^{1/\tau} v_n^* = C_\mathbf{v}^{1/\tau}$, where $\mathbf{v}^* = (v_n^*)_{n \in \mathbb{N}}$ is a non-decreasing rearrangement of $\mathbf{v}$. Furthermore, $\|\mathbf{v}\|_{\ell_\tau^w} := \|\mathbf{v}\| + |\mathbf{v}|_{\ell_\tau^w}$ is a quasi-norm for ℓ_τ^w. Since the continuous embeddings $\ell_\tau \hookrightarrow \ell_\tau^w \hookrightarrow \ell_{\tau+\varepsilon} \hookrightarrow \ell_2$ hold for $\tau < \tau + \varepsilon < 2$, ℓ_τ^w is 'close' to ℓ_τ and is therefore called *weak* ℓ_τ. The following crucial result connects sequences in ℓ_τ^w to best N-term approximation [19].

Proposition 14 *Let positive real numbers s and τ be related by*

$$\frac{1}{\tau} = s + \frac{1}{2}. \tag{2.256}$$

Then $\mathbf{v} \in \ell_\tau^w$ *if and only if* $\|\mathbf{v} - \mathbf{v}_N\| \lesssim N^{-s} \|\mathbf{v}\|_{\ell_\tau^w}$.

The property that an array of wavelet coefficients $\mathbf{v}$ belongs to ℓ_τ is equivalent to the fact that the expansion $\mathbf{v}^T \Psi_H$ in terms of a wavelet basis Ψ_H for a Hilbert space H belongs to a certain *Besov space* which describes a much weaker regularity measure than a Sobolev space of corresponding order, see, e.g., [16, 39]. Thus, Proposition 14 expresses how much loss of regularity can be compensated by judiciously placing the degrees of freedom in a nonlinear way in order to retain a certain optimal order of error decay.

A key criterion for a scheme SOLVE to exhibit an optimal work/accuracy rate can be formulated through the following property of the respective residual approximation. The routine RES is called τ^*-*sparse* for some $0 < \tau^* < 2$ if the following holds: Whenever the solution $\mathbf{q}$ of (2.214) belongs to ℓ_τ^w for some $\tau^* < \tau < 2$, then for any $\mathbf{v}$ with finite support the output $\mathbf{r}_\eta$ of RES $[\eta, \mathbf{Q}, \mathbf{g}, \mathbf{v}]$ satisfies

$$\|\mathbf{r}_\eta\|_{\ell_\tau^w} \lesssim \max\{\|\mathbf{v}\|_{\ell_\tau^w}, \|\mathbf{q}\|_{\ell_\tau^w}\}$$

and

$$\#supp\,\mathbf{r}_\eta \lesssim \eta^{-1/s} \max\{\|\mathbf{v}\|_{\ell_\tau^w}^{1/s}, \|\mathbf{q}\|_{\ell_\tau^w}^{1/s}\}$$

where s and τ are related by (2.256), and the number of floating point operations needed to compute $\mathbf{r}_\eta$ stays proportional to $\#supp\,\mathbf{r}_\eta$.

The analysis in [20] then yields the following result.

Theorem 8 *If* RES *is* τ^*-*sparse and if the exact solution* $\mathbf{q}$ *of (2.214) belongs to* ℓ_τ^w *for some* $\tau > \tau^*$, *then for every* $\varepsilon > 0$ *algorithm* SOLVE $[\varepsilon, \mathbf{Q}, \mathbf{g}]$ *produces after finitely many steps an output* $\overline{\mathbf{q}}_\varepsilon$ *(which, according to Proposition 12, always satisfies* $\|\mathbf{q} - \overline{\mathbf{q}}_\varepsilon\| < \varepsilon$*) with the following properties: For s and τ related by (2.256), one has*

$$\#supp\,\overline{\mathbf{q}}_\varepsilon \lesssim \varepsilon^{-1/s} \|\mathbf{q}\|_{\ell_\tau^w}^{1/s}, \qquad \|\overline{\mathbf{q}}_\varepsilon\|_{\ell_\tau^w} \lesssim \|\mathbf{q}\|_{\ell_\tau^w}, \tag{2.257}$$

and the number of floating point operations needed to compute $\overline{\mathbf{q}}_\varepsilon$ *remains proportional to* $\#supp\,\overline{\mathbf{q}}_\varepsilon$.

Hence, τ^*-sparsity of the routine RES implies for SOLVE asymptotically optimal work/accuracy rates for a certain range of decay rates given by τ^*. We stress that the algorithm itself does *not* require any a-priori knowledge about the solution such as its actual best N-term approximation rate. Theorem 8 also states that controlling the ℓ_τ^w-norm of the quantities generated in the computations is crucial. This finally explains the role of COARSE in Step (II.2) of SOLVE in terms of the following result [19].

Lemma 4 *Let* $\mathbf{v} \in \ell_\tau^w$ *and let* $\mathbf{w}$ *be any finitely supported approximation such that* $\|\mathbf{v} - \mathbf{w}\| \leq \frac{1}{5}\eta$. *Then the output* $\mathbf{w}_\eta$ *of* COARSE $[\frac{4}{5}\eta, \mathbf{w}]$ *satisfies*

$$\#supp\mathbf{w}_\eta \lesssim \|\mathbf{v}\|_{\ell_\tau^w}^{1/\tau} \eta^{-1/s}, \quad \|\mathbf{v} - \mathbf{w}_\eta\| \lesssim \eta, \quad and \quad \|\mathbf{w}_\eta\|_{\ell_\tau^w} \lesssim \|\mathbf{v}\|_{\ell_\tau^w}. \tag{2.258}$$

This can be interpreted as follows. If an error bound for a given finitely supported approximation $\mathbf{w}$ is known, a certain coarsening using only knowledge about $\mathbf{w}$ produces a new approximation to (the possibly unknown) $\mathbf{v}$ which gives rise to a slightly larger error but realizes the optimal relation between support and accuracy up to a uniform constant. In the scheme SOLVE, this means that by the coarsening step the ℓ_τ^w-norms of the iterates $\mathbf{v}^K$ are controlled.

It remains to establish that for SOLVE$_{\text{DCP}}$ the corresponding routine RES$_{\text{DCP}}$ is τ^*-sparse. The following results from [29] reduce this question to the efficiency of APPLY. We say that APPLY $[\cdot, \mathbf{L}, \cdot]$ is τ^*-*efficient* for some $0 < \tau^* < 2$ if for any finitely supported $\mathbf{v} \in \ell_\tau^w$, for $0 < \tau^* < \tau < 2$, the output $\mathbf{w}_\eta$ of APPLY $[\eta, \mathbf{L}, \mathbf{v}]$ satisfies $\|\mathbf{w}_\eta\|_{\ell_\tau^w} \lesssim \|\mathbf{v}\|_{\ell_\tau^w}$ and $\#supp\,\mathbf{w}_\eta \lesssim \eta^{-1/s}\|\mathbf{v}\|_{\ell_\tau^w}^{1/s}$ for $\eta \to 0$. Here the constants depend only on τ as $\tau \to \tau^*$ and s, τ satisfy (2.256). Moreover, the number of floating point operations needed to compute $\mathbf{w}_\eta$ is to remain proportional to $\#supp\,\mathbf{w}_\eta$.

Proposition 15 *If the* APPLY *schemes in* RES$_{\text{PRM}}$ *and* RES$_{\text{ADJ}}$ *are* τ^*-*efficient for some* $\tau^* < 2$, *then* RES$_{\text{DCP}}$ *is* τ^*-*sparse whenever there exists a constant* C *such that* $C\eta \geq \max\{\delta_v, \delta_p\}$ *and*

$$\max\{\|\tilde{\mathbf{p}}\|_{\ell_\tau^w}, \|\tilde{\mathbf{y}}\|_{\ell_\tau^w}, \|\mathbf{v}\|_{\ell_\tau^w}\} \leq C \left(\|\mathbf{y}\|_{\ell_\tau^w} + \|\mathbf{p}\|_{\ell_\tau^w} + \|\mathbf{u}\|_{\ell_\tau^w}\right),$$

where $\mathbf{v}$ *is the current finitely supported input and* $\tilde{\mathbf{y}}, \tilde{\mathbf{p}}$ *are the initial guesses for the exact solution components* $(\mathbf{y}, \mathbf{p})$.

Theorem 9 *If the* APPLY *schemes appearing in* RES$_{\text{PRM}}$ *and* RES$_{\text{ADJ}}$ *are* τ^*-*efficient for some* $\tau^* < 2$ *and the components of the solution* $(\mathbf{y}, \mathbf{p}, \mathbf{u})$ *of (2.207) all belong to the respective space* ℓ_τ^w *for some* $\tau > \tau^*$, *then the approximate solutions* $\mathbf{y}_\varepsilon, \mathbf{p}_\varepsilon, \mathbf{u}_\varepsilon$, *produced by* SOLVE$_{\text{DCP}}$ *for any target accuracy* ε, *satisfy*

$$\|\mathbf{y}_\varepsilon\|_{\ell_\tau^w} + \|\mathbf{p}_\varepsilon\|_{\ell_\tau^w} + \|\mathbf{u}_\varepsilon\|_{\ell_\tau^w} \lesssim \|\mathbf{y}\|_{\ell_\tau^w} + \|\mathbf{p}\|_{\ell_\tau^w} + \|\mathbf{u}\|_{\ell_\tau^w}, \tag{2.259}$$

and

$$(\#supp\,\mathbf{y}_\varepsilon) + (\#supp\,\mathbf{p}_\varepsilon) + (\#supp\,\mathbf{u}_\varepsilon) \lesssim \left(\|\mathbf{y}\|_{\ell_\tau^w}^{1/s} + \|\mathbf{p}\|_{\ell_\tau^w}^{1/s} + \|\mathbf{u}\|_{\ell_\tau^w}^{1/s}\right)\varepsilon^{-1/s}, \tag{2.260}$$

where the constants only depend on τ when τ approaches τ^*. Moreover, the number of floating point operations required during the execution of SOLVE$_{DCP}$ *remains proportional to the right hand side of (2.260).*

Thus, the practical realization of SOLVE$_{DCP}$ providing optimal work/accuracy rates for a possibly large range of decay rates of the error of best N-term approximation hinges on the availability of τ^*-efficient schemes APPLY with possibly small τ^* for the involved operators.

For the approximate application of wavelet representations of a wide class of operators, including differential operators, one can indeed devise efficient schemes which is a consequence of the cancellation properties (CP) together with the norm equivalences (2.89) for the relevant function spaces. For the example considered above, the τ^*-efficiency of $\mathbf{A}$ defined in (2.198) can be shown whenever $\mathbf{A}$ is s^*-compressible where τ^* and s^* are related by (2.256). One knows that s^* is the larger the higher the 'regularity' of the operator and the order of cancellation properties of the wavelets are. Estimates for s^* in terms of these quantities for spline wavelets and the above differential operator A can be found in [5]. These were refined and extended to trace operators in [62]. Hence, Theorem 9 guarantees asymptotically optimal complexity bounds for $\tau > \tau^*$. This means that the scheme SOLVE$_{DCP}$ recovers rates of the error of best N-term approximation of order N^{-s} for $s < s^*$.

When describing the control problem, it has been pointed out that the wavelet framework allows for a flexible choice of norms in the control functional which is reflected by the diagonal matrices $\mathbf{D}_{\mathscr{Y}}$ and $\mathbf{D}_H$ in (DCP), (2.203) together with (2.204). The following result states that multiplication by either $\mathbf{D}_{\mathscr{Y}}^{-1}$ or $\mathbf{D}_H^{-1}$ makes a sequence more compressible, that is, they produce a shift in weak ℓ_τ spaces [29].

Proposition 16 *For $\beta > 0$, $\mathbf{p} \in \ell_\tau^w$ implies $\mathbf{D}^{-\beta}\mathbf{p} \in \ell_{\tau'}^w$, where $\frac{1}{\tau'} := \frac{1}{\tau} + \frac{\beta}{d}$.*

We can conclude the following. Whatever the sparsity class of the adjoint variable $\mathbf{p}$ is, the control $\mathbf{u}$ is in view of (2.207c) even sparser. This means also that although the control $\mathbf{u}$ may be accurately recovered with relatively few degrees of freedom, the overall solution complexity is in the above case bounded from below by the less sparse auxiliary variable $\mathbf{p}$.

The application of these techniques to control problems constrained by parabolic PDEs can be found in [44]. For an extension of these techniques to control problems involving PDEs with possibly infinite stochastic coefficients which introduce a substantial difficulty, one may consult [57, 58].

References

1. J.P. Aubin, *Applied Functional Analysis*, 2nd edn. (Wiley, Hoboken, 2000)
2. I. Babuška, The finite element method with Lagrange multipliers. Numer. Math. **20**, 179–192 (1973)
3. I. Babuška, The finite element method with penalty. Math. Comput. **27**, 221–228 (1973)

4. A. Barinka, Fast evaluation tools for adaptive wavelet schemes, Ph.D. dissertation, RWTH Aachen, 2004
5. A. Barinka, T. Barsch, Ph. Charton, A. Cohen, S. Dahlke, W. Dahmen, K. Urban, Adaptive wavelet schemes for elliptic problems — implementation and numerical experiments. SIAM J. Sci. Comput. **23**, 910–939 (2001)
6. S. Bertoluzza, Wavelet stabilization of the Lagrange multiplier method. Numer. Math. **86**, 1–28 (2000)
7. D. Braess, *Finite Elements: Theory, Fast Solvers and Applications in Solid Mechanics*, 2nd edn. (Cambridge University Press, Cambridge, 2001)
8. J.H. Bramble, J.E. Pasciak, J. Xu, Parallel multilevel preconditioners. Math. Comput. **55**, 1–22 (1990)
9. F. Brezzi, M. Fortin, *Mixed and Hybrid Finite Element Methods* (Springer, New York, 1991)
10. A. Buffa, H. Harbrecht, A. Kunoth, G. Sangalli, Multilevel preconditioning for isogeometric analysis, Comput. Methods Appl. Mech. Eng. **265**, 63–70 (2013)
11. C. Burstedde, Wavelets methods for linear-quadratic, elliptic optimal control problems, Ph.D. dissertation, University of Bonn, Bonn, 2005
12. C. Burstedde, A. Kunoth, Fast iterative solution of elliptic control problems in wavelet discretizations. J. Comput. Appl. Math. **196**(1), 299–319 (2006)
13. C. Canuto, A. Tabacco, K. Urban, The wavelet element method, part I: construction and analysis. Appl. Comput. Harmon. Anal. **6**, 1–52 (1999)
14. J.M. Carnicer, W. Dahmen, J.M. Peña, Local decomposition of refinable spaces. Appl. Comput. Harmon. Anal. **3**, 127–153 (1996)
15. Z. Ciesielski, T. Figiel, Spline bases in classical function spaces on compact C^∞ manifolds: part I and II. Stud. Math. **76**, 1–58 and 95–136 (1983)
16. A. Cohen, *Numerical Analysis of Wavelet Methods*. Studies in Mathematics and Its Applications, vol. 32 (Elsevier, New York, 2003)
17. A. Cohen, R. Masson, Adaptive wavelet methods for second order elliptic problems, preconditioning and adaptivity. SIAM J. Sci. Comput. **21**, 1006–1026 (1999)
18. A. Cohen, I. Daubechies, J.-C. Feauveau, Biorthogonal bases of compactly supported wavelets, Commun. Pure Appl. Math. **45**, 485–560 (1992)
19. A. Cohen, W. Dahmen, R. DeVore, Adaptive wavelet methods for elliptic operator equations – convergence rates. Math. Comput. **70**, 27–75 (2001)
20. A. Cohen, W. Dahmen, R. DeVore, Adaptive wavelet methods II – beyond the elliptic case. Found. Comput. Math. **2**, 203–245 (2002)
21. A. Cohen, W. Dahmen, R. DeVore, Adaptive wavelet schemes for nonlinear variational problems. SIAM J. Numer. Anal. **41**(5), 1785–1823 (2003)
22. S. Dahlke, W. Dahmen, K. Urban, Adaptive wavelet methods for saddle point problems — optimal convergence rates. SIAM J. Numer. Anal. **40**, 1230–1262 (2002)
23. W. Dahmen, Stability of multiscale transformations. J. Fourier Anal. Appl. **2**, 341–361 (1996)
24. W. Dahmen, Wavelet and multiscale methods for operator equations. Acta Numer. **6**, 55–228 (1997)
25. W. Dahmen, Wavelet methods for PDEs – some recent developments. J. Comput. Appl. Math. **128**, 133–185 (2001)
26. W. Dahmen, Multiscale and wavelet methods for operator equations, in *Multiscale Problems and Methods in Numerical Simulation*, ed. by C. Canuto. C.I.M.E. Lecture Notes in Mathematics, vol. 1825 (Springer, Heidelberg, 2003), pp. 31–96
27. W. Dahmen, A. Kunoth, Multilevel preconditioning. Numer. Math. **63**, 315–344 (1992)
28. W. Dahmen, A. Kunoth, Appending boundary conditions by Lagrange multipliers: analysis of the LBB condition. Numer. Math. **88**, 9–42 (2001)
29. W. Dahmen, A. Kunoth, Adaptive wavelet methods for linear-quadratic elliptic control problems: convergence rates. SIAM J. Control. Optim. **43**(5), 1640–1675 (2005)
30. W. Dahmen, R. Schneider, Wavelets with complementary boundary conditions — function spaces on the cube. Results Math. **34**, 255–293 (1998)

31. W. Dahmen, R. Schneider, Composite wavelet bases for operator equations. Math. Comput. **68**, 1533–1567 (1999)
32. W. Dahmen, R. Schneider, Wavelets on manifolds I: construction and domain decomposition. SIAM J. Math. Anal. **31**, 184–230 (1999)
33. W. Dahmen, R. Stevenson, Element-by-element construction of wavelets satisfying stability and moment conditions. SIAM J. Numer. Anal. **37**, 319–325 (1999)
34. W. Dahmen, A. Kunoth, K. Urban, Biorthogonal spline wavelets on the interval – stability and moment conditions. Appl. Comput. Harmon. Anal. **6**, 132–196 (1999)
35. W. Dahmen, A. Kunoth, R. Schneider, Wavelet least squares methods for boundary value problems. SIAM J. Numer. Anal. **39**, 1985–2013 (2002)
36. I. Daubechies, Orthonormal bases of compactly supported wavelets. Commun. Pure Appl. Math. **41**, 909–996 (1988)
37. R. Dautray, J.-L. Lions, *Mathematical Analysis and Numerical Methods for Science and Technology*. Evolution Problems, vol. 5 (Springer, Berlin, 2000)
38. C. de Boor, *A Practical Guide to Splines*, revised edn. (Springer, New York, 2011)
39. R.A. DeVore, Nonlinear approximation. Acta Numer. **7**, 51–150 (1998)
40. L.C. Evans, *Partial Differential Equations* (AMS, Providence, 1998)
41. V. Girault, R. Glowinski, Error analysis of a fictitious domain method applied to a Dirichlet problem. Jpn. J. Ind. Appl. Math. **12**, 487–514 (1995)
42. V. Girault, P.-A. Raviart, *Finite Element Methods for Navier–Stokes Equations* (Springer, Berlin, 1986)
43. P. Grisvard, *Elliptic Problems in Nonsmooth Domains* (Pitman, New York, 1985)
44. M.D. Gunzburger, A. Kunoth, Space-time adaptive wavelet methods for optimal control problems constrained by parabolic evolution equations. SIAM J. Control. Optim. **49**(3), 1150–1170 (2011)
45. M.D. Gunzburger, H.C. Lee, Analysis, approximation, and computation of a coupled solid/fluid temperature control problem. Comput. Methods Appl. Mech. Eng. **118**, 133–152 (1994)
46. J. Haslinger, R.A.E. Mäkinen, *Introduction to Shape Optimization: Theory, Approximation, and Computation* (SIAM, Philadelphia, 2003)
47. S. Jaffard, Wavelet methods for fast resolution of elliptic problems. SIAM J. Numer. Anal. **29**, 965–986 (1992)
48. J. Krumsdorf, Finite element wavelets for the numerical solution of elliptic partial differential equations on polygonal domains, Diploma thesis (in English), Universität Bonn, Bonn, January 2004
49. K. Kunisch, G. Peichl, Shape optimization for mixed boundary value problems based on an embedding domain method. Dyn. Contin. Discret. Impuls. Syst. **4**, 439–478 (1998)
50. A. Kunoth, *Multilevel Preconditioning* (Verlag Shaker, Aachen, 1994)
51. A. Kunoth, Wavelet methods — elliptic boundary value problems and control problems, in *Advances in Numerical Mathematics* (Teubner, Leipzig, 2001)
52. A. Kunoth, Wavelet techniques for the fictitious domain—Lagrange multiplier approach. Numer. Algorithm **27**, 291–316 (2001)
53. A. Kunoth, Fast iterative solution of saddle point problems in optimal control based on wavelets. Comput. Optim. Appl. **22**, 225–259 (2002)
54. A. Kunoth, Adaptive wavelet methods for an elliptic control problem with Dirichlet boundary control. Numer. Algorithm **39**(1–3), 199–220 (2005)
55. A. Kunoth, Multilevel preconditioning for variational problems, in *Isogeometric Analysis and Applications*, ed. by B. Jüttler, B. Simeon. Lecture Notes in Computational Sciences and Engineering (Springer, 2014), pp. 247–281
56. A. Kunoth, J. Sahner, Wavelets on manifolds: an optimized construction. Math. Comput. **75**, 1319–1349 (2006)
57. A. Kunoth, Chr. Schwab, Analytic regularity and GPC approximation for control problems constrained by parametric elliptic and parabolic PDEs. SIAM J. Control. Optim. **51**(3), 2442–2471 (2013)

58. A. Kunoth, Chr. Schwab, Sparse adaptive tensor Galerkin approximations of stochastic pde-constrained control problems. SIAM/ASA J. Uncertain. Quantif. **4** (2016), 1034–1059
59. J.L. Lions, *Optimal Control of Systems Governed by Partial Differential Equations* (Springer, Berlin, 1971)
60. J. Maes, A. Kunoth, A. Bultheel, BPX-type preconditioners for 2nd and 4th order elliptic problems on the sphere. SIAM J. Numer. Anal. **45**(1), 206–222(2007)
61. P. Oswald, On discrete norm estimates related to multilevel preconditioners in the finite element method, in *Constructive Theory of Functions*, ed. by K.G. Ivanov, P. Petrushev, B. Sendov. Proceedings of International Conference Varna 1991 (Bulgarian Academy of Sciences, Sofia, 1992), pp. 203–214
62. R. Pabel, *Wavelet Methods for PDE Constrained Control Problems with Dirichlet Boundary Control* (Shaker, Maastricht, 2007). https://doi.org/10.2370/236_232
63. R. Pinnau, G. Thömmes, Optimal boundary control of glass cooling processes. Math. Methods Appl. Sci. **27**, 1261–1281 (2004)
64. J. Sahner, On the optimized construction of wavelets on manifolds, Diploma thesis (in English), Universität Bonn, Bonn, September 2003
65. L.L. Schumaker, *Spline Functions: Basic Theory*, 3rd edn. (Cambridge Mathematical Library, Cambridge University Press, Cambridge, 2007)
66. Chr. Schwab, R. Stevenson, Space-time adaptive wavelet methods for parabolic evolution equations. Math. Comput. **78**, 1293–1318 (2009)
67. R. Stenberg, On some techniques for approximating boundary conditions in the finite element method. J. Comput. Appl. Math. **63**, 139–148 (1995)
68. R. Stevenson, Locally supported, piecewise polynomial biorthogonal wavelets on non-uniform meshes. Constr. Approx. **19**, 477–508(2003)
69. W. Sweldens, The lifting scheme: a construction of second generation wavelets. SIAM J. Math. Anal. **29**, 511–546 (1998)
70. J. van den Eshof, G.L.G. Sleijpen, Inexact Krylov subspace methods for linear systems. SIAM J. Matrix Anal. Appl. **26**, 125–153 (2004)
71. H. Yserentant, On the multilevel splitting of finite element spaces. Numer. Math. **49**, 379–412 (1986)
72. E. Zeidler, *Nonlinear Functional Analysis and its Applications; III: Variational Methods and Optimization* (Springer, New York, 1985)

Chapter 3
Generalized Locally Toeplitz Sequences: A Spectral Analysis Tool for Discretized Differential Equations

Carlo Garoni and Stefano Serra-Capizzano

Abstract The theory of Generalized Locally Toeplitz (GLT) sequences was developed in order to solve a specific application problem, namely the problem of computing/analyzing the spectral distribution of matrices arising from the numerical discretization of Differential Equations (DEs). A final goal of this spectral analysis is the design of efficient numerical methods for computing the related numerical solutions. The purpose of this contribution is to introduce the reader to the theory of GLT sequences and to present some of its applications to the computation of the spectral distribution of DE discretization matrices. We will mainly focus on the applications, whereas the theory will be presented in a self-contained tool-kit fashion, without entering into technical details.

3.1 Introduction

Origin and Purpose of the Theory of GLT Sequences The theory of Generalized Locally Toeplitz (GLT) sequences stems from Tilli's work on Locally Toeplitz (LT) sequences [56] and from the spectral theory of Toeplitz matrices [2, 10–13, 37, 44, 55, 57–60]. It was then developed by the authors in [29, 30, 50, 51] and has been recently extended by Barbarino in [3]. It was devised in order to solve a specific application problem, namely the problem of computing/analyzing the spectral distribution of matrices arising from the numerical discretization of

C. Garoni
Institute of Computational Science, University of Italian Switzerland, Lugano, Switzerland

Department of Science and High Technology, University of Insubria, Como, Italy
e-mail: carlo.garoni@usi.ch; carlo.garoni@uninsubria.it

S. Serra-Capizzano (✉)
Department of Science and High Technology, University of Insubria, Como, Italy

Department of Information Technology, Uppsala University, Uppsala, Sweden
e-mail: stefano.serrac@uninsubria.it; stefano.serra@it.uu.se

© Springer Nature Switzerland AG 2018
T. Lyche et al. (eds.), *Splines and PDEs: From Approximation Theory to Numerical Linear Algebra*, Lecture Notes in Mathematics 2219,
https://doi.org/10.1007/978-3-319-94911-6_3

Differential Equations (DEs). A final goal of this spectral analysis is the design of efficient numerical methods for computing the related numerical solutions. The theory of GLT sequences finds applications also in other areas of science (see, e.g., [16] and [29, Sections 10.1–10.4]), but the computation of the spectral distribution of DE discretization matrices remains the main application. The next paragraph is therefore devoted to a general description of this application.

Main Application of the Theory of GLT Sequences Suppose a linear DE

$$\mathscr{A}u = g$$

is discretized by a linear numerical method characterized by a mesh fineness parameter n. In this situation, the computation of the numerical solution reduces to solving a linear system of the form

$$A_n \mathbf{u}_n = \mathbf{g}_n,$$

where the size d_n of the matrix A_n increases with n. What is often observed in practice is that A_n enjoys an asymptotic spectral distribution as $n \to \infty$, i.e., as the mesh is progressively refined. More precisely, it often turns out that, for a large class of test functions F,

$$\lim_{n \to \infty} \frac{1}{d_n} \sum_{j=1}^{d_n} F(\lambda_j(A_n)) = \frac{1}{\mu_k(D)} \int_D F(\kappa(\mathbf{y})) d\mathbf{y},$$

where $\lambda_j(A_n)$, $j = 1, \ldots, d_n$, are the eigenvalues of A_n, μ_k is the Lebesgue measure in $\mathbb{R}^k$, and $\kappa : D \subset \mathbb{R}^k \to \mathbb{C}$. In this situation, the function κ is referred to as the *spectral symbol* of the sequence $\{A_n\}_n$. The spectral information contained in κ can be informally summarized as follows: assuming that n is large enough, the eigenvalues of A_n, except possibly for $o(d_n)$ outliers, are approximately equal to the samples of κ over a uniform grid in D. For example, if $k = 1$, $d_n = n$ and $D = [a, b]$, then, assuming we have no outliers, the eigenvalues of A_n are approximately equal to

$$\kappa\left(a + i\,\frac{b - a}{n}\right), \qquad i = 1, \ldots, n,$$

for n large enough. Similarly, if $k = 2$, $d_n = n^2$ and $D = [a_1, b_1] \times [a_2, b_2]$, then, assuming we have no outliers, the eigenvalues of A_n are approximately equal to

$$\kappa\left(a_1 + i_1\,\frac{b_1 - a_1}{n}, \ a_2 + i_2\,\frac{b_2 - a_2}{n}\right), \qquad i_1, i_2 = 1, \ldots, n,$$

for n large enough. It is then clear that the symbol κ provides a 'compact' and quite accurate description of the spectrum of the matrices A_n (for n large enough).

The theory of GLT sequences is a powerful apparatus for computing the spectral symbol κ. Indeed, the sequence of discretization matrices $\{A_n\}_n$ turns out to be a GLT sequence with symbol (or kernel) κ for many classes of DEs and numerical methods, especially if the numerical method belongs to the family of the so-called 'local methods'. Local methods are, for example, Finite Difference (FD) methods, Finite Element (FE) methods with 'locally supported' basis functions, and collocation methods; in short, all standard numerical methods for the approximation of DEs. We refer the reader to Sect. 3.3.2 and [9, 29, 30, 50–52] for applications of the theory of GLT sequences in the context of FD discretizations of DEs; to Sect. 3.3.3 and [5, 9, 25, 26, 29, 30, 51] for the FE and collocation settings; to Sect. 3.3.4 and [22, 28–30, 32–35, 48] for the case of Isogeometric Analysis (IgA) discretizations, both in the collocation and Galerkin frameworks; and to [23] for a further recent application to fractional DEs.

Practical Use of the Spectral Symbol It is worth emphasizing that the knowledge of the spectral symbol κ, which can be attained through the theory of GLT sequences, is not only interesting in itself, but may also be exploited for practical purposes. Let us mention some of them.

(a) Compare the spectrum of A_n, compactly described by κ, with the spectrum of the differential operator $\mathscr{A}$.
(b) Understand whether the numerical method used to discretize the DE $\mathscr{A}u = g$ is appropriate or not to spectrally approximate the operator $\mathscr{A}$.
(c) Analyze the convergence and predict the behavior of iterative methods (especially, multigrid and preconditioned Krylov methods), when they are applied to the matrix A_n.
(d) Design fast iterative solvers (especially, multigrid and preconditioned Krylov methods) for linear systems with coefficient matrix A_n.

The goal (b) can be achieved through the spectral comparison mentioned in (a) and allows one to classify the various numerical methods on the basis of their spectral approximation properties. In this way, it is possible to select the best approximation technique among a set of given methods. In this regard, we point out that the symbol-based analysis carried out in [35] proved that IgA is superior to classical FE methods in the spectral approximation of the underlying differential operator $\mathscr{A}$. The reason for which the spectral symbol κ can be exploited for the purposes (c)–(d) is the following: the convergence properties of iterative solvers in general (and of multigrid and preconditioned Krylov methods in particular) strongly depend on the spectral features of the matrix to which they are applied; hence, the spectral information provided by κ can be conveniently used for designing fast solvers of this kind and/or analyzing their convergence properties. In this respect, we recall that noteworthy estimates on the superlinear convergence of the Conjugate Gradient (CG) method are strictly related to the asymptotic spectral distribution of the matrices to which the CG method is applied; see [4]. We also refer the reader to [20, 21, 24] for recent developments in the IgA framework, where the spectral symbol was exploited to design ad hoc iterative solvers for IgA discretization matrices.

Description of the Present Work The present work is an excerpt of the book [29]. Its purpose is to introduce the reader to the theory of GLT sequences and its applications in the context of DE discretizations. Following [29], we will here consider only unidimensional DEs both for simplicity and because the key 'GLT ideas' are better conveyed in the univariate setting. For the multivariate setting, the reader is referred to the literature cited above and, especially, to the book [30].

3.2   The Theory of GLT Sequences: A Summary

In this section we present a *self-contained* summary of the theory of GLT sequences. Despite its conciseness, our presentation contains *everything one needs to know* in order to understand the applications presented in the next section.

Matrix-Sequences Throughout this work, by a *matrix-sequence* we mean a sequence of the form $\{A_n\}_n$, where A_n is an $n \times n$ matrix. We say that the matrix-sequence $\{A_n\}_n$ is *Hermitian* if each A_n is Hermitian.

Singular Value and Eigenvalue Distribution of a Matrix-Sequence Let μ_k be the Lebesgue measure in $\mathbb{R}^k$. Throughout this work, all the terminology coming from measure theory (such as 'measurable set', 'measurable function', 'almost everywhere (a.e.)', etc.) is always referred to the Lebesgue measure. Let $C_c(\mathbb{R})$ (resp., $C_c(\mathbb{C})$) be the space of continuous complex-valued functions with bounded support defined on $\mathbb{R}$ (resp., $\mathbb{C}$). If A is a square matrix of size n, the singular values and the eigenvalues of A are denoted by $\sigma_1(A), \ldots, \sigma_n(A)$ and $\lambda_1(A), \ldots, \lambda_n(A)$, respectively. The set of the eigenvalues (i.e., the spectrum) of A is denoted by $\Lambda(A)$.

Definition 1 Let $\{A_n\}_n$ be a matrix-sequence and let $f : D \subset \mathbb{R}^k \to \mathbb{C}$ be a measurable function defined on a set D with $0 < \mu_k(D) < \infty$.

- We say that $\{A_n\}_n$ has a singular value distribution described by f, and we write $\{A_n\}_n \sim_\sigma f$, if

$$\lim_{n \to \infty} \frac{1}{n} \sum_{i=1}^{n} F(\sigma_i(A_n)) = \frac{1}{\mu_k(D)} \int_D F(|f(\mathbf{x})|)\mathrm{d}\mathbf{x}, \qquad \forall F \in C_c(\mathbb{R}).$$

  In this case, f is called the *singular value symbol* of $\{A_n\}_n$.
- We say that $\{A_n\}_n$ has a spectral (or eigenvalue) distribution described by f, and we write $\{A_n\}_n \sim_\lambda f$, if

$$\lim_{n \to \infty} \frac{1}{n} \sum_{i=1}^{n} F(\lambda_i(A_n)) = \frac{1}{\mu_k(D)} \int_D F(f(\mathbf{x}))\mathrm{d}\mathbf{x}, \qquad \forall F \in C_c(\mathbb{C}).$$

  In this case, f is called the *spectral (or eigenvalue) symbol* of $\{A_n\}_n$.

When we write a relation such as $\{A_n\}_n \sim_\sigma f$ or $\{A_n\}_n \sim_\lambda f$, it is understood that $\{A_n\}_n$ is a matrix-sequence and f is a measurable function defined on a subset D of some $\mathbb{R}^k$ with $0 < \mu_k(D) < \infty$. If $\{A_n\}_n$ has both a singular value and a spectral distribution described by f, we write $\{A_n\}_n \sim_{\sigma,\lambda} f$.

We report in **S 1** and **S 2** the statements of two useful results concerning the spectral distribution of matrix-sequences. Throughout this work, if A is an $n \times n$ matrix and $1 \le p \le \infty$, we denote by $\|A\|_p$ the Schatten p-norm of A, i.e., the p-norm of the vector $(\sigma_1(A), \ldots, \sigma_n(A))$ formed by the singular values of A; see [7]. The Schatten ∞-norm $\|A\|_\infty$ is the largest singular value of A and coincides with the classical 2-norm $\|A\|$. The Schatten 1-norm $\|A\|_1$ is the sum of all the singular values of A and is often referred to as the trace-norm of A. The (topological) closure of a set S is denoted by $\overline{S}$.

S 1. If $\{A_n\}_n \sim_\lambda f$ and $\Lambda(A_n) \subseteq S$ for all n then $f \in \overline{S}$ a.e.

S 2. If $A_n = X_n + Y_n$ where

- each X_n is Hermitian and $\{X_n\}_n \sim_\lambda f$,
- $\|X_n\|, \|Y_n\| \le C$ for all n, where C is a constant independent of n,
- $n^{-1}\|Y_n\|_1 \to 0$,

then $\{A_n\}_n \sim_\lambda f$.

Informal Meaning Assuming that f is continuous a.e., the spectral distribution $\{A_n\}_n \sim_\lambda f$ has the following informal meaning: all the eigenvalues of A_n, except possibly for $o(n)$ outliers, are approximately equal to the samples of f over a uniform grid in D (for n large enough). For instance, if $k = 1$ and $D = [a, b]$, then, assuming we have no outliers, the eigenvalues of A_n are approximately equal to

$$f\left(a + i\,\frac{b - a}{n}\right), \qquad i = 1, \ldots, n,$$

for n large enough. Similarly, if $k = 2$, $n = m^2$ and $D = [a_1, b_1] \times [a_2, b_2]$, then, assuming we have no outliers, the eigenvalues of A_n are approximately equal to

$$f\left(a_1 + i\,\frac{b_1 - a_1}{m},\ a_2 + j\,\frac{b_2 - a_2}{m}\right), \qquad i, j = 1, \ldots, m,$$

for n large enough. A completely analogous meaning can also be given for the singular value distribution $\{A_n\}_n \sim_\sigma f$.

Zero-Distributed Sequences A matrix-sequence $\{Z_n\}_n$ such that $\{Z_n\}_n \sim_\sigma 0$ is referred to as a zero-distributed sequence. In other words, $\{Z_n\}_n$ is zero-distributed if and only if

$$\lim_{n \to \infty} \frac{1}{n} \sum_{i=1}^{n} F(\sigma_i(Z_n)) = F(0), \qquad \forall\, F \in C_c(\mathbb{R}).$$

Z1–Z2 will provide us with an important characterization of zero-distributed sequences together with a useful sufficient condition for detecting such sequences. For convenience, throughout this work we use the natural convention $1/\infty = 0$.

Z1. $\{Z_n\}_n \sim_\sigma 0$ if and only if $Z_n = R_n + N_n$ with $\lim_{n\to\infty} n^{-1}\mathrm{rank}(R_n) = \lim_{n\to\infty} \|N_n\| = 0$.

Z2. $\{Z_n\}_n \sim_\sigma 0$ if there is a $p \in [1, \infty]$ such that $\lim_{n\to\infty} n^{-1/p}\|Z_n\|_p = 0$.

Sequences of Diagonal Sampling Matrices If $n \in \mathbb{N}$ and $a : [0, 1] \to \mathbb{C}$, the nth diagonal sampling matrix generated by a is the $n \times n$ diagonal matrix given by

$$D_n(a) = \mathop{\mathrm{diag}}_{i=1,\dots,n} a\left(\frac{i}{n}\right).$$

$\{D_n(a)\}_n$ is called the sequence of diagonal sampling matrices generated by a.

Toeplitz Sequences If $n \in \mathbb{N}$ and $f : [-\pi, \pi] \to \mathbb{C}$ is a function in $L^1([-\pi, \pi])$, the nth Toeplitz matrix generated by f is the $n \times n$ matrix

$$T_n(f) = [f_{i-j}]_{i,j=1}^n,$$

where the numbers f_k are the Fourier coefficients of f,

$$f_k = \frac{1}{2\pi} \int_{-\pi}^{\pi} f(\theta)e^{-ik\theta}\,d\theta, \qquad k \in \mathbb{Z}.$$

$\{T_n(f)\}_n$ is called the Toeplitz sequence generated by f.

T1. For every $n \in \mathbb{N}$ the map $T_n(\cdot) : L^1([-\pi, \pi]) \to \mathbb{C}^{n \times n}$

- is linear: $T_n(\alpha f + \beta g) = \alpha T_n(f) + \beta T_n(g)$ for every $\alpha, \beta \in \mathbb{C}$ and every $f, g \in L^1([-\pi, \pi])$;
- satisfies $(T_n(f))^* = T_n(\overline{f})$ for every $f \in L^1([-\pi, \pi])$, so if f is real then $T_n(f)$ is Hermitian for every n.

T2. If $f \in L^1([-\pi, \pi])$ then $\{T_n(f)\}_n \sim_\sigma f$. If $f \in L^1([-\pi, \pi])$ and f is real then $\{T_n(f)\}_n \sim_\lambda f$.

T3. If $n \in \mathbb{N}$, $1 \le p \le \infty$ and $f \in L^p([-\pi, \pi])$, then $\|T_n(f)\|_p \le \frac{n^{1/p}}{(2\pi)^{1/p}}\|f\|_{L^p}$.

Approximating Classes of Sequences The notion of approximating classes of sequences (a.c.s.) is the fundamental concept on which the theory of GLT sequences is based.

Definition 2 Let $\{A_n\}_n$ be a matrix-sequence and let $\{\{B_{n,m}\}_n\}_m$ be a sequence of matrix-sequences. We say that $\{\{B_{n,m}\}_n\}_m$ is an approximating class of sequences (a.c.s.) for $\{A_n\}_n$ if the following condition is met: for every m there exists n_m such that, for $n \ge n_m$,

$$A_n = B_{n,m} + R_{n,m} + N_{n,m}, \qquad \mathrm{rank}(R_{n,m}) \le c(m)n, \qquad \|N_{n,m}\| \le \omega(m),$$

where n_m, $c(m)$, $\omega(m)$ depend only on m, and

$$\lim_{m \to \infty} c(m) = \lim_{m \to \infty} \omega(m) = 0.$$

Throughout this work, we use the abbreviation 'a.c.s.' for both the singular 'approximating class of sequences' and the plural 'approximating classes of sequences'; it will be clear from the context whether 'a.c.s.' is singular or plural. Roughly speaking, $\{\{B_{n,m}\}_n\}_m$ is an a.c.s. for $\{A_n\}_n$ if, for all sufficiently large m, the sequence $\{B_{n,m}\}_n$ approximates $\{A_n\}_n$ in the sense that A_n is eventually equal to $B_{n,m}$ plus a small-rank matrix (with respect to the matrix size n) plus a small-norm matrix. It turns out that the notion of a.c.s. is a notion of convergence in the space of matrix-sequences $\mathscr{E} = \{\{A_n\}_n : \{A_n\}_n$ is a matrix-sequence$\}$, i.e., there exists a topology $\tau_{\text{a.c.s.}}$ on $\mathscr{E}$ such that $\{\{B_{n,m}\}_n\}_m$ is an a.c.s. for $\{A_n\}_n$ if and only if $\{\{B_{n,m}\}_n\}_m$ converges to $\{A_n\}_n$ in $(\mathscr{E}, \tau_{\text{a.c.s.}})$. The theory of a.c.s. may then be interpreted as an approximation theory for matrix-sequences, and for this reason we will use the convergence notation $\{B_{n,m}\}_n \xrightarrow{\text{a.c.s.}} \{A_n\}_n$ to indicate that $\{\{B_{n,m}\}_n\}_m$ is an a.c.s. for $\{A_n\}_n$.

ACS 1. $\{A_n\}_n \sim_\sigma f$ if and only if there exist matrix-sequences $\{B_{n,m}\}_n \sim_\sigma f_m$ such that $\{B_{n,m}\}_n \xrightarrow{\text{a.c.s.}} \{A_n\}_n$ and $f_m \to f$ in measure.

ACS 2. Suppose each A_n is Hermitian. Then, $\{A_n\}_n \sim_\lambda f$ if and only if there exist Hermitian matrix-sequences $\{B_{n,m}\}_n \sim_\lambda f_m$ such that $\{B_{n,m}\}_n \xrightarrow{\text{a.c.s.}} \{A_n\}_n$ and $f_m \to f$ in measure.

ACS 3. Let $p \in [1, \infty]$ and suppose for every m there exists n_m such that, for $n \geq n_m$, $\|A_n - B_{n,m}\|_p \leq \epsilon(m, n) n^{1/p}$, where $\lim_{m \to \infty} \lim \sup_{n \to \infty} \epsilon(m, n) = 0$. Then $\{B_{n,m}\}_n \xrightarrow{\text{a.c.s.}} \{A_n\}_n$.

Generalized Locally Toeplitz Sequences A Generalized Locally Toeplitz (GLT) sequence $\{A_n\}_n$ is a special matrix-sequence equipped with a measurable function $\kappa : [0, 1] \times [-\pi, \pi] \to \mathbb{C}$, the so-called *symbol* (or *kernel*). We use the notation $\{A_n\}_n \sim_{\text{GLT}} \kappa$ to indicate that $\{A_n\}_n$ is a GLT sequence with symbol κ. The symbol of a GLT sequence is unique in the sense that if $\{A_n\}_n \sim_{\text{GLT}} \kappa$ and $\{A_n\}_n \sim_{\text{GLT}} \xi$ then $\kappa = \xi$ a.e. in $[0, 1] \times [-\pi, \pi]$. The main properties of GLT sequences are summarized in the following list. If A is a matrix, we denote by $A^\dagger$ the Moore–Penrose pseudoinverse of A; we recall that $A^\dagger = A^{-1}$ whenever A is invertible and we refer the reader to [8, 36] for more details on the pseudoinverse of a matrix. If A is a Hermitian matrix and f is a function defined at each point of $\Lambda(A)$, we denote by $f(A)$ the unique matrix such that $f(A)\mathbf{v} = f(\lambda)\mathbf{v}$ whenever $A\mathbf{v} = \lambda\mathbf{v}$; for more on matrix functions, we refer the reader to Higham's book [38].

GLT 1. If $\{A_n\}_n \sim_{\text{GLT}} \kappa$ then $\{A_n\}_n \sim_\sigma \kappa$. If $\{A_n\}_n \sim_{\text{GLT}} \kappa$ and the matrices A_n are Hermitian then $\{A_n\}_n \sim_\lambda \kappa$.

GLT 2. If $\{A_n\}_n \sim_{\text{GLT}} \kappa$ and $A_n = X_n + Y_n$, where

- every X_n is Hermitian,
- $\|X_n\|$, $\|Y_n\| \leq C$ for some constant C independent of n,

- $n^{-1}\|Y_n\|_1 \to 0$,

then $\{A_n\}_n \sim_\lambda \kappa$.

GLT 3. We have

- $\{T_n(f)\}_n \sim_{\mathrm{GLT}} \kappa(x,\theta) = f(\theta)$ if $f \in L^1([-\pi,\pi])$,
- $\{D_n(a)\}_n \sim_{\mathrm{GLT}} \kappa(x,\theta) = a(x)$ if $a : [0,1] \to \mathbb{C}$ is continuous a.e.,
- $\{Z_n\}_n \sim_{\mathrm{GLT}} \kappa(x,\theta) = 0$ if and only if $\{Z_n\}_n \sim_\sigma 0$.

GLT 4. If $\{A_n\}_n \sim_{\mathrm{GLT}} \kappa$ and $\{B_n\}_n \sim_{\mathrm{GLT}} \xi$ then

- $\{A_n^*\}_n \sim_{\mathrm{GLT}} \overline{\kappa}$,
- $\{\alpha A_n + \beta B_n\}_n \sim_{\mathrm{GLT}} \alpha\kappa + \beta\xi$ for all $\alpha, \beta \in \mathbb{C}$,
- $\{A_n B_n\}_n \sim_{\mathrm{GLT}} \kappa\xi$.

GLT 5. If $\{A_n\}_n \sim_{\mathrm{GLT}} \kappa$ and $\kappa \neq 0$ a.e. then $\{A_n^\dagger\}_n \sim_{\mathrm{GLT}} \kappa^{-1}$.

GLT 6. If $\{A_n\}_n \sim_{\mathrm{GLT}} \kappa$ and each A_n is Hermitian, then $\{f(A_n)\}_n \sim_{\mathrm{GLT}} f(\kappa)$ for every continuous function $f : \mathbb{C} \to \mathbb{C}$.

GLT 7. $\{A_n\}_n \sim_{\mathrm{GLT}} \kappa$ if and only if there exist GLT sequences $\{B_{n,m}\}_n \sim_{\mathrm{GLT}} \kappa_m$ such that $\{B_{n,m}\}_n \xrightarrow{\mathrm{a.c.s.}} \{A_n\}_n$ and $\kappa_m \to \kappa$ in measure.

3.3 Applications

In this section we present several applications of the theory of GLT sequences to the spectral analysis of DE discretization matrices. Our aim is to show how to compute the singular value and eigenvalue distribution of matrix-sequences arising from a DE discretization through the 'GLT tools' presented in the previous section. We begin by considering FD discretizations, then we will move to FE discretizations, and finally we will focus on IgA discretizations. Before starting, we collect below some auxiliary results.

3.3.1 Preliminaries

3.3.1.1 Matrix-Norm Inequalities

If $1 \leq p \leq \infty$, the symbol $|\cdot|_p$ denotes both the p-norm of vectors and the associated operator norm for matrices:

$$|\mathbf{x}|_p = \begin{cases} \left(\sum_{i=1}^m |x_i|^p\right)^{1/p}, & \text{if } 1 \leq p < \infty, \\ \max_{i=1,\dots,m} |x_i|, & \text{if } p = \infty, \end{cases} \quad \mathbf{x} \in \mathbb{C}^m,$$

$$|X|_p = \max_{\substack{\mathbf{x}\in\mathbb{C}^m \\ \mathbf{x}\neq\mathbf{0}}} \frac{|X\mathbf{x}|_p}{|\mathbf{x}|_p}, \quad X \in \mathbb{C}^{m\times m}.$$

The 2-norm $|\cdot|_2$ is also known as the spectral (or Euclidean) norm and it is preferably denoted by $\|\cdot\|$. Important inequalities involving the p-norms with $p = 1, 2, \infty$ are the following:

$$\|X\| \leq \sqrt{|X|_1|X|_\infty}, \qquad X \in \mathbb{C}^{m \times m}, \qquad (3.1)$$

$$\|X\| \geq |x_{ij}|, \qquad i, j = 1, \ldots, m, \qquad X \in \mathbb{C}^{m \times m}; \qquad (3.2)$$

see [8, 36]. Since it is known that $|X|_1 = \max_{j=1,\ldots,m} \sum_{i=1}^{m} |x_{ij}|$ and $|X|_\infty = \max_{i=1,\ldots,m} \sum_{j=1}^{m} |x_{ij}|$, the inequalities (3.1)–(3.2) are particularly useful to estimate the spectral norm of a matrix when we have bounds for its components.

As mentioned in Sect. 3.2, the Schatten p-norm of an $n \times n$ matrix A is defined as the p-norm of the vector $(\sigma_1(A), \ldots, \sigma_n(A))$ formed by the singular values of A. The Schatten ∞-norm $\|A\|_\infty$ is the largest singular value $\sigma_{\max}(A)$ and coincides with the spectral norm $\|A\|$. The Schatten 1-norm $\|A\|_1$ is the sum of all the singular values of A and is often referred to as the trace-norm of A. The Schatten p-norms are deeply studied in Bhatia's book [7]. Here, we just recall a couple of basic trace-norm inequalities that we shall need in what follows:

$$\|X\|_1 \leq \text{rank}(X)\|X\| \leq m\|X\|, \qquad X \in \mathbb{C}^{m \times m}, \qquad (3.3)$$

$$\|X\|_1 \leq \sum_{i,j=1}^{m} |x_{ij}|, \qquad X \in \mathbb{C}^{m \times m}. \qquad (3.4)$$

The inequality (3.3) follows from the equation $\sigma_{\max}(X) = \|X\|$ and the definition $\|X\|_1 = \sum_{i=1}^{m} \sigma_i(X) = \sum_{i=1}^{\text{rank}(X)} \sigma_i(X)$. For the proof of the inequality (3.4), see, e.g., [29, Section 2.4.3].

3.3.1.2 GLT Preconditioning

The next theorem is an important result in the context of GLT preconditioning, but it will be used only in Sect. 3.3.4.3. The reader may then decide to skip it on first reading and come back here afterwards, just before going into Sect. 3.3.4.3.

Theorem 1 *Let $\{A_n\}_n$ be a sequence of Hermitian matrices such that $\{A_n\}_n \sim_{\text{GLT}} \kappa$, and let $\{P_n\}_n$ be a sequence of Hermitian Positive Definite (HPD) matrices such that $\{P_n\}_n \sim_{\text{GLT}} \xi$ with $\xi \neq 0$ a.e. Then, the sequence of preconditioned matrices $P_n^{-1} A_n$ satisfies*

$$\{P_n^{-1} A_n\}_n \sim_{\text{GLT}} \xi^{-1}\kappa,$$

and

$$\{P_n^{-1} A_n\}_n \sim_{\sigma, \lambda} \xi^{-1}\kappa.$$

Proof The GLT relation $\{P_n^{-1} A_n\}_n \sim_{\text{GLT}} \xi^{-1} \kappa$ is a direct consequence of **GLT 4**–**GLT 5**. The singular value distribution $\{P_n^{-1} A_n\}_n \sim_\sigma \xi^{-1} \kappa$ follows immediately from **GLT 1**. The only difficult part is the spectral distribution $\{P_n^{-1} A_n\}_n \sim_\lambda \xi^{-1} \kappa$, which does not follow from **GLT 1** because $P_n^{-1} A_n$ is not Hermitian in general.

Since P_n is HPD, the eigenvalues of P_n are positive and the matrices $P_n^{1/2}$, $P_n^{-1/2}$ are well-defined. Moreover,

$$P_n^{-1} A_n \sim P_n^{-1/2} A_n P_n^{-1/2}, \tag{3.5}$$

where $X \sim Y$ means that X is similar to Y. The good news is that $P_n^{-1/2} A_n P_n^{-1/2}$ is Hermitian and, moreover, by **GLT 4**–**GLT 6** (with **GLT 6** applied to $f(z) = |z|^{1/2}$), we have

$$\{P_n^{-1/2} A_n P_n^{-1/2}\}_n \sim_{\text{GLT}} |\xi|^{-1/2} \kappa |\xi|^{-1/2} = |\xi|^{-1} \kappa = \xi^{-1} \kappa;$$

note that the latter equation follows from the fact that $\xi \geq 0$ a.e. by **S 1**, since P_n is HPD and $\{P_n\}_n \sim_\lambda \xi$ by **GLT 1**. Since $P_n^{-1/2} A_n P_n^{-1/2}$ is Hermitian, **GLT 1** yields

$$\{P_n^{-1/2} A_n P_n^{-1/2}\}_n \sim_\lambda \xi^{-1} \kappa.$$

Thus, by the similarity (3.5), $\{P_n^{-1} A_n\}_n \sim_\lambda \xi^{-1} \kappa$.   □

3.3.1.3 Arrow-Shaped Sampling Matrices

If $n \in \mathbb{N}$ and $a : [0, 1] \to \mathbb{C}$, the nth arrow-shaped sampling matrix generated by a is denoted by $S_n(a)$ and is defined as the following symmetric matrix of size n:

$$(S_n(a))_{i,j} = (D_n(a))_{\min(i,j),\min(i,j)} = a\left(\frac{\min(i,j)}{n}\right), \qquad i, j = 1, \ldots, n, \tag{3.6}$$

that is,

$$S_n(a) = \begin{bmatrix} a(\tfrac{1}{n}) & a(\tfrac{1}{n}) & a(\tfrac{1}{n}) & \cdots & \cdots & a(\tfrac{1}{n}) \\ a(\tfrac{1}{n}) & a(\tfrac{2}{n}) & a(\tfrac{2}{n}) & \cdots & \cdots & a(\tfrac{2}{n}) \\ a(\tfrac{1}{n}) & a(\tfrac{2}{n}) & a(\tfrac{3}{n}) & \cdots & \cdots & a(\tfrac{3}{n}) \\ \vdots & \vdots & \vdots & \ddots & & \vdots \\ \vdots & \vdots & \vdots & & \ddots & \vdots \\ a(\tfrac{1}{n}) & a(\tfrac{2}{n}) & a(\tfrac{3}{n}) & \cdots & \cdots & a(1) \end{bmatrix}.$$

The name is due to the fact that, if we imagine to color the matrix $S_n(a)$ by assigning the color i to the entries $a(\tfrac{i}{n})$, the resulting picture looks like a sort of arrow pointing toward the upper left corner. Throughout this work, if $X, Y \in \mathbb{C}^{m \times m}$, we denote by $X \circ Y$ the componentwise (Hadamard) product of X and Y:

$$(X \circ Y)_{ij} = x_{ij} y_{ij}, \qquad i, j = 1, \ldots, m.$$

Moreover, if $g : D \to \mathbb{C}$ is continuous over D, with $D \subseteq \mathbb{C}^k$ for some k, we denote by $\omega_g(\cdot)$ the modulus of continuity of g,

$$\omega_g(\delta) = \sup_{\substack{\mathbf{x}, \mathbf{y} \in D \\ \|\mathbf{x} - \mathbf{y}\| \leq \delta}} |g(\mathbf{x}) - g(\mathbf{y})|, \qquad \delta > 0.$$

If we need/want to specify D, we will say that $\omega_g(\cdot)$ is the modulus of continuity of g over D.

Theorem 2 *Let $a : [0, 1] \to \mathbb{C}$ be continuous and let f be a trigonometric polynomial of degree $\leq r$. Then, we have*

$$\|S_n(a) \circ T_n(f) - D_n(a) T_n(f)\| \leq (2r + 1) \|f\|_\infty \, \omega_a\left(\frac{r}{n}\right) \tag{3.7}$$

for every $n \in \mathbb{N}$,

$$\|S_n(a) \circ T_n(f)\| \leq C \tag{3.8}$$

for every $n \in \mathbb{N}$ and for some constant C independent of n, and

$$\{S_n(a) \circ T_n(f)\}_n \sim_{\text{GLT}} a(x) f(\theta). \tag{3.9}$$

Proof For all $i, j = 1, \ldots, n$,

- if $|i - j| > r$, then the Fourier coefficient f_{i-j} is zero and, consequently,

$$(S_n(a) \circ T_n(f))_{ij} = (S_n(a))_{ij} (T_n(f))_{ij} = a\left(\frac{\min(i, j)}{n}\right) f_{i-j} = 0,$$

$$(D_n(a) T_n(f))_{ij} = (D_n(a))_{ii} (T_n(f))_{ij} = a\left(\frac{i}{n}\right) f_{i-j} = 0;$$

- if $|i - j| \leq r$, then, using (3.2) and **T 3**, we obtain

$$|(S_n(a) \circ T_n(f))_{ij} - (D_n(a) T_n(f))_{ij}| = |(S_n(a))_{ij} (T_n(f))_{ij} - (D_n(a))_{ii} (T_n(f))_{ij}|$$

$$= |(S_n(a))_{ij} - (D_n(a))_{ii}| \, |(T_n(f))_{ij}|$$

$$\leq \left| a\left(\frac{\min(i, j)}{n}\right) - a\left(\frac{i}{n}\right) \right| \|T_n(f)\|$$

$$\leq \|f\|_\infty \, \omega_a\left(\left|\frac{\min(i, j)}{n} - \frac{i}{n}\right|\right).$$

Since $|i - j| \leq r$, we have

$$\left| \frac{\min(i, j)}{n} - \frac{i}{n} \right| \leq \frac{|j - i|}{n} \leq \frac{r}{n},$$

hence

$$\left| (S_n(a) \circ T_n(f))_{ij} - (D_n(a)T_n(f))_{ij} \right| \leq \|f\|_\infty \, \omega_a\left(\frac{r}{n}\right).$$

It follows from the first item that the nonzero entries in each row and column of $S_n(a) \circ T_n(f) - D_n(a)T_n(f)$ are at most $2r + 1$. Hence, from the second item we infer that the 1-norm and the ∞-norm of $S_n(a) \circ T_n(f) - D_n(a)T_n(f)$ are bounded by $(2r + 1)\|f\|_\infty \, \omega_a(\frac{r}{n})$. The application of (3.1) yields (3.7). Using (3.7) and **T 3** we obtain

$$\|S_n(a) \circ T_n(f)\| \leq \|S_n(a) \circ T_n(f) - D_n(a)T_n(f)\| + \|D_n(a)\| \, \|T_n(f)\|$$

$$\leq (2r + 1)\|f\|_\infty \, \omega_a\left(\frac{r}{n}\right) + \|a\|_\infty \|f\|_\infty,$$

which implies (3.8). Finally, since $\omega_a(\frac{r}{n}) \to 0$ as $n \to \infty$, the matrix-sequence $\{S_n(a) \circ T_n(f) - D_n(a)T_n(f)\}_n$ is zero-distributed by (3.7) and **Z 1** (or **Z 2**). Thus, (3.9) follows from **GLT 3–GLT 4**. □

3.3.2 FD Discretization of Differential Equations

3.3.2.1 FD Discretization of Diffusion Equations

Consider the following second-order differential problem:

$$\begin{cases} -(a(x)u'(x))' = f(x), & x \in (0, 1), \\ u(0) = \alpha, \quad u(1) = \beta, \end{cases} \tag{3.10}$$

where $a \in C([0, 1])$ and f is a given function. To ensure the well-posedness of this problem, further conditions on a and f should be imposed; for example, $f \in L^2([0, 1])$ and $a \in C^1([0, 1])$ with $a(x) > 0$ for every $x \in [0, 1]$, so that problem (3.10) is elliptic (see Chapter 8 of [14], especially the Sturm-Liouville problem on page 223). However, we here only assume that $a \in C([0, 1])$ as the GLT analysis presented herein does not require any other assumption.

FD Discretization We consider the discretization of (3.10) by the classical second-order central FD scheme on a uniform grid. In the case where $a(x)$ is constant, this is also known as the $(-1, 2, -1)$ scheme. Let us describe it shortly; for more details on FD methods, we refer the reader to the available literature (see, e.g., [53] or any good book on FDs). Choose a discretization parameter $n \in \mathbb{N}$, set $h = \frac{1}{n+1}$ and $x_j = jh$ for all $j \in [0, n+1]$. For $j = 1, \ldots, n$ we approximate $-(a(x)u'(x))'|_{x=x_j}$ by the classical second-order central FD formula:

$$- (a(x)u'(x))'|_{x=x_j} \approx - \frac{a(x_{j+\frac{1}{2}})u'(x_{j+\frac{1}{2}}) - a(x_{j-\frac{1}{2}})u'(x_{j-\frac{1}{2}})}{h}$$

$$\approx - \frac{a(x_{j+\frac{1}{2}})\dfrac{u(x_{j+1}) - u(x_j)}{h} - a(x_{j-\frac{1}{2}})\dfrac{u(x_j) - u(x_{j-1})}{h}}{h}$$

$$= \frac{-a(x_{j+\frac{1}{2}})u(x_{j+1}) + \left(a(x_{j+\frac{1}{2}}) + a(x_{j-\frac{1}{2}})\right)u(x_j) - a(x_{j-\frac{1}{2}})u(x_{j-1})}{h^2}.$$

$$(3.11)$$

This means that the nodal values of the solution u satisfy (approximately) the following linear system:

$$- a(x_{j+\frac{1}{2}})u(x_{j+1}) + \left(a(x_{j+\frac{1}{2}}) + a(x_{j-\frac{1}{2}})\right)u(x_j) - a(x_{j-\frac{1}{2}})u(x_{j-1}) = h^2 f(x_j),$$

$$j = 1, \ldots, n.$$

We then approximate the solution by the piecewise linear function that takes the value u_j in x_j for $j = 0, \ldots, n+1$, where $u_0 = \alpha$, $u_{n+1} = \beta$, and $\mathbf{u} = (u_1, \ldots, u_n)^T$ solves

$$- a(x_{j+\frac{1}{2}})u_{j+1} + \left(a(x_{j+\frac{1}{2}}) + a(x_{j-\frac{1}{2}})\right)u_j - a(x_{j-\frac{1}{2}})u_{j-1} = h^2 f(x_j),$$

$$j = 1, \ldots, n.$$

$$(3.12)$$

The matrix of the linear system (3.12) is the $n \times n$ tridiagonal symmetric matrix given by

$$A_n = \begin{bmatrix} a_{\frac{1}{2}} + a_{\frac{3}{2}} & -a_{\frac{3}{2}} & & & & \\ -a_{\frac{3}{2}} & a_{\frac{3}{2}} + a_{\frac{5}{2}} & -a_{\frac{5}{2}} & & & \\ & -a_{\frac{5}{2}} & \ddots & \ddots & & \\ & & \ddots & \ddots & -a_{n-\frac{1}{2}} & \\ & & & -a_{n-\frac{1}{2}} & a_{n-\frac{1}{2}} + a_{n+\frac{1}{2}} \end{bmatrix}, \qquad (3.13)$$

where $a_i = a(x_i)$ for all $i \in [0, n+1]$.

GLT Analysis of the FD Discretization Matrices We are going to see that the theory of GLT sequences allows one to compute the singular value and spectral distribution of the sequence of FD discretization matrices $\{A_n\}_n$. Actually, this is the fundamental example that led to the birth of the theory of LT sequences and, subsequently, of GLT sequences.

Theorem 3 *If* $a \in C([0, 1])$ *then*

$$\{A_n\}_n \sim_{GLT} a(x)(2 - 2\cos\theta) \qquad (3.14)$$

and

$$\{A_n\}_n \sim_{\sigma,\lambda} a(x)(2 - 2\cos\theta). \qquad (3.15)$$

Proof It suffices to prove (3.14) because (3.15) follows from (3.14) and **GLT 1** as the matrices A_n are symmetric. Consider the matrix

$$D_n(a)T_n(2 - 2\cos\theta) = \begin{bmatrix} 2a(\frac{1}{n}) & -a(\frac{1}{n}) & & & \\ -a(\frac{2}{n}) & 2a(\frac{2}{n}) & -a(\frac{2}{n}) & & \\ & -a(\frac{3}{n}) & \ddots & \ddots & \\ & & \ddots & \ddots & -a(\frac{n-1}{n}) \\ & & & -a(1) & 2a(1) \end{bmatrix}. \qquad (3.16)$$

In view of the inequalities $\left|x_j - \frac{j}{n}\right| \le \frac{1}{n+1} = h$, $j = 1, \ldots, n$, a direct comparison between (3.16) and (3.13) shows that the modulus of each diagonal entry of the matrix $A_n - D_n(a)T_n(2 - 2\cos\theta)$ is bounded by $2\,\omega_a(3h/2)$, and the modulus of each off-diagonal entry of $A_n - D_n(a)T_n(2 - 2\cos\theta)$ is bounded by $\omega_a(3h/2)$. Therefore, the 1-norm and the ∞-norm of $A_n - D_n(a)T_n(2 - 2\cos\theta)$ are bounded by $4\,\omega_a(3h/2)$, and so, by (3.1),

$$\|A_n - D_n(a)T_n(2 - 2\cos\theta)\| \le 4\,\omega_a(3h/2) \to 0 \text{ as } n \to \infty.$$

Setting $Z_n = A_n - D_n(a)T_n(2 - 2\cos\theta)$, we have $\{Z_n\}_n \sim_\sigma 0$ by **Z 1** (or **Z 2**). Since

$$A_n = D_n(a)T_n(2 - 2\cos\theta) + Z_n,$$

GLT 3 and **GLT 4** yield (3.14). □

Remark 1 (Formal Structure of the Symbol) From a formal viewpoint (i.e., disregarding the regularity of $a(x)$ and $u(x)$), problem (3.10) can be rewritten in the form

$$\begin{cases} -a(x)u''(x) - a'(x)u'(x) = f(x), & x \in (0, 1), \\ u(0) = \alpha, & u(1) = \beta. \end{cases}$$

From this reformulation, it appears more clearly that the symbol $a(x)(2 - 2\cos\theta)$ consists of the two 'ingredients':

- The coefficient of the higher-order differential operator, namely $a(x)$, in the physical variable x. To make a parallelism with Hörmander's theory [39], the higher-order differential operator $-a(x)u''(x)$ is the so-called principal symbol of the complete differential operator $-a(x)u''(x) - a'(x)u'(x)$ and $a(x)$ is then the coefficient of the principal symbol.
- The trigonometric polynomial associated with the FD formula $(-1, 2, -1)$ used to approximate the higher-order derivative $-u''(x)$, namely $2 - 2\cos\theta = -e^{i\theta} + 2 - e^{-i\theta}$, in the Fourier variable θ. To see that $(-1, 2, -1)$ is precisely the FD formula used to approximate $-u''(x)$, simply imagine $a(x) = 1$ and note that in this case the FD scheme (3.11) becomes

$$-u''(x_j) \approx \frac{-u(x_{j+1}) + 2u(x_j) - u(x_{j-1})}{h^2},$$

i.e., the FD formula $(-1, 2, -1)$ to approximate $-u''(x_j)$.

We observe that the term $-a'(x)u'(x)$, which only depends on lower-order derivatives of $u(x)$, does not enter the expression of the symbol.

Remark 2 (Nonnegativity and Order of the Zero at $\theta = 0$) The trigonometric polynomial $2 - 2\cos\theta$ is nonnegative on $[-\pi, \pi]$ and it has a unique zero of order 2 at $\theta = 0$, because

$$\lim_{\theta \to 0} \frac{2 - 2\cos\theta}{\theta^2} = 1.$$

This reflects the fact that the associated FD formula $(-1, 2, -1)$ approximates $-u''(x)$, which is a differential operator of order 2 (it is also nonnegative on the space of functions $v \in C^2([0, 1])$ such that $v(0) = v(1) = 0$, in the sense that $\int_0^1 -v''(x)v(x)dx = \int_0^1 (v'(x))^2 dx \geq 0$ for all such v).

3.3.2.2 FD Discretization of Convection-Diffusion-Reaction Equations

1st Part

Suppose we add to the diffusion equation (3.10) a convection and a reaction term. In this way, we obtain the following convection-diffusion-reaction equation in divergence form with Dirichlet boundary conditions:

$$\begin{cases} -(a(x)u'(x))' + b(x)u'(x) + c(x)u(x) = f(x), & x \in (0, 1), \\ u(0) = \alpha, \quad u(1) = \beta, \end{cases} \tag{3.17}$$

where $a : [0, 1] \to \mathbb{R}$ is continuous as before and we assume that $b, c : [0, 1] \to \mathbb{R}$ are bounded. Based on Remark 1, we expect that the term $b(x)u'(x) + c(x)u(x)$, which only involves lower-order derivatives of $u(x)$, does not enter the expression of the symbol. In other words, if we discretize the higher-order term $-(a(x)u'(x))'$ as in (3.11), the symbol of the resulting FD discretization matrices B_n should be again $a(x)(2 - 2\cos\theta)$. We are going to show that this is in fact the case.

FD Discretization Let $n \in \mathbb{N}$, set $h = \frac{1}{n+1}$ and $x_j = jh$ for all $j \in [0, n + 1]$. Consider the discretization of (3.17) by the FD scheme defined as follows.

- To approximate the higher-order (diffusion) term $-(a(x)u'(x))'$, use again the FD formula (3.11), i.e.,

$$- (a(x)u'(x))'|_{x=x_j}$$

$$\approx \frac{-a(x_{j+\frac{1}{2}})u(x_{j+1}) + \left(a(x_{j+\frac{1}{2}}) + a(x_{j-\frac{1}{2}})\right)u(x_j) - a(x_{j-\frac{1}{2}})u(x_{j-1})}{h^2}. \tag{3.18}$$

- To approximate the convection term $b(x)u'(x)$, use any (consistent) FD formula; to fix the ideas, here we use the second-order central formula

$$b(x)u'(x)|_{x=x_j} \approx b(x_j)\frac{u(x_{j+1}) - u(x_{j-1})}{2h}. \tag{3.19}$$

- To approximate the reaction term $c(x)u(x)$, use the obvious equation

$$c(x)u(x)|_{x=x_j} = c(x_j)u(x_j). \tag{3.20}$$

The resulting FD discretization matrix B_n admits a natural decomposition as

$$B_n = A_n + Z_n, \tag{3.21}$$

where A_n is the matrix coming from the discretization of the higher-order (diffusion) term $-(a(x)u'(x))$, while Z_n is the matrix coming from the discretization of the lower-order (convection and reaction) terms $b(x)u'(x)$ and $c(x)u(x)$. Note that A_n

is given by (3.13) and Z_n is given by

$$
Z_n = \frac{h}{2}
\begin{bmatrix}
0 & b_1 & & & \\
-b_2 & 0 & b_2 & & \\
& \ddots & \ddots & \ddots & \\
& & -b_{n-1} & 0 & b_{n-1} \\
& & & -b_n & 0
\end{bmatrix}
+ h^2
\begin{bmatrix}
c_1 & & & & \\
& c_2 & & & \\
& & \ddots & & \\
& & & c_{n-1} & \\
& & & & c_n
\end{bmatrix},
\tag{3.22}
$$

where $b_i = b(x_i)$ and $c_i = c(x_i)$ for all $i = 1, \ldots, n$.

GLT Analysis of the FD Discretization Matrices We now prove that Theorem 3 holds unchanged with B_n in place of A_n. This highlights a general aspect: *lower-order terms such as $b(x)u'(x) + c(x)u(x)$ do not enter the expression of the symbol and do not affect in any way the asymptotic singular value and spectral distribution of DE discretization matrices.*

Theorem 4 *If $a \in C([0, 1])$ and $b, c : [0, 1] \to \mathbb{R}$ are bounded then*

$$
\{B_n\}_n \sim_{\mathrm{GLT}} a(x)(2 - 2\cos\theta)
\tag{3.23}
$$

and

$$
\{B_n\}_n \sim_{\sigma, \lambda} a(x)(2 - 2\cos\theta).
\tag{3.24}
$$

Proof By (3.1), the matrix Z_n in (3.22) satisfies

$$
\|Z_n\| \leq h\|b\|_\infty + h^2\|c\|_\infty \leq C/n
\tag{3.25}
$$

for some constant C independent of n. As a consequence, $\{Z_n\}_n$ is zero-distributed by **Z 1** (or **Z 2**), hence $\{Z_n\}_n \sim_{\mathrm{GLT}} 0$ by **GLT 3**. Since $\{A_n\}_n \sim_{\mathrm{GLT}} a(x)(2 - 2\cos\theta)$ by Theorem 3, the decomposition (3.21) and **GLT 4** imply (3.23).

Now, if the convection term is not present, i.e., $b(x) = 0$ identically, then B_n is symmetric and (3.24) follows from (3.23) and **GLT 1**. If $b(x)$ is not identically 0, then B_n is not symmetric in general and so (3.23) and **GLT 1** only imply the singular value distribution $\{B_n\}_n \sim_\sigma a(x)(2 - 2\cos\theta)$. Nevertheless, in view of the decomposition (3.21), since A_n is symmetric, since $\|Z_n\|_1 = O(1)$ by the inequalities (3.25) and (3.3), and since $\|A_n\| \leq 4\|a\|_\infty$ by (3.1), the spectral distribution $\{B_n\}_n \sim_\lambda a(x)(2 - 2\cos\theta)$ holds (by **GLT 2**) even if $b(x)$ is an arbitrary bounded function. $\square$

2nd Part

So far, we only considered differential equations with Dirichlet boundary
conditions. A natural question is the following: if we change the boundary
conditions in (3.17), does the expression of the symbol change? The answer is 'no':
*boundary conditions do not affect the singular value and eigenvalue distribution
because they only produce a small-rank perturbation in the resulting discretization
matrices.* To understand better this point, we consider problem (3.17) with Neumann
boundary conditions:

$$\begin{cases} -(a(x)u'(x))' + b(x)u'(x) + c(x)u(x) = f(x), & x \in (0, 1), \\ u'(0) = \alpha, \quad u'(1) = \beta. \end{cases} \tag{3.26}$$

FD Discretization We discretize (3.26) by the same FD scheme considered in the
1st part, which is defined by the FD formulas (3.18)–(3.20). In this way, we arrive
at the linear system

$$- a(x_{j+\frac{1}{2}})u_{j+1} + \big(a(x_{j+\frac{1}{2}}) + a(x_{j-\frac{1}{2}})\big)u_j - a(x_{j-\frac{1}{2}})u_{j-1}$$

$$+ \frac{h}{2}\big(b(x_j)u_{j+1} - b(x_j)u_{j-1}\big) + h^2 c(x_j)u_j = h^2 f(x_j), \qquad j = 1, \ldots, n, \tag{3.27}$$

which is formed by n equations in the $n + 2$ unknowns $u_0, u_1, \ldots, u_n, u_{n+1}$. Note
that u_0 and u_{n+1} should now be considered as unknowns, because they are not
specified by the Dirichlet boundary conditions. However, as it is common in the
FD context, u_0 and u_{n+1} are expressed in terms of $u_1, \ldots, u_n$ by exploiting the
Neumann boundary conditions. The simplest choice is to express u_0 and u_{n+1} as a
function of u_1 and u_n, respectively, by imposing the conditions

$$\frac{u_1 - u_0}{h} = \alpha, \qquad \frac{u_{n+1} - u_n}{h} = \beta, \tag{3.28}$$

which yield $u_0 = u_1 - \alpha h$ and $u_{n+1} = u_n + \beta h$. Substituting into (3.27), we obtain
a linear system with n equations and n unknowns $u_1, \ldots, u_n$. Setting $a_i = a(x_i)$,
$b_i = b(x_i)$, $c_i = c(x_i)$ for all $i \in [0, n + 1]$, the matrix of this system is

$$C_n = B_n + R_n = A_n + Z_n + R_n, \tag{3.29}$$

where A_n, B_n, Z_n are given by (3.13), (3.21), (3.22), respectively, and

$$
R_n = \begin{bmatrix} -a_{\frac{1}{2}} - \dfrac{h}{2}b_1 & & \\ & \ddots & \\ & & -a_{n+\frac{1}{2}} + \dfrac{h}{2}b_n \end{bmatrix}
$$

is a small-rank correction coming from the discretization (3.28) of the boundary conditions.

GLT Analysis of the FD Discretization Matrices We prove that Theorems 3 and 4 hold unchanged with C_n in place of A_n and B_n, respectively.

Theorem 5 *If $a \in C([0, 1])$ and $b, c : [0, 1] \to \mathbb{R}$ are bounded then*

$$
\{C_n\}_n \sim_{\text{GLT}} a(x)(2 - 2\cos\theta) \tag{3.30}
$$

and

$$
\{C_n\}_n \sim_{\sigma, \lambda} a(x)(2 - 2\cos\theta). \tag{3.31}
$$

Proof Let C denote a generic constant independent of n. It is clear that $\|R_n\| \leq \|a\|_\infty + (h/2)\|b\|_\infty \leq C$. Moreover, since $\|R_n\|_1 \leq \text{rank}(R_n)\|R_n\| \leq C$, the matrix-sequence $\{R_n\}_n$ is zero-distributed by **Z 2**. Note that $\{Z_n\}_n$ is zero-distributed as well because $\|Z_n\| \leq C/n$ by (3.25). In view of the decomposition (3.29), Theorem 3 and **GLT 3–GLT 4** imply (3.30).

If the matrices C_n are symmetric (this happens if $b(x) = 0$), from (3.30) and **GLT 1** we immediately obtain (3.31). If the matrices C_n are not symmetric, from (3.30) and **GLT 1** we only obtain the singular value distribution in (3.31). However, in view of (3.29), since $\|R_n + Z_n\|_1 = o(n)$ and $\|R_n + Z_n\|$, $\|A_n\| \leq C$, the spectral distribution in (3.31) holds (by **GLT 2**) even if the matrices C_n are not symmetric.   □

3rd Part

Consider the following convection-diffusion-reaction problem:

$$
\begin{cases} -a(x)u''(x) + b(x)u'(x) + c(x)u(x) = f(x), & x \in (0, 1), \\ u(0) = \alpha, \quad u(1) = \beta, \end{cases} \tag{3.32}
$$

where $a : [0, 1] \to \mathbb{R}$ is continuous and $b, c : [0, 1] \to \mathbb{R}$ are bounded. The difference with respect to problem (3.17) is that the higher-order differential operator now appears in non-divergence form, i.e., we have $-a(x)u''(x)$ instead of

$-(a(x)u'(x))'$. Nevertheless, based on Remark 1, if we use again the FD formula $(-1, 2, -1)$ to discretize the second derivative $-u''(x)$, the symbol of the resulting FD discretization matrices should be again $a(x)(2 - 2\cos\theta)$. We are going to show that this is in fact the case.

FD Discretization Let $n \in \mathbb{N}$, set $h = \frac{1}{n+1}$ and $x_j = jh$ for all $j = 0, \ldots, n + 1$. We discretize again (3.32) by the central second-order FD scheme, which in this case is defined by the following formulas:

$$-a(x)u''(x)|_{x=x_j} \approx a(x_j)\frac{-u(x_{j+1}) + 2u(x_j) - u(x_{j-1})}{h^2}, \qquad j = 1, \ldots, n,$$

$$b(x)u'(x)|_{x=x_j} \approx b(x_j)\frac{u(x_{j+1}) - u(x_{j-1})}{2h}, \qquad j = 1, \ldots, n,$$

$$c(x)u(x)|_{x=x_j} = c(x_j)u(x_j), \qquad j = 1, \ldots, n.$$

Then, we approximate the solution of (3.32) by the piecewise linear function that takes the value u_j at the point x_j for $j = 0, \ldots, n + 1$, where $u_0 = \alpha$, $u_{n+1} = \beta$, and $\mathbf{u} = (u_1, \ldots, u_n)^T$ solves the linear system

$$a(x_j)(-u_{j+1} + 2u_j - u_{j-1}) + \frac{h}{2}b(x_j)(u_{j+1} - u_{j-1}) + h^2c(x_j)u_j = h^2 f(x_j),$$

$$j = 1, \ldots, n.$$

The matrix E_n of this linear system can be decomposed according to the diffusion, convection and reaction term, as follows:

$$E_n = K_n + Z_n, \tag{3.33}$$

where Z_n is the sum of the convection and reaction matrix and is given by (3.22), while

$$K_n = \begin{bmatrix} 2a_1 & -a_1 & & & & \\ -a_2 & 2a_2 & -a_2 & & & \\ & \ddots & \ddots & \ddots & & \\ & & -a_{n-1} & 2a_{n-1} & -a_{n-1} \\ & & & -a_n & 2a_n \end{bmatrix} \tag{3.34}$$

is the diffusion matrix ($a_i = a(x_i)$ for all $i = 1, \ldots, n$).

GLT Analysis of the FD Discretization Matrices Despite the nonsymmetry of the diffusion matrix, which is due to the non-divergence form of the higher-order (diffusion) operator $-a(x)u''(x)$, we will prove that Theorems 3–5 hold unchanged with E_n in place of A_n, B_n, C_n, respectively.

Theorem 6 *If $a \in C([0, 1])$ and $b, c : [0, 1] \to \mathbb{R}$ are bounded then*

$$\{E_n\}_n \sim_{\mathrm{GLT}} a(x)(2 - 2\cos\theta) \tag{3.35}$$

and

$$\{E_n\}_n \sim_{\sigma, \lambda} a(x)(2 - 2\cos\theta). \tag{3.36}$$

Proof Throughout this proof, the letter C will denote a generic constant independent of n. By (3.25),

$$\|Z_n\| \le C/n,$$

hence $\{Z_n\}_n$ is zero-distributed. We prove that

$$\{K_n\}_n \sim_{\mathrm{GLT}} a(x)(2 - 2\cos\theta), \tag{3.37}$$

after which (3.35) will follow from **GLT 3–GLT 4** and the decomposition (3.33). It is clear from (3.34) that

$$K_n = \operatorname*{diag}_{i=1,\dots,n} (a_i)\, T_n(2 - 2\cos\theta).$$

By **T 3** applied with $p = \infty$, we obtain

$$\|K_n - D_n(a)T_n(2 - 2\cos\theta)\| \le \left\| \operatorname*{diag}_{i=1,\dots,n} (a_i) - D_n(a) \right\| \|T_n(2 - 2\cos\theta)\|$$

$$\le \omega_a(h)\, \|2 - 2\cos\theta\|_\infty = 4\,\omega_a(h),$$

which tends to 0 as $n \to \infty$. We conclude that $\{K_n - D_n(a)T_n(2 - 2\cos\theta)\}_n$ is zero-distributed, and so (3.37) follows from **GLT 3–GLT 4**.

From (3.35) and **GLT 1** we obtain the singular value distribution in (3.36). To obtain the spectral distribution, the idea is to exploit the fact that K_n is 'almost' symmetric, because $a(x)$ varies continuously when x ranges in $[0, 1]$, and so $a(x_j) \approx a(x_{j+1})$ for all $j = 1, \dots, n - 1$ (when n is large enough). Therefore, by replacing K_n with one of its symmetric approximations $\tilde{K}_n$, we can write

$$E_n = \tilde{K}_n + (K_n - \tilde{K}_n) + Z_n, \tag{3.38}$$

and in view of the decomposition (3.38) we want to obtain the spectral distribution in (3.36) from **GLT 2** applied with $X_n = \tilde{K}_n$ and $Y_n = (K_n - \tilde{K}_n) + Z_n$. Let

$$
\tilde{K}_n = \begin{bmatrix}
2a_1 & -a_1 \\
-a_1 & 2a_2 & -a_2 \\
& \ddots & \ddots & \ddots \\
& & -a_{n-2} & 2a_{n-1} & -a_{n-1} \\
& & & -a_{n-1} & 2a_n
\end{bmatrix}. \tag{3.39}
$$

Since

$$
\|K_n - \tilde{K}_n\| \le \sqrt{|K_n - \tilde{K}_n|_1 \, |K_n - \tilde{K}_n|_\infty} \le \max_{i=1,\dots,n-1} |a_{i+1} - a_i| \le \omega_a(h) \to 0,
$$

$$
\|K_n\| \le \sqrt{|K_n|_1 |K_n|_\infty} \le 4\|a\|_\infty \le C,
$$

$$
\|Z_n\| \to 0,
$$

it follows from **GLT 2** that $\{E_n\}_n \sim_\lambda a(x)(2 - 2\cos\theta)$.   □

Remark 3 In the proof of Theorem 6 we could also choose

$$
\tilde{K}_n = S_n(a) \circ T_n(2 - 2\cos\theta) = \begin{bmatrix}
2\tilde{a}_1 & -\tilde{a}_1 \\
-\tilde{a}_1 & 2\tilde{a}_2 & -\tilde{a}_2 \\
& \ddots & \ddots & \ddots \\
& & -\tilde{a}_{n-2} & 2\tilde{a}_{n-1} & -\tilde{a}_{n-1} \\
& & & -\tilde{a}_{n-1} & 2\tilde{a}_n
\end{bmatrix},
$$

where $\tilde{a}_i = a(\frac{i}{n})$ for all $i = 1, \dots, n$ and $S_n(a)$ is the arrow-shaped sampling matrix defined in (3.6). With this choice of $\tilde{K}_n$, nothing changes in the proof of Theorem 6 except for the bound of $\|K_n - \tilde{K}_n\|$, which becomes $\|K_n - \tilde{K}_n\| \le 4\omega_a(h)$.

4th Part

Based on Remark 1, if we change the FD scheme to discretize the differential problem (3.32), the symbol should become $a(x)p(\theta)$, where $p(\theta)$ is the trigonometric polynomial associated with the new FD formula used to approximate the second derivative $-u''(x)$ (the higher-order differential operator). We are going to show through an example that this is indeed the case.

FD Discretization Consider the convection-diffusion-reaction problem (3.32). Instead of the second-order central FD scheme $(-1, 2, -1)$, this time we use the fourth-order central FD scheme $\frac{1}{12}(1, -16, 30, -16, 1)$ to approximate the second derivative $-u''(x)$. In other words, for $j = 2, \ldots, n-1$ we approximate the higher-order term $-a(x)u''(x)$ by the FD formula

$$-a(x)u''(x)|_{x=x_j} \approx a(x_j)\frac{u(x_{j+2})-16u(x_{j+1})+30u(x_j)-16u(x_{j-1})+u(x_{j-2})}{12h^2},$$

while for $j = 1, n$ we use again the FD scheme $(-1, 2, -1)$,

$$-a(x)u''(x)|_{x=x_j} \approx a(x_j)\frac{-u(x_{j+1})+2u(x_j)-u(x_{j-1})}{h^2}.$$

From a numerical viewpoint, this is not a good choice because the FD formula $\frac{1}{12}(1, -16, 30, -16, 1)$ is a very accurate fourth-order formula, and in order not to destroy the accuracy one would gain from this formula, one should use a fourth-order scheme also for $j = 1, n$ instead of the classical $(-1, 2, -1)$. However, in this work we are not concerned with this kind of issues and we use the classical $(-1, 2, -1)$ because it is simpler and allows us to better illustrate the GLT analysis without introducing useless technicalities. As already observed before, the FD schemes used to approximate the lower-order terms $b(x)u'(x)$ and $c(x)u(x)$ do not affect the symbol, as well as the singular value and eigenvalue distribution, of the resulting sequence of discretization matrices. To illustrate once again this point, in this example we assume to approximate $b(x)u'(x)$ and $c(x)u(x)$ by the following 'strange' FD formulas: for $j = 1, \ldots, n$,

$$b(x)u'(x)|_{x=x_j} \approx b(x_j)\frac{u(x_j)-u(x_{j-1})}{h},$$

$$c(x)u(x)|_{x=x_j} \approx c(x_j)\frac{u(x_{j+1})+u(x_j)+u(x_{j-1})}{3}.$$

Setting $a_i = a(x_i)$, $b_i = b(x_i)$, $c_i = c(x_i)$ for all $i = 1, \ldots, n$, the resulting FD discretization matrix P_n can be decomposed according to the diffusion, convection and reaction term, as follows:

$$P_n = K_n + Z_n,$$

where Z_n is the sum of the convection and reaction matrix,

$$Z_n = h\begin{bmatrix} b_1 & & & & \\ -b_2 & b_2 & & & \\ & \ddots & \ddots & & \\ & & -b_{n-1} & b_{n-1} & \\ & & & -b_n & b_n \end{bmatrix} + \frac{h^2}{3}\begin{bmatrix} c_1 & c_1 & & & \\ c_2 & c_2 & c_2 & & \\ & \ddots & \ddots & \ddots & \\ & & c_{n-1} & c_{n-1} & c_{n-1} \\ & & & c_n & c_n \end{bmatrix},$$

while K_n is the diffusion matrix,

$$K_n = \frac{1}{12}\begin{bmatrix} 24a_1 & -12a_1 \\ -16a_2 & 30a_2 & -16a_2 & a_2 \\ & a_3 & -16a_3 & 30a_3 & -16a_3 & a_3 \\ & & \ddots & \ddots & \ddots & \ddots & \ddots \\ & & & a_{n-2} & -16a_{n-2} & 30a_{n-2} & -16a_{n-2} & a_{n-2} \\ & & & & a_{n-1} & -16a_{n-1} & 30a_{n-1} & -16a_{n-1} \\ & & & & & & -12a_n & 24a_n \end{bmatrix}.$$

GLT Analysis of the FD Discretization Matrices Let $p(\theta)$ be the trigonometric polynomial associated with the FD formula $\frac{1}{12}(1, -16, 30, -16, 1)$ used to approximate the second derivative $-u''(x)$, i.e.,

$$p(\theta) = \frac{1}{12}(e^{-2i\theta} - 16e^{-i\theta} + 30 - 16e^{i\theta} + e^{2i\theta}) = \frac{1}{12}(30 - 32\cos\theta + 2\cos(2\theta)).$$

Based on Remark 1, the following result is not unexpected.

Theorem 7 *If $a \in C([0, 1])$ and $b, c : [0, 1] \to \mathbb{R}$ are bounded then*

$$\{P_n\}_n \sim_{\text{GLT}} a(x)p(\theta) \tag{3.40}$$

and

$$\{P_n\}_n \sim_{\sigma, \lambda} a(x)p(\theta). \tag{3.41}$$

Proof Throughout this proof, the letter C will denote a generic constant independent of n. To simultaneously obtain (3.40) and (3.41), we consider the following decomposition of P_n:

$$P_n = \tilde{K}_n + (K_n - \tilde{K}_n) + Z_n,$$

where $\tilde{K}_n$ is the symmetric approximation of K_n given by

$$\tilde{K}_n = S_n(a) \circ T_n(p)$$

$$= \frac{1}{12} \begin{bmatrix} 30\tilde{a}_1 & -16\tilde{a}_1 & \tilde{a}_1 \\ -16\tilde{a}_1 & 30\tilde{a}_2 & -16\tilde{a}_2 & \tilde{a}_2 \\ \tilde{a}_1 & -16\tilde{a}_2 & 30\tilde{a}_3 & -16\tilde{a}_3 & \tilde{a}_3 \\ & \ddots & \ddots & \ddots & \ddots & \ddots \\ & & \tilde{a}_{n-4} & -16\tilde{a}_{n-3} & 30\tilde{a}_{n-2} & -16\tilde{a}_{n-2} & \tilde{a}_{n-2} \\ & & & \tilde{a}_{n-3} & -16\tilde{a}_{n-2} & 30\tilde{a}_{n-1} & -16\tilde{a}_{n-1} \\ & & & & \tilde{a}_{n-2} & -16\tilde{a}_{n-1} & 30\tilde{a}_n \end{bmatrix}$$

($\tilde{a}_i = a(\frac{i}{n})$ for all $i = 1, \ldots, n$). We show that:

(a) $\{\tilde{K}_n\}_n \sim_{\mathrm{GLT}} a(x)p(\theta)$;
(b) $\|K_n\|$, $\|\tilde{K}_n\| \le C$ and $\|Z_n\| \to 0$;
(c) $\|K_n - \tilde{K}_n\|_1 = o(n)$.

Note that (b)–(c) imply that $\{(K_n - \tilde{K}_n) + Z_n\}_n \sim_\sigma 0$ by **Z 2**. Once we have proved (a)–(c), the GLT relation (3.40) follows from **GLT 4**, the singular value distribution in (3.41) follows from (3.40) and **GLT 1**, and the spectral distribution in (3.41) follows from **GLT 2** applied with $X_n = \tilde{K}_n$ and $Y_n = (K_n - \tilde{K}_n) + Z_n$.

Proof of (a) See Theorem 2.

Proof of (b) We have

$$\|Z_n\| \le \sqrt{|Z_n|_1 \, |Z_n|_\infty} \le 2h\|b\|_\infty + h^2\|c\|_\infty \to 0,$$

$$\|K_n\| \le \sqrt{|K_n|_1 \, |K_n|_\infty} \le \frac{64}{12}\|a\|_\infty,$$

$$\|\tilde{K}_n\| \le \sqrt{|\tilde{K}_n|_1 \, |\tilde{K}_n|_\infty} \le \frac{64}{12}\|a\|_\infty.$$

Note that the uniform boundedness of $\|\tilde{K}_n\|$ with respect to n was already known from Theorem 2.

Proof of (c) A direct comparison between K_n and $\tilde{K}_n$ shows that

$$K_n = \tilde{K}_n + R_n + N_n,$$

where $N_n = K_n - \tilde{K}_n - R_n$ and R_n is the matrix whose rows are all zeros except for the first and the last one, which are given by

$$\frac{1}{12}\left[\, 24a_1 - 30\tilde{a}_1 \qquad -12a_1 + 16\tilde{a}_1 \qquad -\tilde{a}_1 \qquad 0 \qquad \cdots \qquad 0 \,\right]$$

and

$$\frac{1}{12}\left[\, 0 \qquad \cdots \qquad 0 \qquad -\tilde{a}_{n-2} \qquad -12a_n + 16\tilde{a}_{n-1} \qquad 24a_n - 30\tilde{a}_n \,\right],$$

respectively. We have

$$\|R_n\| \le \frac{83}{12}\|a\|_\infty, \qquad \text{rank}(R_n) \le 2, \qquad \|N_n\| \le \frac{64}{12}\omega_a\!\left(\frac{2}{n}\right)$$

and

$$\|K_n - \tilde{K}_n\|_1 \le \|R_n\|_1 + \|N_n\|_1 \le \text{rank}(R_n)\|R_n\| + n\|N_n\|,$$

hence $\|K_n - \tilde{K}_n\|_1 = o(n)$. $\qquad\qquad\qquad\qquad\qquad\qquad\qquad\qquad\qquad\qquad\square$

Remark 4 (Nonnegativity and Order of the Zero at $\theta = 0$) Despite we have changed the FD scheme to approximate the second derivative $-u''(x)$, the resulting trigonometric polynomial $p(\theta)$ retains some properties of $2 - 2\cos\theta$. In particular, $p(\theta)$ is nonnegative over $[-\pi, \pi]$ and it has a unique zero of order 2 at $\theta = 0$, because

$$\lim_{\theta\to 0}\frac{p(\theta)}{\theta^2} = 1 = \lim_{\theta\to 0}\frac{2 - 2\cos\theta}{\theta^2}.$$

This reflects the fact the associated FD formula $\frac{1}{12}(1, -16, 30, -16, 1)$ approximates $-u''(x)$, which is a differential operator of order 2 and it is also nonnegative on $\{v \in C^2([0, 1]) : v(0) = v(1) = 0\}$; cf. Remark 2.

3.3.2.3 FD Discretization of Higher-Order Equations

So far we only considered the FD discretization of second-order differential equations. In order to show that the GLT analysis is not limited to second-order equations, in this section we deal with an higher-order problem. For simplicity, we focus on the following fourth-order problem with homogeneous Dirichlet–Neumann boundary conditions:

$$\begin{cases} a(x)u^{(4)}(x) = f(x), & x \in (0, 1), \\ u(0) = 0, & u(1) = 0, \\ u'(0) = 0, & u'(1) = 0, \end{cases} \qquad (3.42)$$

where $a \in C([0, 1])$ and f is a given function. We do not consider more complicated boundary conditions, and we do not include terms with lower-order derivatives, because we know from Remark 1 and the experience gained from the previous section that both these ingredients only serve to complicate things, but ultimately they do not affect the symbol, as well as the singular value and eigenvalue distribution, of the resulting discretization matrices. Based on Remark 1, the symbol of the matrix-sequence arising from the FD discretization of (3.42) should be $a(x)q(\theta)$, where $q(\theta)$ is the trigonometric polynomial associated with the FD formula used to discretize $u^{(4)}(x)$. We will see that this is in fact the case.

FD Discretization We approximate the fourth derivative $u^{(4)}(x)$ by the second-order central FD scheme $(1, -4, 6, -4, 1)$, which yields the approximation

$$a(x)u^{(4)}(x)|_{x=x_j} \approx a(x_j)\frac{u(x_{j+2}) - 4u(x_{j+1}) + 6u(x_j) - 4u(x_{j-1}) + u(x_{j-2})}{h^4},$$

for all $j = 2, \ldots, n + 1$; here, $h = \frac{1}{n+3}$ and $x_j = jh$ for $j = 0, \ldots, n + 3$. Taking into account the homogeneous boundary conditions, we approximate the solution of (3.42) by the piecewise linear function that takes the value u_j in x_j for $j = 0, \ldots, n + 3$, where $u_0 = u_1 = u_{n+2} = u_{n+3} = 0$ and $\mathbf{u} = (u_2, \ldots, u_{n+1})^T$ is the solution of the linear system

$$a(x_j)(u_{j+2} - 4u_{j+1} + 6u_j - 4u_{j-1} + u_{j-2}) = h^4 f(x_j), \qquad j = 2, \ldots, n + 1.$$

The matrix A_n of this linear system is given by

$$A_n = \begin{bmatrix} 6a_2 & -4a_2 & a_2 \\ -4a_3 & 6a_3 & -4a_3 & a_3 \\ a_4 & -4a_4 & 6a_4 & -4a_4 & a_4 \\ & \ddots & \ddots & \ddots & \ddots & \ddots \\ & & a_{n-1} & -4a_{n-1} & 6a_{n-1} & -4a_{n-1} & a_{n-1} \\ & & & a_n & -4a_n & 6a_n & -4a_n \\ & & & & a_{n+1} & -4a_{n+1} & 6a_{n+1} \end{bmatrix},$$

where $a_i = a(x_i)$ for all $i = 2, \ldots, n + 1$.

GLT Analysis of the FD Discretization Matrices Let $q(\theta)$ be the trigonometric polynomial associated with the FD formula $(1, -4, 6, -4, 1)$, i.e.,

$$q(\theta) = e^{-2i\theta} - 4e^{-i\theta} + 6 - 4e^{i\theta} + e^{2i\theta} = 6 - 8\cos\theta + 2\cos(2\theta).$$

Theorem 8 *If* $a \in C([0, 1])$ *then*

$$\{A_n\}_n \sim_{\text{GLT}} a(x)q(\theta) \tag{3.43}$$

and

$$\{A_n\}_n \sim_{\sigma, \lambda} a(x)q(\theta). \tag{3.44}$$

Proof We show that

$$\|A_n - S_n(a) \circ T_n(q)\| \to 0. \tag{3.45}$$

Once this is proved, since $\{S_n(a) \circ T_n(q)\}_n \sim_{\text{GLT}} a(x)q(\theta)$ and $\|S_n(a) \circ T_n(q)\|$ is uniformly bounded with respect to n (by Theorem 2), and since $\|A_n\| \leq 16\|a\|_\infty$ by (3.1), the relations (3.43)–(3.44) follow from the decomposition

$$A_n = S_n(a) \circ T_n(q) + (A_n - S_n(a) \circ T_n(q))$$

and from **GLT 1–GLT 4**, taking into account that $S_n(a) \circ T_n(p)$ is symmetric and $\{A_n - S_n(a) \circ T_n(q)\}_n$ is zero-distributed by (3.45) and **Z 1** (or **Z 2**). Let us then prove (3.45). The matrices A_n and $S_n(a) \circ T_n(q)$ are banded (pentadiagonal) and, for all $i, j = 1, \ldots, n$ with $|i - j| \leq 2$, a crude estimates gives

$$\left|(A_n)_{ij} - (S_n(a) \circ T_n(q))_{ij}\right| = \left|a_{i+1}(T_n(q))_{ij} - a\left(\frac{\min(i, j)}{n}\right)(T_n(q))_{ij}\right|$$

$$= \left|a\left(\frac{i+1}{n+3}\right) - a\left(\frac{\min(i, j)}{n}\right)\right| |(T_n(q))_{ij}|$$

$$\leq 6\omega_a\left(\frac{6}{n}\right).$$

Hence, by (3.1), $\|A_n - S_n(a) \circ T_n(q)\| \leq 5 \cdot 6\omega_a(\frac{6}{n}) \to 0.$ $\qquad \square$

Remark 5 (Nonnegativity and Order of the Zero at $\theta = 0$) The polynomial $q(\theta)$ is nonnegative over $[-\pi, \pi]$ and has a unique zero of order 4 at $\theta = 0$, because

$$\lim_{\theta \to 0} \frac{q(\theta)}{\theta^4} = 1.$$

This reflects the fact that the FD formula $(1, -4, 6, -4, 1)$ associated with $q(\theta)$ approximates the fourth derivative $u^{(4)}(x)$, which is a differential operator of order 4 (it is also nonnegative on the space of functions $v \in C^4([0, 1])$ such that $v(0) = v(1) = 0$ and $v'(0) = v'(1) = 0$, in the sense that $\int_0^1 v^{(4)}(x)v(x)\mathrm{d}x = \int_0^1 (v''(x))^2\mathrm{d}x \geq 0$ for all such v); see also Remarks 2 and 4.

3.3.2.4   Non-uniform FD Discretizations

All the FD discretizations considered in the previous sections are based on uniform grids. It is natural to ask whether the theory of GLT sequences finds applications also in the context of non-uniform FD discretizations. The answer to this question is affirmative, at least in the case where the non-uniform grid is obtained as the mapping of a uniform grid through a fixed function G, independent of the mesh size. In this section we illustrate this claim by means of a simple example.

FD Discretization   Consider the diffusion equation (3.10) with $a \in C([0, 1])$. Take a discretization parameter $n \in \mathbb{N}$, fix a set of grid points $0 = x_0 < x_1 < \ldots < x_{n+1} = 1$ and define the corresponding stepsizes $h_j = x_j - x_{j-1}$, $j = 1, \ldots, n+1$. For each $j = 1, \ldots, n$, we approximate $-(a(x)u'(x))'|_{x=x_j}$ by the FD formula

$$- (a(x)u'(x))'|_{x=x_j} \approx -\frac{a(x_j + \frac{h_{j+1}}{2})u'(x_j + \frac{h_{j+1}}{2}) - a(x_j - \frac{h_j}{2})u'(x_j - \frac{h_j}{2})}{\frac{h_{j+1}}{2} + \frac{h_j}{2}}$$

$$\approx -\frac{a(x_j + \frac{h_{j+1}}{2})\frac{u(x_{j+1}) - u(x_j)}{h_{j+1}} - a(x_j - \frac{h_j}{2})\frac{u(x_j) - u(x_{j-1})}{h_j}}{\frac{h_{j+1}}{2} + \frac{h_j}{2}}$$

which is equal to $\dfrac{2}{h_j + h_{j+1}}$ times

$$-\frac{a(x_j - \frac{h_j}{2})}{h_j}u(x_{j-1}) + \left(\frac{a(x_j - \frac{h_j}{2})}{h_j} + \frac{a(x_j + \frac{h_{j+1}}{2})}{h_{j+1}}\right)u(x_j)$$

$$-\frac{a(x_j + \frac{h_{j+1}}{2})}{h_{j+1}}u(x_{j+1}).$$

This means that the nodal values of the solution u satisfy (approximately) the following linear system:

$$-\frac{a(x_j - \frac{h_j}{2})}{h_j}u(x_{j-1}) + \left(\frac{a(x_j - \frac{h_j}{2})}{h_j} + \frac{a(x_j + \frac{h_{j+1}}{2})}{h_{j+1}}\right)u(x_j)$$

$$-\frac{a(x_j + \frac{h_{j+1}}{2})}{h_{j+1}}u(x_{j+1})$$

$$= \frac{h_j + h_{j+1}}{2}f(x_j), \qquad j = 1, \ldots, n.$$

We then approximate the solution by the piecewise linear function that takes the value u_j in x_j for $j = 0, \ldots, n + 1$, where $u_0 = \alpha$, $u_{n+1} = \beta$, and $\mathbf{u} =$

$(u_1, \ldots, u_n)^T$ solves

$$
-\frac{a(x_j - \frac{h_j}{2})}{h_j} u_{j-1} + \left(\frac{a(x_j - \frac{h_j}{2})}{h_j} + \frac{a(x_j + \frac{h_{j+1}}{2})}{h_{j+1}} \right) u_j - \frac{a(x_j + \frac{h_{j+1}}{2})}{h_{j+1}} u_{j+1}
$$

$$
= \frac{h_j + h_{j+1}}{2} f(x_j), \qquad j = 1, \ldots, n.
$$

The matrix of this linear system is the $n \times n$ tridiagonal symmetric matrix given by

$$
\text{tridiag}_n \left[-\frac{a(x_j - \frac{h_j}{2})}{h_j}, \ \frac{a(x_j - \frac{h_j}{2})}{h_j} + \frac{a(x_j + \frac{h_{j+1}}{2})}{h_{j+1}}, \ -\frac{a(x_j + \frac{h_{j+1}}{2})}{h_{j+1}} \right]. \qquad (3.46)
$$

GLT Analysis of the FD Discretization Matrices Let $h = \frac{1}{n+1}$ and $\hat{x}_j = jh$, $j = 0, \ldots, n + 1$. In the following, we assume that the set of points $\{x_0, x_1, \ldots, x_{n+1}\}$ is obtained as the mapping of the uniform grid $\{\hat{x}_0, \hat{x}_1, \ldots, \hat{x}_{n+1}\}$ through a fixed function G, i.e., $x_j = G(\hat{x}_j)$ for $j = 0, \ldots, n + 1$, where $G : [0, 1] \to [0, 1]$ is an increasing and bijective map, independent of the mesh parameter n. The resulting FD discretization matrix (3.46) will be denoted by $A_{G,n}$ in order to emphasize its dependence on G. In formulas,

$$
A_{G,n} = \text{tridiag}_n \left[-\frac{a(G(\hat{x}_j) - \frac{h_j}{2})}{h_j}, \ \frac{a(G(\hat{x}_j) - \frac{h_j}{2})}{h_j} \right. \qquad (3.47)
$$

$$
\left. + \frac{a(G(\hat{x}_j) + \frac{h_{j+1}}{2})}{h_{j+1}}, \ -\frac{a(G(\hat{x}_j) + \frac{h_{j+1}}{2})}{h_{j+1}} \right]
$$

with

$$
h_j = G(\hat{x}_j) - G(\hat{x}_{j-1}), \qquad j = 1, \ldots, n + 1.
$$

Theorem 9 Let $a \in C([0, 1])$. Suppose $G : [0, 1] \to [0, 1]$ is an increasing bijective map in $C^1([0, 1])$ and there exist at most finitely many points $\hat{x}$ such that $G'(\hat{x}) = 0$. Then

$$
\left\{ \frac{1}{n+1} A_{G,n} \right\}_n \sim_{\text{GLT}} \frac{a(G(\hat{x}))}{G'(\hat{x})} (2 - 2\cos\theta) \qquad (3.48)
$$

and

$$
\left\{ \frac{1}{n+1} A_{G,n} \right\}_n \sim_{\sigma, \lambda} \frac{a(G(\hat{x}))}{G'(\hat{x})} (2 - 2\cos\theta). \qquad (3.49)
$$

Proof We only prove (3.48) because (3.49) follows immediately from (3.48) and **GLT 1** as the matrices $A_{G,n}$ are symmetric. Since $G \in C^1([0, 1])$, for every $j = 1, \ldots, n$ there exist $\alpha_j \in [\hat{x}_{j-1}, \hat{x}_j]$ and $\beta_j \in [\hat{x}_j, \hat{x}_{j+1}]$ such that

$$h_j = G(\hat{x}_j) - G(\hat{x}_{j-1}) = G'(\alpha_j)h = (G'(\hat{x}_j) + \delta_j)h, \qquad (3.50)$$

$$h_{j+1} = G(\hat{x}_{j+1}) - G(\hat{x}_j) = G'(\beta_j)h = (G'(\hat{x}_j) + \varepsilon_j)h, \qquad (3.51)$$

where

$$\delta_j = G'(\alpha_j) - G'(\hat{x}_j),$$

$$\varepsilon_j = G'(\beta_j) - G'(\hat{x}_j).$$

Note that

$$|\delta_j|, |\varepsilon_j| \le \omega_{G'}(h), \qquad j = 1, \ldots, n,$$

where $\omega_{G'}$ is the modulus of continuity of G'. In view of (3.50) and (3.51), we have, for each $j = 1, \ldots, n$,

$$a\left(G(\hat{x}_j) - \frac{h_j}{2}\right) = a\left(G(\hat{x}_j) - \frac{h}{2}(G'(\hat{x}_j) + \delta_j)\right) = a(G(\hat{x}_j)) + \mu_j, \qquad (3.52)$$

$$a\left(G(\hat{x}_j) + \frac{h_{j+1}}{2}\right) = a\left(G(\hat{x}_j) + \frac{h}{2}(G'(\hat{x}_j) + \varepsilon_j)\right) = a(G(\hat{x}_j)) + \eta_j, \qquad (3.53)$$

where

$$\mu_j = a\left(G(\hat{x}_j) - \frac{h}{2}(G'(\hat{x}_j) + \delta_j)\right) - a(G(\hat{x}_j)),$$

$$\eta_j = a\left(G(\hat{x}_j) + \frac{h}{2}(G'(\hat{x}_j) + \varepsilon_j)\right) - a(G(\hat{x}_j)).$$

This time

$$|\mu_j|, |\eta_j| \le C_G \omega_a(h), \qquad j = 1, \ldots, n,$$

where ω_a is the modulus of continuity of a and C_G is a constant depending only on G. Substituting (3.50)–(3.53) in (3.47), we obtain

$$\frac{1}{n+1}A_{G,n} = h\, A_{G,n} \qquad (3.54)$$

$$= \text{tridiag}_n\left[-\frac{a(G(\hat{x}_j)) + \mu_j}{G'(\hat{x}_j) + \delta_j}, \frac{a(G(\hat{x}_j)) + \mu_j}{G'(\hat{x}_j) + \delta_j}\right.$$

$$\left. + \frac{a(G(\hat{x}_j)) + \eta_j}{G'(\hat{x}_j) + \varepsilon_j}, -\frac{a(G(\hat{x}_j)) + \eta_j}{G'(\hat{x}_j) + \varepsilon_j}\right].$$

Let $\tilde{x}_j = \frac{j}{n}$ for $j = 1, \ldots, n$ and consider the matrix

$$D_n\left(\frac{a(G(\hat{x}))}{G'(\hat{x})}\right)T_n(2-2\cos\theta) = \text{tridiag}_n\left[-\frac{a(G(\tilde{x}_j))}{G'(\tilde{x}_j)}, 2\frac{a(G(\tilde{x}_j))}{G'(\tilde{x}_j)}, -\frac{a(G(\tilde{x}_j))}{G'(\tilde{x}_j)}\right].$$
(3.55)

In view of the inequality $|\hat{x}_j - \tilde{x}_j| \leq h$, which is satisfied for all $j = 1, \ldots, n$, the matrix (3.55) seems to be an 'approximation' of $\frac{1}{n+1}A_{G,n}$; cf. (3.54) and (3.55). Since the function $a(G(\hat{x}))/G'(\hat{x})$ is continuous a.e., **GLT 3** and **GLT 4** yield

$$\left\{D_n\left(\frac{a(G(\hat{x}))}{G'(\hat{x})}\right)T_n(2-2\cos\theta)\right\}_n \sim_{\text{GLT}} \frac{a(G(\hat{x}))}{G'(\hat{x})}(2-2\cos\theta).$$

We are going to show that

$$\left\{D_n\left(\frac{a(G(\hat{x}))}{G'(\hat{x})}\right)T_n(2-2\cos\theta)\right\}_n \xrightarrow{\text{a.c.s.}} \left\{\frac{1}{n+1}A_{G,n}\right\}_n.$$
(3.56)

Once this is proved, (3.48) follows immediately from **GLT 7**.

We first prove (3.56) in the case where $G'(\hat{x})$ does not vanish over $[0, 1]$, so that

$$m_{G'} = \min_{\hat{x}\in[0,1]} G'(\hat{x}) > 0.$$

In this case, we will show directly that $\|Z_n\| \to 0$, where

$$Z_n = \frac{1}{n+1}A_{G,n} - D_n\left(\frac{a(G(\hat{x}))}{G'(\hat{x})}\right)T_n(2-2\cos\theta).$$
(3.57)

The matrix Z_n in (3.57) is tridiagonal and a straightforward computation based on (3.54)–(3.55) shows that all its components are bounded in modulus by a quantity that depends only on n, G, a and that converges to 0 as $n \to \infty$. For example, if $j = 2, \ldots, n$, then

$$|(Z_n)_{j,j-1}| = \left|\frac{a(G(\hat{x}_j)) + \mu_j}{G'(\hat{x}_j) + \delta_j} - \frac{a(G(\tilde{x}_j))}{G'(\tilde{x}_j)}\right|$$

$$\leq \left|\frac{a(G(\hat{x}_j)) + \mu_j}{G'(\hat{x}_j) + \delta_j} - \frac{a(G(\hat{x}_j))}{G'(\hat{x}_j) + \delta_j}\right| + \left|\frac{a(G(\hat{x}_j))}{G'(\hat{x}_j) + \delta_j} - \frac{a(G(\hat{x}_j))}{G'(\hat{x}_j)}\right|$$

$$+ \left|\frac{a(G(\hat{x}_j))}{G'(\hat{x}_j)} - \frac{a(G(\tilde{x}_j))}{G'(\tilde{x}_j)}\right|$$

$$= \left|\frac{\mu_j}{G'(\hat{x}_j) + \delta_j}\right| + \left|\frac{a(G(\hat{x}_j))\delta_j}{G'(\hat{x}_j)(G'(\hat{x}_j) + \delta_j)}\right| + \left|\frac{a(G(\hat{x}_j))}{G'(\hat{x}_j)} - \frac{a(G(\tilde{x}_j))}{G'(\tilde{x}_j)}\right|$$

$$\leq \frac{C_G\omega_a(h)}{m_{G'}} + \frac{\|a\|_\infty\omega_{G'}(h)}{m_{G'}^2} + \omega_{a(G)/G'}(h),$$
(3.58)

where in the last inequality we used the fact that $G'(\hat{x}_j) + \delta_j = G'(\alpha_j)$ by (3.50). Thus, $\|Z_n\| \to 0$ as $n \to \infty$ by (3.1).

Now we consider the case where G has a finite number of points $\hat{x}$ where $G'(\hat{x}) = 0$. In this case, the previous argument does not work because $m_{G'} = 0$. However, we can still prove (3.56) in the following way. Let $\hat{x}^{(1)}, \ldots, \hat{x}^{(s)}$ be the points where G' vanishes, and consider the balls (intervals) $B(\hat{x}^{(k)}, \frac{1}{m}) = \{\hat{x} \in [0, 1] : |\hat{x} - \hat{x}^{(k)}| < \frac{1}{m}\}$. The function G' is continuous and positive on the complement of the union $\bigcup_{k=1}^{s} B(\hat{x}^{(k)}, \frac{1}{m})$, so

$$m_{G', m} = \min_{\hat{x} \in [0,1] \setminus \bigcup_{k=1}^{s} B(\hat{x}^{(k)}, \frac{1}{m})} G'(\hat{x}) > 0.$$

For all indices $j = 1, \ldots, n$ such that $\hat{x}_j \in [0, 1] \setminus \bigcup_{k=1}^{s} B(\hat{x}^{(k)}, \frac{1}{m})$, the components in the jth row of the matrix (3.57) are bounded in modulus by a quantity that depends only on n, m, G, a and that converges to 0 as $n \to \infty$. This becomes immediately clear if we note that, for such indices j, the inequality (3.58) holds unchanged with $m_{G'}$ replaced by $m_{G', m}$ and with $\omega_{a(G)/G'}$ replaced by $\omega_{a(G)/G', m}$, the modulus of continuity of $a(G)/G'$ over $[0, 1] \setminus \bigcup_{k=1}^{s} B(\hat{x}^{(k)}, \frac{1}{m})$. The number of remaining rows of Z_n (the rows corresponding to indices j such that $\hat{x}_j \in \bigcup_{k=1}^{s} B(\hat{x}^{(k)}, \frac{1}{m})$) is at most $2s(n+1)/m + s$. Indeed, each interval $B(\hat{x}^{(k)}, \frac{1}{m})$ has length $2/m$ (at most) and can contain at most $2(n + 1)/m + 1$ grid points $\hat{x}_j$. Thus, for every n, m we can split the matrix Z_n into the sum of two terms, i.e.,

$$Z_n = R_{n,m} + N_{n,m},$$

where $N_{n,m}$ is obtained from Z_n by setting to zero all the rows corresponding to indices j such that $\hat{x}_j \in \bigcup_{k=1}^{s} B(\hat{x}^{(k)}, \frac{1}{m})$ and $R_{n,m} = Z_n - N_{n,m}$ is obtained from Z_n by setting to zero all the rows corresponding to indices j such that $\hat{x}_j \in [0, 1] \setminus \bigcup_{k=1}^{s} B(\hat{x}^{(k)}, \frac{1}{m})$. From the above discussion we have

$$\lim_{n \to \infty} \|N_{n,m}\| = 0$$

for all m, and

$$\text{rank}(R_{n,m}) \leq \frac{2s(n + 1)}{m} + s$$

for all m, n. In particular, for each m we can choose n_m such that, for $n \geq n_m$, $\text{rank}(R_{n,m}) \leq 3sn/m$ and $\|N_{n,m}\| \leq 1/m$. The convergence (3.56) now follows from the definition of a.c.s. $\square$

An increasing bijective map $G : [0, 1] \to [0, 1]$ in $C^1([0, 1])$ is said to be regular if $G'(\hat{x}) \neq 0$ for all $\hat{x} \in [0, 1]$ and is said to be singular otherwise, i.e., if $G'(\hat{x}) = 0$ for some $\hat{x} \in [0, 1]$. If G is singular, any point $\hat{x} \in [0, 1]$ such that $G'(\hat{x}) = 0$ is referred to as a singularity point (or simply a singularity) of G. The choice of a map G with one or more singularity points corresponds to adopting a local refinement strategy, according to which the grid points x_j rapidly accumulate at the G-images of the singularities as n increases. For example, if

$$G(\hat{x}) = \hat{x}^q, \qquad q > 1, \tag{3.59}$$

then 0 is a singularity of G (because $G'(0) = 0$) and the grid points

$$x_j = G(\hat{x}_j) = \left(\frac{j}{n+1}\right)^q, \quad j = 0, \ldots, n+1,$$

rapidly accumulate at $G(0) = 0$ as $n \to \infty$. We note that, whenever G is singular, the symbol in (3.48) is unbounded (except in some rare cases where $a(G(\hat{x}))$ and $G'(\hat{x})$ vanish simultaneously).

3.3.3 FE Discretization of Differential Equations

3.3.3.1 FE Discretization of Convection-Diffusion-Reaction Equations

Consider the following convection-diffusion-reaction problem in divergence form with Dirichlet boundary conditions:

$$\begin{cases} -(a(x)u'(x))' + b(x)u'(x) + c(x)u(x) = f(x), & x \in (0, 1), \\ u(0) = u(1) = 0, \end{cases} \tag{3.60}$$

where $f \in L^2([0, 1])$ and the coefficients a, b, c are only assumed to be in $L^\infty([0, 1])$. These sole assumptions are enough to perform the GLT analysis of the matrices arising from the FE discretization of (3.60). In this sense, we are going to see that the theory of GLT sequences allows one to derive the singular value and spectral distribution of DE discretization matrices under very weak hypotheses on the DE coefficients.

FE Discretization We consider the approximation of (3.60) by classical linear FEs on a uniform mesh in [0, 1] with stepsize $h = \frac{1}{n+1}$. We briefly describe here this approximation technique and for more details we refer the reader to [45, Chapter 4] or to any other good book on FEs. We first recall from [14, Chapter 8] that, if $\Omega \subset \mathbb{R}$ is a bounded interval whose endpoints are, say, α and β, $H^1(\Omega)$ denotes the (Sobolev) space of functions $v \in L^2(\Omega)$ possessing a weak (Sobolev) derivative in $L^2(\Omega)$. We also recall that each $v \in H^1(\Omega)$ coincides a.e. with a continuous function in $C(\overline{\Omega})$, and $H^1(\Omega)$ can also be defined as the following subspace of $C(\overline{\Omega})$:

$$H^1(\Omega) = \left\{ v \in C(\overline{\Omega}) : v \text{ is differentiable a.e. with } v' \in L^2(\Omega), \right.$$

$$\left. v(x) = v(\alpha) + \int_\alpha^x v'(y)dy \text{ for all } x \in \overline{\Omega} \right\}.$$

In this definition, the weak derivative of a $v \in H^1(\Omega)$ is just the classical derivative v' (which exists a.e.). Let

$$H_0^1(\Omega) = \{v \in H^1(\Omega) : v(\alpha) = v(\beta) = 0\}.$$

The weak form of (3.60) reads as follows [14, Chapter 8]: find $u \in H_0^1([0, 1])$ such that

$$a(u, w) = f(w), \qquad \forall w \in H_0^1([0, 1]),$$

where

$$a(u, w) = \int_0^1 a(x) u'(x) w'(x) \mathrm{d}x + \int_0^1 b(x) u'(x) w(x) \mathrm{d}x + \int_0^1 c(x) u(x) w(x) \mathrm{d}x,$$

$$f(w) = \int_0^1 f(x) w(x) \mathrm{d}x.$$

Let $h = \frac{1}{n+1}$ and $x_i = ih$, $i = 0, \ldots, n+1$. In the linear FE approach based on the uniform mesh $\{x_0, \ldots, x_{n+1}\}$, we fix the subspace $\mathscr{W}_n = \mathrm{span}(\varphi_1, \ldots, \varphi_n) \subset H_0^1([0, 1])$, where $\varphi_1, \ldots, \varphi_n$ are the so-called hat-functions:

$$\varphi_i(x) = \frac{x - x_{i-1}}{x_i - x_{i-1}} \chi_{[x_{i-1}, x_i)}(x) + \frac{x_{i+1} - x}{x_{i+1} - x_i} \chi_{[x_i, x_{i+1})}(x), \qquad i = 1, \ldots, n;$$

(3.61)

see Fig. 3.1. Note that $\mathscr{W}_n$ is the space of piecewise linear functions corresponding to the sequence of points $0 = x_0 < x_1 < \ldots < x_{n+1} = 1$ and vanishing on the boundary of the domain $[0, 1]$. In formulas,

$$\mathscr{W}_n = \left\{ s : [0, 1] \to \mathbb{R} : s\big|_{\left[\frac{i}{n+1}, \frac{i+1}{n+1}\right)} \in \mathbb{P}_1, \ i = 0, \ldots, n, \ s(0) = s(1) = 0 \right\},$$

where $\mathbb{P}_1$ is the space of polynomials of degree less than or equal to 1. We look for an approximation $u_{\mathscr{W}_n}$ of u by solving the following (Galerkin) problem: find $u_{\mathscr{W}_n} \in \mathscr{W}_n$ such that

$$a(u_{\mathscr{W}_n}, w) = f(w), \qquad \forall w \in \mathscr{W}_n.$$

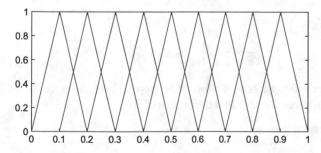

Fig. 3.1 Graph of the hat-functions $\varphi_1, \ldots, \varphi_n$ for $n = 9$

Since $\{\varphi_1, \ldots, \varphi_n\}$ is a basis of $\mathcal{W}_n$, we can write $u_{\mathcal{W}_n} = \sum_{j=1}^{n} u_j \varphi_j$ for a unique vector $\mathbf{u} = (u_1, \ldots, u_n)^T$. By linearity, the computation of $u_{\mathcal{W}_n}$ (i.e., of $\mathbf{u}$) reduces to solving the linear system

$$A_n \mathbf{u} = \mathbf{f},$$

where $\mathbf{f} = (\mathrm{f}(\varphi_1), \ldots, \mathrm{f}(\varphi_n))^T$ and A_n is the stiffness matrix,

$$A_n = [a(\varphi_j, \varphi_i)]_{i,j=1}^{n}.$$

Note that A_n admits the following decomposition:

$$A_n = K_n + Z_n, \tag{3.62}$$

where

$$K_n = \left[\int_0^1 a(x)\varphi_j'(x)\varphi_i'(x)\mathrm{d}x \right]_{i,j=1}^{n} \tag{3.63}$$

is the (symmetric) diffusion matrix and

$$Z_n = \left[\int_0^1 b(x)\varphi_j'(x)\varphi_i(x)\mathrm{d}x \right]_{i,j=1}^{n} + \left[\int_0^1 c(x)\varphi_j(x)\varphi_i(x)\mathrm{d}x \right]_{i,j=1}^{n} \tag{3.64}$$

is the sum of the convection and reaction matrix.

GLT Analysis of the FE Discretization Matrices Using the theory of GLT sequences we now derive the spectral and singular value distribution of the sequence of normalized stiffness matrices $\{\frac{1}{n+1} A_n\}_n$.

Theorem 10 *If $a, b, c \in L^{\infty}([0, 1])$ then*

$$\left\{ \frac{1}{n+1} A_n \right\}_n \sim_{\mathrm{GLT}} a(x)(2 - 2\cos\theta) \tag{3.65}$$

and

$$\left\{ \frac{1}{n+1} A_n \right\}_n \sim_{\sigma, \lambda} a(x)(2 - 2\cos\theta). \tag{3.66}$$

Proof The proof consists of the following steps. Throughout the proof, the letter C will denote a generic constant independent of n.

Step 1 We show that

$$\left\| \frac{1}{n+1} K_n \right\| \leq C \tag{3.67}$$

and

$$\left\|\frac{1}{n+1}Z_n\right\| \le C/n. \tag{3.68}$$

To prove (3.67), we note that K_n is a banded (tridiagonal) matrix, due to the local support property $\text{supp}(\varphi_i) = [x_{i-1}, x_{i+1}]$, $i = 1, \ldots, n$. Moreover, by the inequality $|\varphi_i'(x)| \le n + 1$, for all $i, j = 1, \ldots, n$ we have

$$|(K_n)_{ij}| = \left|\int_0^1 a(x)\varphi_j'(x)\varphi_i'(x)dx\right| = \left|\int_{x_{i-1}}^{x_{i+1}} a(x)\varphi_j'(x)\varphi_i'(x)dx\right|$$

$$\le (n+1)^2\|a\|_{L^\infty}\int_{x_{i-1}}^{x_{i+1}} dx = 2(n+1)\|a\|_{L^\infty}.$$

Thus, the components of the tridiagonal matrix $\frac{1}{n+1}K_n$ are bounded (in modulus) by $2\|a\|_{L^\infty}$, and (3.67) follows from (3.1).

To prove (3.68), we follow the same argument as for the proof of (3.67). Due to the local support property of the hat-functions, Z_n is tridiagonal. Moreover, by the inequalities $|\varphi_i(x)| \le 1$ and $|\varphi_i'(x)| \le n + 1$, for all $i, j = 1, \ldots, n$ we have

$$|(Z_n)_{ij}| = \left|\int_{x_{i-1}}^{x_{i+1}} b(x)\varphi_j'(x)\varphi_i(x)dx + \int_{x_{i-1}}^{x_{i+1}} c(x)\varphi_j(x)\varphi_i(x)dx\right|$$

$$\le 2\|b\|_{L^\infty} + \frac{2\|c\|_{L^\infty}}{n+1},$$

and (3.68) follows from (3.1).

Step 2 Consider the linear operator $K_n(\cdot) : L^1([0, 1]) \to \mathbb{R}^{n \times n}$,

$$K_n(g) = \left[\int_0^1 g(x)\varphi_j'(x)\varphi_i'(x)dx\right]_{i,j=1}^n.$$

By (3.63), we have $K_n = K_n(a)$. The next three steps are devoted to show that

$$\left\{\frac{1}{n+1}K_n(g)\right\}_n \sim_{\text{GLT}} g(x)(2 - 2\cos\theta), \qquad \forall g \in L^1([0, 1]). \tag{3.69}$$

Once this is done, the theorem is proved. Indeed, by applying (3.69) with $g = a$ we immediately get $\{\frac{1}{n+1}K_n\}_n \sim_{\text{GLT}} a(x)(2 - 2\cos\theta)$. Since $\{\frac{1}{n+1}Z_n\}_n$ is zero-distributed by Step 1, (3.65) follows from the decomposition

$$\frac{1}{n+1}A_n = \frac{1}{n+1}K_n + \frac{1}{n+1}Z_n \tag{3.70}$$

and from **GLT 3**–**GLT 4**; and the singular value distribution in (3.66) follows from
GLT 1. If $b(x) = 0$ identically, then $\frac{1}{n+1}A_n$ is symmetric and also the spectral
distribution in (3.66) follows from **GLT 1**. If $b(x)$ is not identically 0, the spectral
distribution in (3.66) follows from **GLT 2** applied to the decomposition (3.70),
taking into account what we have proved in Step 1.

Step 3 We first prove (3.69) in the constant-coefficient case where $g = 1$ identically.
In this case, a direct computation based on (3.61) shows that

$$
K_n(1) = \left[\int_0^1 \varphi_j'(x)\varphi_i'(x)\mathrm{d}x \right]_{i,j=1}^n = \frac{1}{h}
\begin{bmatrix}
2 & -1 & & & \\
-1 & 2 & -1 & & \\
& \ddots & \ddots & \ddots & \\
& & -1 & 2 & -1 \\
& & & -1 & 2
\end{bmatrix}
= \frac{1}{h} T_n(2 - 2\cos\theta),
$$

and the desired relation $\{\frac{1}{n+1} K_n(1)\}_n \sim_{\mathrm{GLT}} 2 - 2\cos\theta$ follows from **GLT 3**. Note
that it is precisely the analysis of the constant-coefficient case considered in this step
that allows one to realize what is the correct normalization factor. In our case, this is
$\frac{1}{n+1}$, which removes the $\frac{1}{h}$ from $K_n(1)$ and yields a normalized matrix $\frac{1}{n+1}K_n(1) =$
$T_n(2 - 2\cos\theta)$, whose components are *bounded away from* 0 *and* ∞ (actually, in
the present case they are even constant).

Step 4 Now we prove (3.69) in the case where $g \in C([0, 1])$. We first illustrate
the idea, and then we go into the details. The proof is based on the fact that the
hat-functions (3.61) are 'locally supported'. Indeed, the support $[x_{i-1}, x_{i+1}]$ of the
ith hat-function $\varphi_i(x)$ is located near the point $\frac{i}{n} \in [x_i, x_{i+1}]$, and the amplitude of
the support tends to 0 as $n \to \infty$. In this sense, the linear FE method considered
herein belongs to the family of the so-called 'local' methods. Since $g(x)$ varies
continuously over $[0, 1]$, the (i, j) entry of $K_n(g)$ can be approximated as follows,
for every $i, j = 1, \ldots, n$:

$$
(K_n(g))_{ij} = \int_0^1 g(x)\varphi_j'(x)\varphi_i'(x)\mathrm{d}x = \int_{x_{i-1}}^{x_{i+1}} g(x)\varphi_j'(x)\varphi_i'(x)\mathrm{d}x
$$

$$
\approx g\left(\frac{i}{n}\right) \int_{x_{i-1}}^{x_{i+1}} \varphi_j'(x)\varphi_i'(x)\mathrm{d}x = g\left(\frac{i}{n}\right) \int_0^1 \varphi_j'(x)\varphi_i'(x)\mathrm{d}x
$$

$$
= g\left(\frac{i}{n}\right)(K_n(1))_{ij}.
$$

This approximation can be rewritten in matrix form as

$$
K_n(g) \approx D_n(g)K_n(1). \tag{3.71}
$$

We will see that (3.71) implies that $\{\frac{1}{n+1}K_n(g) - \cdot\frac{1}{n+1}D_n(g)K_n(1)\}_n \sim_\sigma 0$, and (3.69) will then follow from Step 3 and **GLT 3–GLT 4**.

Let us now go into the details. Since $\text{supp}(\varphi_i) = [x_{i-1}, x_{i+1}]$ and $|\varphi_i'(x)| \le n+1$, for all $i, j = 1, \ldots, n$ we have

$$
\left|(K_n(g))_{ij} - (D_n(g)K_n(1))_{ij}\right| = \left|\int_0^1 \left[g(x) - g\left(\frac{i}{n}\right)\right]\varphi_j'(x)\varphi_i'(x)dx\right|
$$

$$
\le (n+1)^2 \int_{x_{i-1}}^{x_{i+1}} \left|g(x) - g\left(\frac{i}{n}\right)\right| dx
$$

$$
\le 2(n+1)\,\omega_g\left(\frac{2}{n+1}\right).
$$

It follows that each entry of the matrix $Y_n = \frac{1}{n+1}K_n(g) - \frac{1}{n+1}D_n(g)K_n(1)$ is bounded in modulus by $2\,\omega_g(\frac{2}{n+1})$. Moreover, Y_n is banded (tridiagonal), because of the local support property of the hat-functions. Thus, both the 1-norm and the ∞-norm of Y_n are bounded by $C\,\omega_g(\frac{2}{n+1})$, and (3.1) yields $\|Y_n\| \le C\,\omega_g(\frac{2}{n+1}) \to 0$ as $n \to \infty$. Hence, $\{Y_n\}_n \sim_\sigma 0$, which implies (3.69) by Step 3 and **GLT 3–GLT 4**.

Step 5 Finally, we prove (3.69) in the general case where $g \in L^1([0, 1])$. By the density of $C([0, 1])$ in $L^1([0, 1])$, there exist continuous functions $g_m \in C([0, 1])$ such that $g_m \to g$ in $L^1([0, 1])$. By Step 4,

$$
\left\{\frac{1}{n+1}K_n(g_m)\right\}_n \sim_{\text{GLT}} g_m(x)(2 - 2\cos\theta). \tag{3.72}
$$

Moreover,

$$
g_m(x)(2 - 2\cos\theta) \to g(x)(2 - 2\cos\theta) \quad \text{in measure.} \tag{3.73}
$$

We show that

$$
\left\{\frac{1}{n+1}K_n(g_m)\right\}_n \xrightarrow{\text{a.c.s.}} \left\{\frac{1}{n+1}K_n(g)\right\}_n. \tag{3.74}
$$

Since $\sum_{i=1}^n |\varphi_i'(x)| \le 2(n+1)$ for all $x \in [0, 1]$, by (3.4) we obtain

$$
\|K_n(g) - K_n(g_m)\|_1 \le \sum_{i,j=1}^n |(K_n(g))_{ij} - (K_n(g_m))_{ij}|
$$

$$
= \sum_{i,j=1}^n \left|\int_0^1 [g(x) - g_m(x)]\varphi_j'(x)\varphi_i'(x)dx\right|
$$

$$\leq \int_0^1 |g(x) - g_m(x)| \sum_{i,j=1}^n |\varphi'_j(x)| \, |\varphi'_i(x)| dx$$

$$\leq 4(n+1)^2 \|g - g_m\|_{L^1}$$

and

$$\left\| \frac{1}{n+1} K_n(g) - \frac{1}{n+1} K_n(g_m) \right\|_1 \leq Cn \|g - g_m\|_{L^1}.$$

Thus, $\{\frac{1}{n+1} K_n(g_m)\}_n \xrightarrow{\text{a.c.s.}} \{\frac{1}{n+1} K_n(g)\}_n$ by **ACS 3**. In view of (3.72)–(3.74), the relation (3.69) follows from **GLT 7**.   □

Remark 6 (Formal Structure of the Symbol) Problem (3.60) can be formally rewritten as follows:

$$\begin{cases} -a(x)u''(x) + (b(x) - a'(x))u'(x) + c(x)u(x) = f(x), & x \in (0,1), \\ u(0) = u(1) = 0. \end{cases}$$

(3.75)

It is then clear that the symbol $a(x)(2 - 2\cos\theta)$ has the same formal structure of the higher-order differential operator $-a(x)u''(x)$ associated with (3.75) (as in the FD case; see Remark 1). The formal analogy becomes even more evident if we note that $2 - 2\cos\theta$ is the trigonometric polynomial in the Fourier variable coming from the FE discretization of the (negative) second derivative $-u''(x)$. Indeed, as we have seen in Step 3 of the proof of Theorem 10, $2 - 2\cos\theta$ is the symbol of the sequence of FE diffusion matrices $\{\frac{1}{n+1} K_n(1)\}_n$, which arises from the FE approximation of the Poisson problem

$$\begin{cases} -u''(x) = f(x), & x \in (0,1), \\ u(0) = u(1) = 0, \end{cases}$$

that is, problem (3.60) in the case where $a(x) = 1$ and $b(x) = c(x) = 0$ identically.

3.3.3.2 FE Discretization of a System of Equations

In this section we consider the linear FE approximation of a system of differential equations, namely

$$\begin{cases} -(a(x)u'(x))' + v'(x) = f(x), & x \in (0,1), \\ -u'(x) - \rho v(x) = g(x), & x \in (0,1), \\ u(0) = 0, \quad u(1) = 0, \\ v(0) = 0, \quad v(1) = 0, \end{cases}$$

(3.76)

where ρ is a constant and a is only assumed to be in $L^1([0, 1])$. As we shall see, the resulting discretization matrices appear in the so-called saddle point form [6, p. 3], and we will illustrate the way to compute the asymptotic spectral and singular value distribution of their Schur complements using the theory of GLT sequences. It is worth noting that the Schur complement is a key tool for the numerical treatment of the related linear systems [6, Section 5]. The analysis of this section is similar to the analysis in [25, Section 2], but the discretization technique considered herein is a pure FE approximation, whereas in [25, Section 2] the authors adopted a mixed FD/FE technique.

FE Discretization We consider the approximation of (3.76) by linear FEs on a uniform mesh in $[0, 1]$ with stepsize $h = \frac{1}{n+1}$. Let us describe it shortly. The weak form of (3.76) reads as follows[1]: find $u, v \in H_0^1([0, 1])$ such that, for all $w \in H_0^1([0, 1])$,

$$\begin{cases} \int_0^1 a(x)u'(x)w'(x)\mathrm{d}x + \int_0^1 v'(x)w(x)\mathrm{d}x = \int_0^1 f(x)w(x)\mathrm{d}x, \\ -\int_0^1 u'(x)w(x)\mathrm{d}x - \rho \int_0^1 v(x)w(x)\mathrm{d}x = \int_0^1 g(x)w(x)\mathrm{d}x. \end{cases} \quad (3.77)$$

Let $h = \frac{1}{n+1}$ and $x_i = ih$, $i = 0, \ldots, n+1$. In the linear FE approach based on the mesh $\{x_0, \ldots, x_{n+1}\}$, we fix the subspace $\mathscr{W}_n = \mathrm{span}(\varphi_1, \ldots, \varphi_n) \subset H_0^1([0, 1])$, where $\varphi_1, \ldots, \varphi_n$ are the hat-functions in (3.61) (see also Fig. 3.1). Then, we look for approximations $u_{\mathscr{W}_n}, v_{\mathscr{W}_n}$ of u, v by solving the following (Galerkin) problem: find $u_{\mathscr{W}_n}, v_{\mathscr{W}_n} \in \mathscr{W}_n$ such that, for all $w \in \mathscr{W}_n$,

$$\begin{cases} \int_0^1 a(x)u'_{\mathscr{W}_n}(x)w'(x)\mathrm{d}x + \int_0^1 v'_{\mathscr{W}_n}(x)w(x)\mathrm{d}x = \int_0^1 f(x)w(x)\mathrm{d}x, \\ -\int_0^1 u'_{\mathscr{W}_n}(x)w(x)\mathrm{d}x - \rho \int_0^1 v_{\mathscr{W}_n}(x)w(x)\mathrm{d}x = \int_0^1 g(x)w(x)\mathrm{d}x. \end{cases}$$

Since $\{\varphi_1, \ldots, \varphi_n\}$ is a basis of $\mathscr{W}_n$, we can write $u_{\mathscr{W}_n} = \sum_{j=1}^n u_j \varphi_j$ and $v_{\mathscr{W}_n} = \sum_{j=1}^n v_j \varphi_j$ for unique vectors $\mathbf{u} = (u_1, \ldots, u_n)^T$ and $\mathbf{v} = (v_1, \ldots, v_n)^T$. By linearity, the computation of $u_{\mathscr{W}_n}, v_{\mathscr{W}_n}$ (i.e., of $\mathbf{u}, \mathbf{v}$) reduces to solving the linear system

$$A_{2n} \begin{bmatrix} \mathbf{u} \\ \mathbf{v} \end{bmatrix} = \begin{bmatrix} \mathbf{f} \\ \mathbf{g} \end{bmatrix},$$

---

[1]We are proceeding formally here, because the assumption $a \in L^1([0, 1])$ is too weak to ensure that the weak form (3.77) is well-defined. Keep in mind, however, that our formal derivation is correct if $a \in L^\infty([0, 1])$.

where $\mathbf{f} = \left[\int_0^1 f(x)\varphi_i(x)dx\right]_{i=1}^n$, $\mathbf{g} = \left[\int_0^1 g(x)\varphi_i(x)dx\right]_{i=1}^n$ and A_{2n} is the stiffness matrix, which possesses the following saddle point structure:

$$A_{2n} = \begin{bmatrix} K_n & H_n \\ H_n^T & -\rho M_n \end{bmatrix}.$$

Here, the blocks K_n, H_n, M_n are square matrices of size n, and precisely

$$K_n = \left[\int_0^1 a(x)\varphi_j'(x)\varphi_i'(x)dx\right]_{i,j=1}^n,$$

$$H_n = \left[\int_0^1 \varphi_j'(x)\varphi_i(x)dx\right]_{i,j=1}^n = \frac{1}{2}\begin{bmatrix} 0 & 1 & & & \\ -1 & 0 & 1 & & \\ & \ddots & \ddots & \ddots & \\ & & -1 & 0 & 1 \\ & & & -1 & 0 \end{bmatrix} = -i\,T_n(\sin\theta),$$

$$M_n = \left[\int_0^1 \varphi_j(x)\varphi_i(x)dx\right]_{i,j=1}^n = \frac{h}{6}\begin{bmatrix} 4 & 1 & & & \\ 1 & 4 & 1 & & \\ & \ddots & \ddots & \ddots & \\ & & 1 & 4 & 1 \\ & & & 1 & 4 \end{bmatrix} = \frac{h}{3}\,T_n(2+\cos\theta).$$

Note that K_n is exactly the matrix appearing in (3.63). Note also that the matrices K_n, M_n are symmetric, while H_n is skew-symmetric: $H_n^T = -H_n = i\,T_n(\sin\theta)$.

GLT Analysis of the Schur Complements of the FE Discretization Matrices
Assume that the matrices K_n are invertible. This is satisfied, for example, if $a > 0$ a.e., in which case the matrices K_n are positive definite. The (negative) Schur complement of A_{2n} is the symmetric matrix given by

$$S_n = \rho M_n + H_n^T\,K_n^{-1}\,H_n = \frac{\rho h}{3}\,T_n(2+\cos\theta) + T_n(\sin\theta)\,K_n^{-1}\,T_n(\sin\theta). \quad (3.78)$$

In the following, we perform the GLT analysis of the sequence of normalized Schur complements $\{(n+1)S_n\}_n$, and we compute its asymptotic spectral and singular value distribution under the additional necessary assumption that $a \neq 0$ a.e.

Theorem 11 *Let $\rho \in \mathbb{R}$ and $a \in L^1([0,1])$. Suppose that the matrices K_n are invertible and that $a \neq 0$ a.e. Then*

$$\{(n+1)S_n\}_n \sim_{\mathrm{GLT}} \varsigma(x,\theta) \quad (3.79)$$

and

$${(n + 1)S_n}_n \sim_{\sigma, \lambda} \varsigma(x, \theta), \tag{3.80}$$

where

$$\varsigma(x, \theta) = \frac{\rho}{3}(2 + \cos \theta) + \frac{\sin^2 \theta}{a(x)(2 - 2 \cos \theta)}.$$

Proof In view of (3.78), we have

$$(n + 1)S_n = \frac{\rho}{3} T_n(2 + \cos \theta) + T_n(\sin \theta) \left(\frac{1}{n + 1} K_n \right)^{-1} T_n(\sin \theta).$$

Moreover, by (3.69),

$$\left\{ \frac{1}{n + 1} K_n \right\}_n = \left\{ \frac{1}{n + 1} K_n(a) \right\}_n \sim_{\text{GLT}} a(x)(2 - 2 \cos \theta).$$

Therefore, under the assumption that $a \neq 0$ a.e., the GLT relation (3.79) follows from **GLT 3–GLT 5**. The singular value and spectral distributions in (3.80) follow from (3.79) and **GLT 1** as the Schur complements S_n are symmetric. □

3.3.4 IgA Discretization of Differential Equations

Isogeometric Analysis (IgA) is a modern and successful paradigm introduced in [18, 40] for analyzing problems governed by DEs. Its goal is to improve the connection between numerical simulation and Computer-Aided Design (CAD) systems. The main idea in IgA is to use directly the geometry provided by CAD systems and to approximate the solutions of DEs by the same type of functions (usually, B-splines or NURBS). In this way, it is possible to save about 80% of the CPU time, which is normally employed in the translation between two different languages (e.g., between FEs and CAD or between FDs and CAD). In its original formulation [18, 40], IgA employs Galerkin discretizations, which are typical of the FE approach. In the Galerkin framework an efficient implementation requires special numerical quadrature rules when constructing the resulting system of equations; see, e.g., [42]. To avoid this issue, isogeometric collocation methods have been recently introduced in [1]. Detailed comparisons with IgA Galerkin have shown the advantages of IgA collocation in terms of accuracy versus computational cost, in particular when higher-order approximation degrees are adopted [49]. Within the framework of IgA collocation, many applications have been successfully tackled, showing its potential and flexibility. Interested readers are referred to the recent review [47] and references therein. Section 3.3.4.1 is devoted to the isogeometric collocation approach, whereas the more traditional isogeometric Galerkin methods will be addressed in Sects. 3.3.4.2–3.3.4.3.

3.3.4.1 B-Spline IgA Collocation Discretization of Convection-Diffusion-Reaction Equations

Consider the convection-diffusion-reaction problem

$$
\begin{cases}
-(a(x)u'(x))' + b(x)u'(x) + c(x)u(x) = f(x), & x \in \Omega, \\
u(x) = 0, & x \in \partial\Omega,
\end{cases}
\tag{3.81}
$$

where Ω is a bounded open interval of $\mathbb{R}$, $a : \overline{\Omega} \to \mathbb{R}$ is a function in $C^1(\overline{\Omega})$ and $b, c, f : \overline{\Omega} \to \mathbb{R}$ are functions in $C(\overline{\Omega})$. We consider the isogeometric collocation approximation of (3.81) based on uniform B-splines of degree $p \geq 2$. Since this approximation technique is not as known as FDs or FEs, we describe it below in some detail. For more on IgA collocation methods, see [1, 47].

Isogeometric Collocation Approximation Problem (3.81) can be reformulated as follows:

$$
\begin{cases}
-a(x)u''(x) + s(x)u'(x) + c(x)u(x) = f(x), & x \in \Omega, \\
u(x) = 0, & x \in \partial\Omega,
\end{cases}
\tag{3.82}
$$

where $s(x) = b(x) - a'(x)$. In the standard collocation method, we choose a finite dimensional vector space $\mathcal{W}$, consisting of sufficiently smooth functions defined on $\overline{\Omega}$ and vanishing on the boundary $\partial\Omega$; we call $\mathcal{W}$ the approximation space. Then, we introduce a set of $N = \dim \mathcal{W}$ collocation points $\{\tau_1, \ldots, \tau_N\} \subset \Omega$ and we look for a function $u_{\mathcal{W}} \in \mathcal{W}$ satisfying the differential equation (3.82) at the points τ_i, i.e.,

$$
-a(\tau_i)u''_{\mathcal{W}}(\tau_i) + s(\tau_i)u'_{\mathcal{W}}(\tau_i) + c(\tau_i)u_{\mathcal{W}}(\tau_i) = f(\tau_i), \qquad i = 1, \ldots, N.
$$

The function $u_{\mathcal{W}}$ is taken as an approximation to the solution u of (3.82). If we fix a basis $\{\varphi_1, \ldots, \varphi_N\}$ for $\mathcal{W}$, then we have $u_{\mathcal{W}} = \sum_{j=1}^N u_j \varphi_j$ for a unique vector $\mathbf{u} = (u_1, \ldots, u_N)^T$, and, by linearity, the computation of $u_{\mathcal{W}}$ (i.e., of $\mathbf{u}$) reduces to solving the linear system

$$
A\mathbf{u} = \mathbf{f},
$$

where $\mathbf{f} = \left[f(\tau_i)\right]_{i=1}^N$ and

$$
\begin{aligned}
A &= \left[-a(\tau_i)\varphi_j''(\tau_i) + s(\tau_i)\varphi_j'(\tau_i) + c(\tau_i)\varphi_j(\tau_i)\right]_{i,j=1}^N \\
&= \left(\operatorname*{diag}_{i=1,\ldots,N} a(\tau_i)\right)\left[-\varphi_j''(\tau_i)\right]_{i,j=1}^N + \left(\operatorname*{diag}_{i=1,\ldots,N} s(\tau_i)\right)\left[\varphi_j'(\tau_i)\right]_{i,j=1}^N \\
&\quad + \left(\operatorname*{diag}_{i=1,\ldots,N} c(\tau_i)\right)\left[\varphi_j(\tau_i)\right]_{i,j=1}^N
\end{aligned}
\tag{3.83}
$$

is the collocation matrix.

Now, suppose that the physical domain Ω can be described by a global geometry function $G : [0, 1] \to \overline{\Omega}$, which is invertible and satisfies $G(\partial([0, 1])) = \partial\overline{\Omega}$. Let

$$\{\hat{\varphi}_1, \ldots, \hat{\varphi}_N\} \tag{3.84}$$

be a set of basis functions defined on the parametric (or reference) domain $[0, 1]$ and vanishing on the boundary $\partial([0, 1])$. Let

$$\{\hat{\tau}_1, \ldots, \hat{\tau}_N\} \tag{3.85}$$

be a set of N collocation points in $(0, 1)$. In the isogeometric collocation approach, we find an approximation $u_{\mathscr{W}}$ of u by using the standard collocation method described above, in which

- the approximation space is chosen as $\mathscr{W} = \mathrm{span}(\varphi_1, \ldots \varphi_N)$, with

$$\varphi_i(x) = \hat{\varphi}_i(G^{-1}(x)) = \hat{\varphi}_i(\hat{x}), \qquad x = G(\hat{x}), \qquad i = 1, \ldots, N, \tag{3.86}$$

- the collocation points in the physical domain Ω are defined as

$$\tau_i = G(\hat{\tau}_i), \qquad i = 1, \ldots, N. \tag{3.87}$$

The resulting collocation matrix A is given by (3.83), with the basis functions φ_i and the collocation points τ_i defined as in (3.86)–(3.87).

Assuming that G and $\hat{\varphi}_i$, $i = 1, \ldots, N$, are sufficiently regular, we can apply standard differential calculus to express A in terms of G and $\hat{\varphi}_i$, $\hat{\tau}_i$, $i = 1, \ldots, N$. Let us work out this expression. For any $u : \overline{\Omega} \to \mathbb{R}$, consider the corresponding function $\hat{u} : [0, 1] \to \mathbb{R}$, which is defined on the parametric domain by

$$\hat{u}(\hat{x}) = u(x), \qquad x = G(\hat{x}). \tag{3.88}$$

In other words, $\hat{u}(\hat{x}) = u(G(\hat{x}))$.[2] Then, u satisfies (3.82) if and only if $\hat{u}$ satisfies the corresponding transformed problem

$$\begin{cases} -a_G(\hat{x})\hat{u}''(\hat{x}) + s_G(\hat{x})\hat{u}'(\hat{x}) + c_G(\hat{x})\hat{u}(\hat{x}) = f(G(\hat{x})), & \hat{x} \in (0, 1), \\ \hat{u}(\hat{x}) = 0, & \hat{x} \in \partial((0, 1)), \end{cases} \tag{3.89}$$

where a_G, s_G, c_G are, respectively, the transformed diffusion, convection, reaction coefficient. They are given by

$$a_G(\hat{x}) = \frac{a(G(\hat{x}))}{(G'(\hat{x}))^2}, \tag{3.90}$$

---

[2]Note that $\hat{\varphi}_i(\hat{x}) = \varphi_i(G(\hat{x}))$ for $i = 1, \ldots, N$, so $\hat{\varphi}_1, \ldots, \hat{\varphi}_N$ are obtained from $\varphi_1, \ldots, \varphi_N$ by the rule (3.88). Moreover, the equation $\tau_i = G(\hat{\tau}_i)$ is the same as the relation $x = G(\hat{x})$ in (3.88).

$$s_G(\hat{x}) = \frac{a(G(\hat{x}))G''(\hat{x})}{(G'(\hat{x}))^3} + \frac{s(G(\hat{x}))}{G'(\hat{x})}, \tag{3.91}$$

$$c_G(\hat{x}) = c(G(\hat{x})), \tag{3.92}$$

for $\hat{x} \in [0, 1]$. The collocation matrix A in (3.83) can be expressed in terms of G and $\hat{\varphi}_i$, $\hat{\tau}_i$, $i = 1, \ldots, N$, as follows:

$$
\begin{aligned}
A &= \left[-a_G(\hat{\tau}_i)\hat{\varphi}_j''(\hat{\tau}_i) + s_G(\hat{\tau}_i)\hat{\varphi}_j'(\hat{\tau}_i) + c_G(\hat{\tau}_i)\hat{\varphi}_j(\hat{\tau}_i) \right]_{i,j=1}^{N} \\
&= \left(\operatorname*{diag}_{i=1,\ldots,N} a_G(\hat{\tau}_i) \right) \left[-\hat{\varphi}_j''(\hat{\tau}_i) \right]_{i,j=1}^{N} + \left(\operatorname*{diag}_{i=1,\ldots,N} s_G(\hat{\tau}_i) \right) \left[\hat{\varphi}_j'(\hat{\tau}_i) \right]_{i,j=1}^{N} \\
&\quad + \left(\operatorname*{diag}_{i=1,\ldots,N} c_G(\hat{\tau}_i) \right) \left[\hat{\varphi}_j(\hat{\tau}_i) \right]_{i,j=1}^{N}.
\end{aligned} \tag{3.93}
$$

In the IgA context, the geometry map G is expressed in terms of the functions $\hat{\varphi}_i$, in accordance with the isoparametric approach [18, Section 3.1]. Moreover, the functions $\hat{\varphi}_i$ themselves are usually B-splines or their rational versions, the so-called NURBS. In this section, the role of the $\hat{\varphi}_i$ will be played by B-splines over uniform knot sequences. Furthermore, we do not limit ourselves to the isoparametric approach, but we allow the geometry map G to be any sufficiently regular function from $[0, 1]$ to $\overline{\Omega}$, not necessarily expressed in terms of B-splines. Finally, following [1], the collocation points $\hat{\tau}_i$ will be chosen as the Greville abscissae corresponding to the B-splines $\hat{\varphi}_i$.

B-Splines and Greville Abscissae For $p, n \geq 1$, consider the uniform knot sequence

$$t_1 = \cdots = t_{p+1} = 0 < t_{p+2} < \cdots < t_{p+n} < 1 = t_{p+n+1} = \cdots = t_{2p+n+1}, \tag{3.94}$$

where

$$t_{i+p+1} = \frac{i}{n}, \qquad i = 0, \ldots, n. \tag{3.95}$$

The B-splines of degree p on this knot sequence are denoted by

$$N_{i,[p]} : [0, 1] \to \mathbb{R}, \qquad i = 1, \ldots, n + p, \tag{3.96}$$

and are defined recursively as follows [19]: for $1 \leq i \leq n + 2p$,

$$N_{i,[0]}(t) = \chi_{[t_i, t_{i+1})}(t), \qquad t \in [0, 1]; \tag{3.97}$$

for $1 \leq k \leq p$ and $1 \leq i \leq n + 2p - k$,

$$N_{i,[k]}(t) = \frac{t - t_i}{t_{i+k} - t_i} N_{i,[k-1]}(t) + \frac{t_{i+k+1} - t}{t_{i+k+1} - t_{i+1}} N_{i+1,[k-1]}(t), \qquad t \in [0, 1], \tag{3.98}$$

where we assume that a fraction with zero denominator is zero. The Greville abscissa $\xi_{i,[p]}$ associated with the B-spline $N_{i,[p]}$ is defined by

$$\xi_{i,[p]} = \frac{t_{i+1} + t_{i+2} + \ldots + t_{i+p}}{p}, \qquad i = 1, \ldots, n + p. \tag{3.99}$$

We know from [19] that the functions $N_{1,[p]}, \ldots, N_{n+p,[p]}$ belong to $C^{p-1}([0, 1])$ and form a basis for the spline space

$$\left\{ s \in C^{p-1}([0, 1]) : \ s|_{\left[\frac{i}{n}, \frac{i+1}{n}\right)} \in \mathbb{P}_p, \ i = 0, \ldots, n - 1 \right\},$$

where $\mathbb{P}_p$ is the space of polynomials of degree less than or equal to p. Moreover, $N_{1,[p]}, \ldots, N_{n+p,[p]}$ possess the following properties [19].

- Local support property:

$$\mathrm{supp}(N_{i,[p]}) = [t_i, t_{i+p+1}], \qquad i = 1, \ldots, n + p. \tag{3.100}$$

- Vanishment on the boundary:

$$N_{i,[p]}(0) = N_{i,[p]}(1) = 0, \qquad i = 2, \ldots, n + p - 1. \tag{3.101}$$

- Nonnegative partition of unity:

$$N_{i,[p]}(t) \geq 0, \qquad t \in [0, 1], \qquad i = 1, \ldots, n + p, \tag{3.102}$$

$$\sum_{i=1}^{n+p} N_{i,[p]}(t) = 1, \qquad t \in [0, 1]. \tag{3.103}$$

- Bounds for derivatives:

$$\sum_{i=1}^{n+p} |N'_{i,[p]}(t)| \leq 2pn, \qquad t \in [0, 1], \tag{3.104}$$

$$\sum_{i=1}^{n+p} |N''_{i,[p]}(t)| \leq 4p(p - 1)n^2, \qquad t \in [0, 1]. \tag{3.105}$$

Note that the derivatives $N'_{1,[p]}(t), \ldots, N'_{n+p,[p]}(t)$ (resp., $N''_{1,[p]}(t), \ldots, N''_{n+p,[p]}(t)$) may not be defined at some of the points $\frac{1}{n}, \ldots, \frac{n-1}{n}$ when $p = 1$ (resp., $p = 1, 2$). In the summations (3.104)–(3.105), it is understood that the undefined values are counted as 0.

Let $\phi_{[q]}$ be the cardinal B-spline of degree $q \geq 0$ over the uniform knot sequence $\{0, 1, \ldots, q+1\}$, which is defined recursively as follows [19]:

$$\phi_{[0]}(t) = \chi_{[0,1)}(t), \qquad t \in \mathbb{R}, \tag{3.106}$$

$$\phi_{[q]}(t) = \frac{t}{q}\phi_{[q-1]}(t) + \frac{q+1-t}{q}\phi_{[q-1]}(t-1), \qquad t \in \mathbb{R}, \qquad q \geq 1. \tag{3.107}$$

It is known from [15, 19] that $\phi_{[q]} \in C^{q-1}(\mathbb{R})$ and

$$\mathrm{supp}(\phi_{[q]}) = [0, q+1]. \tag{3.108}$$

Moreover, the following symmetry property holds by [31, Lemma 3] (see also [15, p. 86]):

$$\phi_{[q]}^{(r)}\left(\frac{q+1}{2}+t\right) = (-1)^r\phi_{[q]}^{(r)}\left(\frac{q+1}{2}-t\right), \qquad t \in \mathbb{R}, \qquad r, q \geq 0, \tag{3.109}$$

where $\phi_{[q]}^{(r)}$ is the rth derivative of $\phi_{[q]}$. Note that $\phi_{[q]}^{(r)}(t)$ is defined for all $t \in \mathbb{R}$ if $r < q$, and for all $t \in \mathbb{R}\backslash\{0, 1, \ldots, q+1\}$ if $r \geq q$. Nevertheless, (3.109) holds for all $t \in \mathbb{R}$, because when the left-hand side is not defined, the right-hand side is not defined as well. Concerning the L^2 inner products of derivatives of cardinal B-splines, it was proved in [31, Lemma 4] that

$$\int_{\mathbb{R}} \phi_{[q_1]}^{(r_1)}(t)\phi_{[q_2]}^{(r_2)}(t+\tau)\mathrm{d}t = (-1)^{r_1}\phi_{[q_1+q_2+1]}^{(r_1+r_2)}(q_1+1+\tau)$$
$$= (-1)^{r_2}\phi_{[q_1+q_2+1]}^{(r_1+r_2)}(q_2+1-\tau) \tag{3.110}$$

for every $\tau \in \mathbb{R}$ and every $q_1, q_2, r_1, r_2 \geq 0$. Equation (3.110) is a property of the more general family of box splines [54] and generalizes the result appearing in [15, p. 89]. Cardinal B-splines are of interest herein, because the so-called central basis functions $N_{i,[p]}$, $i = p+1, \ldots, n$, are uniformly shifted and scaled versions of the cardinal B-spline $\phi_{[p]}$. This is illustrated in Figs. 3.2 and 3.3 for $p = 3$. In formulas, we have

$$N_{i,[p]}(t) = \phi_{[p]}(nt-i+p+1), \qquad t \in [0,1], \qquad i = p+1, \ldots, n, \tag{3.111}$$

and, consequently,

$$N'_{i,[p]}(t) = n\,\phi'_{[p]}(nt-i+p+1), \qquad t \in [0,1], \qquad i = p+1, \ldots, n, \tag{3.112}$$

$$N''_{i,[p]}(t) = n^2\phi''_{[p]}(nt-i+p+1), \qquad t \in [0,1], \qquad i = p+1, \ldots, n. \tag{3.113}$$

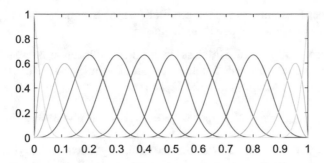

Fig. 3.2 Graph of the B-splines $N_{i,[p]}$, $i = 1, \ldots, n + p$, for $p = 3$ and $n = 10$; the central basis functions $N_{i,[p]}$, $i = p + 1, \ldots, n$, are depicted in blue

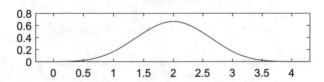

Fig. 3.3 Graph of the cubic cardinal B-spline $\phi_{[3]}$

Remark 7 For degree $p = 1$, the central B-spline basis functions $N_{2,[1]}, \ldots, N_{n,[1]}$ are the hat-functions $\varphi_1, \ldots, \varphi_{n-1}$ corresponding to the grid points

$$x_i = ih, \qquad i = 0, \ldots, n, \qquad h = \frac{1}{n}.$$

To see this, simply write (3.98) for $p = 1$ and compare it with (3.61). The graph of $N_{2,[1]}, \ldots, N_{n,[1]}$ for $n = 10$ is depicted in Fig. 3.1.

In view of (3.99) and (3.100), the Greville abscissa $\xi_{i,[p]}$ lies in the support of $N_{i,[p]}$,

$$\xi_{i,[p]} \in \mathrm{supp}(N_{i,[p]}) = [t_i, t_{i+p+1}], \qquad i = 1, \ldots, n + p. \qquad (3.114)$$

The central Greville abscissae $\xi_{i,[p]}$, $i = p + 1, \ldots, n$, which are the Greville abscissae associated with the central basis functions (3.111), simplify to

$$\xi_{i,[p]} = \frac{i}{n} - \frac{p+1}{2n}, \qquad i = p + 1, \ldots, n. \qquad (3.115)$$

The Greville abscissae are somehow equivalent, in an asymptotic sense, to the uniform knots in $[0, 1]$. More precisely,

$$\left| \xi_{i,[p]} - \frac{i}{n+p} \right| \le \frac{C_p}{n}, \qquad i = 1, \ldots, n + p, \qquad (3.116)$$

where C_p depends only on p. The proof of (3.116) is a matter of straightforward computations; we leave the details to the reader.

B-Spline IgA Collocation Matrices In the IgA collocation approach based on (uniform) B-splines, the basis functions $\hat{\varphi}_1, \ldots, \hat{\varphi}_N$ in (3.84) are chosen as the B-splines $N_{2,[p]}, \ldots, N_{n+p-1,[p]}$ in (3.96), i.e.,

$$\hat{\varphi}_i = N_{i+1,[p]}, \qquad i = 1, \ldots, n + p - 2. \tag{3.117}$$

In this setting, $N = n + p - 2$. Note that the boundary functions $N_{1,[p]}$ and $N_{n+p,[p]}$ are excluded because they do not vanish on the boundary $\partial([0, 1])$; see also Fig. 3.2. As for the collocation points $\hat{\tau}_1 \ldots, \hat{\tau}_N$ in (3.85), they are chosen as the Greville abscissae $\xi_{2,[p]}, \ldots, \xi_{n+p-1,[p]}$ in (3.99), i.e.,

$$\hat{\tau}_i = \xi_{i+1,[p]}, \qquad i = 1, \ldots, n + p - 2. \tag{3.118}$$

In what follows we assume $p \geq 2$, so as to ensure that $N''_{j+1,[p]}(\xi_{i+1,[p]})$ is defined for all $i, j = 1, \ldots, n + p - 2$. The collocation matrix (3.93) resulting from the choices of $\hat{\varphi}_i, \hat{\tau}_i$ as in (3.117)–(3.118) will be denoted by $A^{[p]}_{G,n}$, in order to emphasize its dependence on the geometry map G and the parameters n, p:

$$A^{[p]}_{G,n} = \Big[-a_G(\xi_{i+1,[p]})N''_{j+1,[p]}(\xi_{i+1,[p]}) + s_G(\xi_{i+1,[p]})N'_{j+1,[p]}(\xi_{i+1,[p]})$$
$$+ c_G(\xi_{i+1,[p]})N_{j+1,[p]}(\xi_{i+1,[p]}) \Big]^{n+p-2}_{i,j=1}$$
$$= D^{[p]}_n(a_G) K^{[p]}_n + D^{[p]}_n(s_G) H^{[p]}_n + D^{[p]}_n(c_G) M^{[p]}_n,$$

where

$$D^{[p]}_n(v) = \operatorname*{diag}_{i=1,\ldots,n+p-2} v(\xi_{i+1,[p]})$$

is the diagonal sampling matrix containing the samples of the function $v : [0, 1] \to \mathbb{R}$ at the Greville abscissae, and

$$K^{[p]}_n = \big[-N''_{j+1,[p]}(\xi_{i+1,[p]}) \big]^{n+p-2}_{i,j=1},$$
$$H^{[p]}_n = \big[N'_{j+1,[p]}(\xi_{i+1,[p]}) \big]^{n+p-2}_{i,j=1},$$
$$M^{[p]}_n = \big[N_{j+1,[p]}(\xi_{i+1,[p]}) \big]^{n+p-2}_{i,j=1}.$$

Note that $A^{[p]}_{G,n}$ can be decomposed as follows:

$$A^{[p]}_{G,n} = K^{[p]}_{G,n} + Z^{[p]}_{G,n},$$

where

$$K_{G,n}^{[p]} = \left[-a_G(\xi_{i+1,[p]})N''_{j+1,[p]}(\xi_{i+1,[p]}) \right]_{i,j=1}^{n+p-2} = D_n^{[p]}(a_G)\, K_n^{[p]}$$

is the diffusion matrix, i.e., the matrix resulting from the discretization of the higher-order (diffusion) term in (3.82), and

$$Z_{G,n}^{[p]} = \left[s_G(\xi_{i+1,[p]})N'_{j+1,[p]}(\xi_{i+1,[p]}) + c_G(\xi_{i+1,[p]})N_{j+1,[p]}(\xi_{i+1,[p]}) \right]_{i,j=1}^{n+p-2}$$
$$= D_n^{[p]}(s_G)\, H_n^{[p]} + D_n^{[p]}(c_G)\, M_n^{[p]}$$

is the matrix resulting from the discretization of the terms in (3.82) with lower-order derivatives (i.e., the convection and reaction terms). As already noticed in the previous sections about FD and FE discretizations, the matrix $Z_{G,n}^{[p]}$ can be regarded as a 'residual term', since it comes from the discretization of the lower-order differential operators. Indeed, we shall see that the norm of $Z_{G,n}^{[p]}$ is negligible with respect to the norm of the diffusion matrix $K_{G,n}^{[p]}$ when the discretization parameter n is large, because, after normalization by n^2, it will turn out that $\|n^{-2}Z_{G,n}^{[p]}\|$ tends to 0 as $n \to \infty$ (contrary to $\|n^{-2}K_{G,n}^{[p]}\|$, which remains bounded away from 0 and ∞).

Let us now provide an approximate construction of $K_n^{[p]}$, $M_n^{[p]}$, $H_n^{[p]}$. This is necessary for the GLT analysis of this section. We only construct the submatrices

$$\left[(K_n^{[p]})_{ij}\right]_{i,j=p}^{n-1}, \qquad \left[(H_n^{[p]})_{ij}\right]_{i,j=p}^{n-1}, \qquad \left[(M_n^{[p]})_{ij}\right]_{i,j=p}^{n-1}, \qquad (3.119)$$

which are determined by the central basis functions (3.111) and by the central Greville abscissae (3.115). Note that the submatrix $[(K_n^{[p]})_{ij}]_{i,j=p}^{n-1}$, when embedded in any matrix of size $n + p - 2$ at the right place (identified by the row and column indices $p, \ldots, n-1$), provides an approximation of $K_n^{[p]}$ up to a low-rank correction. A similar consideration also applies to the submatrices $[(H_n^{[p]})_{ij}]_{i,j=p}^{n-1}$ and $[(M_n^{[p]})_{ij}]_{i,j=p}^{n-1}$. A direct computation based on (3.109), (3.111)–(3.113) and (3.115) shows that, for $i, j = p, \ldots, n-1$,

$$(K_n^{[p]})_{ij} = -n^2\phi''_{[p]}\left(\frac{p+1}{2} + i - j\right) = -n^2\phi''_{[p]}\left(\frac{p+1}{2} - i + j\right),$$

$$(H_n^{[p]})_{ij} = n\,\phi'_{[p]}\left(\frac{p+1}{2} + i - j\right) = -n\,\phi'_{[p]}\left(\frac{p+1}{2} - i + j\right),$$

$$(M_n^{[p]})_{ij} = \phi_{[p]}\left(\frac{p+1}{2} + i - j\right) = \phi_{[p]}\left(\frac{p+1}{2} - i + j\right).$$

Since their entries depend only on the difference $i - j$, the submatrices (3.119) are Toeplitz matrices, and precisely

$$\left[(K_n^{[p]})_{ij}\right]_{i,j=p}^{n-1} = n^2\left[-\phi_{[p]}''\left(\frac{p+1}{2} - i + j\right)\right]_{i,j=p}^{n-1} = n^2\, T_{n-p}(f_p), \qquad (3.120)$$

$$\left[(H_n^{[p]})_{ij}\right]_{i,j=p}^{n-1} = n\left[-\phi_{[p]}'\left(\frac{p+1}{2} - i + j\right)\right]_{i,j=p}^{n-1} = n\,\mathrm{i}\, T_{n-p}(g_p), \qquad (3.121)$$

$$\left[(M_n^{[p]})_{ij}\right]_{i,j=p}^{n-1} = \left[\phi_{[p]}\left(\frac{p+1}{2} - i + j\right)\right]_{i,j=p}^{n-1} = T_{n-p}(h_p), \qquad (3.122)$$

where

$$f_p(\theta) = \sum_{k\in\mathbb{Z}} -\phi_{[p]}''\left(\frac{p+1}{2} - k\right) \mathrm{e}^{\mathrm{i}k\theta}$$

$$= -\phi_{[p]}''\left(\frac{p+1}{2}\right) - 2\sum_{k=1}^{\lfloor p/2\rfloor} \phi_{[p]}''\left(\frac{p+1}{2} - k\right)\cos(k\theta), \qquad (3.123)$$

$$g_p(\theta) = -\mathrm{i}\sum_{k\in\mathbb{Z}} -\phi_{[p]}'\left(\frac{p+1}{2} - k\right) \mathrm{e}^{\mathrm{i}k\theta}$$

$$= -2\sum_{k=1}^{\lfloor p/2\rfloor} \phi_{[p]}'\left(\frac{p+1}{2} - k\right)\sin(k\theta), \qquad (3.124)$$

$$h_p(\theta) = \sum_{k\in\mathbb{Z}} \phi_{[p]}\left(\frac{p+1}{2} - k\right) \mathrm{e}^{\mathrm{i}k\theta}$$

$$= \phi_{[p]}\left(\frac{p+1}{2}\right) + 2\sum_{k=1}^{\lfloor p/2\rfloor} \phi_{[p]}\left(\frac{p+1}{2} - k\right)\cos(k\theta); \qquad (3.125)$$

note that we used (3.108)–(3.109) to simplify the expressions of $f_p(\theta)$, $g_p(\theta)$, $h_p(\theta)$. It follows from (3.120) that $T_{n-p}(f_p)$ is the principal submatrix of both $n^{-2}K_n^{[p]}$ and $T_{n+p-2}(f_p)$ corresponding to the set of indices $p, \ldots, n-1$. Similar results follow from (3.121)–(3.122), and so we obtain

$$n^{-2}K_n^{[p]} = T_{n+p-2}(f_p) + R_n^{[p]}, \qquad \mathrm{rank}(R_n^{[p]}) \leq 4(p-1), \qquad (3.126)$$

$$-\mathrm{i}\,n^{-1}H_n^{[p]} = T_{n+p-2}(g_p) + S_n^{[p]}, \qquad \mathrm{rank}(S_n^{[p]}) \leq 4(p-1), \qquad (3.127)$$

$$M_n^{[p]} = T_{n+p-2}(h_p) + V_n^{[p]}, \qquad \mathrm{rank}(V_n^{[p]}) \leq 4(p-1). \qquad (3.128)$$

To better appreciate the above construction, let us see two examples. We only consider the case of the matrix $K_n^{[p]}$ because for $H_n^{[p]}$ and $M_n^{[p]}$ the situation is the same. In the first example, we fix $p = 3$. The matrix $K_n^{[3]}$ is given by

$$K_n^{[3]} = \frac{n^2}{6} \begin{bmatrix} 33 & -7 & -2 & & & & & & \\ -9 & 15 & -6 & & & & & & \\ & -6 & 12 & -6 & & & & & \\ & & -6 & 12 & -6 & & & & \\ & & & \ddots & \ddots & \ddots & & & \\ & & & & -6 & 12 & -6 & & \\ & & & & & -6 & 12 & -6 & \\ & & & & & & -6 & 15 & -9 \\ & & & & & & -2 & -7 & 33 \end{bmatrix}.$$

The submatrix $T_{n-2}(f_3)$ appears in correspondence of the highlighted box and we have

$$f_3(\theta) = \frac{1}{6}(-6e^{i\theta} + 12 - 6e^{-i\theta}) = 2 - 2\cos\theta,$$

as given by (3.123) for $p = 3$. In the second example, we fix $p = 4$. The matrix $K_n^{[4]}$ is given by

$$K_n^{[4]} = \frac{n^2}{96} \begin{bmatrix} 855 & -133 & -71 & -3 & & & & & & & \\ -81 & 243 & -63 & -27 & & & & & & & \\ -36 & -36 & 132 & -48 & -12 & & & & & & \\ -16 & -44 & 120 & -48 & -12 & & & & & & \\ & -12 & -48 & 120 & -48 & -12 & & & & & \\ & -12 & -48 & 120 & -48 & -12 & & & & & \\ & & & \ddots & \ddots & \ddots & \ddots & \ddots & & & \\ & & & & -12 & -48 & 120 & -48 & -12 & & \\ & & & & & -12 & -48 & 120 & -48 & -12 & \\ & & & & & & -12 & -48 & 120 & -44 & -16 \\ & & & & & & & -12 & -48 & 132 & -36 & -36 \\ & & & & & & & & -27 & -63 & 243 & -81 \\ & & & & & & & & -3 & -71 & -133 & 855 \end{bmatrix}.$$

The submatrix $T_{n-3}(f_4)$ appears in correspondence of the highlighted box and we have

$$f_4(\theta) = \frac{1}{96}(-12e^{2i\theta} - 48e^{i\theta} + 120 - 48e^{-i\theta} - 12e^{-2i\theta}) = \frac{5}{4} - \cos\theta - \frac{1}{4}\cos(2\theta),$$

as given by (3.123) for $p = 4$.

Before passing to the GLT analysis of the collocation matrices $A_{G,n}^{[p]}$, we prove the existence of an n-independent bound for the spectral norms of $n^{-2}K_n^{[p]}$, $n^{-1}H_n^{[p]}$, $M_n^{[p]}$. Actually, one could also prove that the components of $n^{-2}K_n^{[p]}$, $n^{-1}H_n^{[p]}$, $M_n^{[p]}$ do not depend on n as illustrated above for the matrix $n^{-2}K_n^{[p]}$ in the cases $p = 3, 4$. However, for our purposes it suffices to show that, for every $p \geq 2$, there exists a constant $C^{[p]}$ such that, for all n,

$$\|n^{-2}K_n^{[p]}\| \leq C^{[p]}, \qquad \|n^{-1}H_n^{[p]}\| \leq C^{[p]}, \qquad \|M_n^{[p]}\| \leq C^{[p]}. \qquad (3.129)$$

To prove (3.129), we note that $K_n^{[p]}$, $H_n^{[p]}$, $M_n^{[p]}$ are banded, with bandwidth bounded by $2p+1$. Indeed, if $|i - j| > p$, one can show that $(K_n^{[p]})_{ij} = (H_n^{[p]})_{ij} = (M_n^{[p]})_{ij} = 0$ by using (3.114), which implies that $\xi_{i+1,[p]}$ lies outside or on the border of supp$(N_{j+1,[p]})$, whose intersection with supp$(N_{i+1,[p]})$ consists of at most one of the knots t_k. Moreover, by (3.102)–(3.105), for all $i, j = 1, \ldots, n + p - 2$ we have

$$|(K_n^{[p]})_{ij}| = |N_{j+1,[p]}''(\xi_{i+1,[p]})| \leq 4p(p - 1)n^2,$$

$$|(H_n^{[p]})_{ij}| = |N_{j+1,[p]}'(\xi_{i+1,[p]})| \leq 2pn,$$

$$|(M_n^{[p]})_{ij}| = |N_{j+1,[p]}(\xi_{i+1,[p]})| \leq 1.$$

Hence, (3.129) follows from (3.1).

GLT Analysis of the B-Spline IgA Collocation Matrices Assuming that the geometry map G possesses some regularity properties, we show that, for any $p \geq 2$, the sequence of normalized IgA collocation matrices $\{n^{-2}A_{G,n}^{[p]}\}_n$ is a GLT sequence whose symbol describes both its singular value and spectral distribution.

Theorem 12 *Let Ω be a bounded open interval of $\mathbb{R}$, let $a \in C^1(\overline{\Omega})$ and $b, c \in C(\overline{\Omega})$. Let $p \geq 2$ and let $G : [0, 1] \to \overline{\Omega}$ be such that $G \in C^2([0, 1])$ and $G'(\hat{x}) \neq 0$ for all $\hat{x} \in [0, 1]$. Then*

$$\{n^{-2}A_{G,n}^{[p]}\}_n \sim_{\text{GLT}} f_{G,p} \qquad (3.130)$$

and

$$\{n^{-2}A_{G,n}^{[p]}\}_n \sim_{\sigma, \lambda} f_{G,p}, \qquad (3.131)$$

where

$$f_{G,p}(\hat{x}, \theta) = a_G(\hat{x})f_p(\theta) = \frac{a(G(\hat{x}))}{(G'(\hat{x}))^2}f_p(\theta) \qquad (3.132)$$

and $f_p(\theta)$ is defined in (3.123).

Proof The proof consists of the following steps. Throughout the proof, the letter C will denote a generic constant independent of n.

Step 1 We show that

$$\|n^{-2} K_{G,n}^{[p]}\| \leq C \tag{3.133}$$

and

$$\|n^{-2} Z_{G,n}^{[p]}\| \leq C/n. \tag{3.134}$$

To prove (3.133), it suffices to use the regularity of G and (3.129):

$$\|n^{-2} K_{G,n}^{[p]}\| = \|n^{-2} D_n^{[p]}(a_G) K_n^{[p]}\| \leq \|a_G\|_\infty C^{[p]} \leq \frac{C^{[p]} \|a\|_\infty}{\min_{\hat{x} \in [0,1]} |G'(\hat{x})|^2}.$$

The proof of (3.134) is similar. It suffices to use the fact that $G \in C^2([0,1])$ and (3.129):

$$\|n^{-2} Z_{G,n}^{[p]}\| = \|n^{-2} D_n^{[p]}(s_G) H_n^{[p]} + n^{-2} D_n^{[p]}(c_G) M_n^{[p]}\|$$

$$\leq n^{-1} C^{[p]} \left(\frac{\|a\|_\infty \|G''\|_\infty}{\min_{\hat{x} \in [0,1]} |G'(\hat{x})|^3} + \frac{\|a'\|_\infty + \|b\|_\infty}{\min_{\hat{x} \in [0,1]} |G'(\hat{x})|} \right) + n^{-2} C^{[p]} \|c\|_\infty.$$

Step 2 Define the symmetric matrix

$$\tilde{K}_{G,n}^{[p]} = S_{n+p-2}(a_G) \circ n^2 T_{n+p-2}(f_p), \tag{3.135}$$

where we recall that $S_m(v)$ is the mth arrow-shaped sampling matrix generated by v (see (3.6)), and consider the following decomposition of $n^{-2} A_{G,n}^{[p]}$:

$$n^{-2} A_{G,n}^{[p]} = n^{-2} \tilde{K}_{G,n}^{[p]} + \left(n^{-2} K_{G,n}^{[p]} - n^{-2} \tilde{K}_{G,n}^{[p]} \right) + n^{-2} Z_{G,n}^{[p]}. \tag{3.136}$$

We know from Theorem 2 that $\|n^{-2} \tilde{K}_{G,n}^{[p]}\| \leq C$ and $\{n^{-2} \tilde{K}_{G,n}^{[p]}\}_n \sim_{\text{GLT}} f_{G,p}(\hat{x}, \theta)$.

Step 3 We show that

$$\|n^{-2} K_{G,n}^{[p]} - n^{-2} \tilde{K}_{G,n}^{[p]}\|_1 = o(n). \tag{3.137}$$

Once this is done, the thesis is proved. Indeed, from (3.137) and (3.134) we obtain

$$\|(n^{-2} K_{G,n}^{[p]} - n^{-2} \tilde{K}_{G,n}^{[p]}) + n^{-2} Z_{G,n}^{[p]}\|_1 \leq \|n^{-2} K_{G,n}^{[p]} - n^{-2} \tilde{K}_{G,n}^{[p]}\|_1$$

$$+ \|n^{-2} Z_{G,n}^{[p]}\|_n = o(n),$$

hence $\{(n^{-2}K_{G,n}^{[p]} - n^{-2}\tilde{K}_{G,n}^{[p]}) + n^{-2}Z_{G,n}^{[p]}\}_n$ is zero-distributed by **Z 2**. Thus, the GLT relation (3.130) follows from the decomposition (3.136) and **GLT 3**–**GLT 4**, the singular value distribution in (3.131) follows from **GLT 1**, and the eigenvalue distribution in (3.131) follows from **GLT 2**.

To prove (3.137), we decompose the difference $n^{-2}K_{G,n}^{[p]} - n^{-2}\tilde{K}_{G,n}^{[p]}$ as follows:

$$n^{-2}K_{G,n}^{[p]} - n^{-2}\tilde{K}_{G,n}^{[p]} = n^{-2}D_n^{[p]}(a_G)\, K_n^{[p]} - S_{n+p-2}(a_G) \circ T_{n+p-2}(f_p)$$

$$= n^{-2}D_n^{[p]}(a_G)\, K_n^{[p]} - D_n^{[p]}(a_G)\, T_{n+p-2}(f_p) \tag{3.138}$$

$$+ D_n^{[p]}(a_G)\, T_{n+p-2}(f_p) - D_{n+p-2}(a_G)\, T_{n+p-2}(f_p) \tag{3.139}$$

$$+ D_{n+p-2}(a_G)\, T_{n+p-2}(f_p) - S_{n+p-2}(a_G) \circ T_{n+p-2}(f_p). \tag{3.140}$$

We consider separately the three matrices in (3.138)–(3.140) and we show that their trace-norms are $o(n)$.

- By (3.126), the rank of the matrix (3.138) is bounded by $4(p-1)$. By the regularity of G, the inequality (3.129) and **T 3**, the spectral norm of (3.138) is bounded by C. Thus, the trace-norm of (3.138) is $o(n)$ (actually, $O(1)$) by (3.3).
- By (3.116), the continuity of a_G and **T 3**, the spectral norm of the matrix (3.139) is bounded by $C\omega_{a_G}(n^{-1})$, so it tends to 0. Hence, the trace-norm of (3.139) is $o(n)$ by (3.3).
- By Theorem 2, the spectral norm of the matrix (3.140) is bounded by $C\omega_{a_G}(n^{-1})$, so it tends to 0. Hence, the trace-norm of (3.140) is $o(n)$ by (3.3).

In conclusion, $\|n^{-2}K_{G,n}^{[p]} - n^{-2}\tilde{K}_{G,n}^{[p]}\|_1 = o(n)$. □

Remark 8 (Formal Structure of the Symbol) We invite the reader to compare the symbol (3.132) with the transformed problem (3.89). It is clear that the higher-order operator $-a_G(\hat{x})\hat{u}''(\hat{x})$ has a discrete spectral counterpart $a_G(\hat{x})f_p(\theta)$ which looks formally the same (as in the FD and FE cases; see Remarks 1 and 6). To better appreciate the formal analogy, note that $f_p(\theta)$ is the trigonometric polynomial in the Fourier variable coming from the B-spline IgA collocation discretization of the second derivative $-\hat{u}''(\hat{x})$ on the parametric domain $[0, 1]$. Indeed, $f_p(\theta)$ is the symbol of the sequence of B-spline IgA collocation diffusion matrices $\{n^{-2}K_n^{[p]}\}_n$, which arises from the B-spline IgA collocation approximation of (3.82) in the case where $a(x) = 1, b(x) = c(x) = 0$ identically, $\Omega = (0, 1)$ and G is the identity map over $[0, 1]$; note that in this case (3.82) is the same as (3.89), $x = G(\hat{x}) = \hat{x}$ and $u = \hat{u}$.

Remark 9 (Nonnegativity and Order of the Zero at $\theta = 0$) Figure 3.4 shows the graph of $f_p(\theta)$ normalized by its maximum $M_{f_p} = \max_{\theta \in [-\pi, \pi]} f_p(\theta)$ for $p = 2, \ldots, 10$. Note that $f_2(\theta) = f_3(\theta) = 2 - 2\cos\theta$. We see from the figure (and it was proved in [22]) that $f_p(\theta)$ is nonnegative over $[-\pi, \pi]$ and has a unique zero of order 2 at $\theta = 0$ because

$$\lim_{\theta \to 0} \frac{f_p(\theta)}{\theta^2} = 1.$$

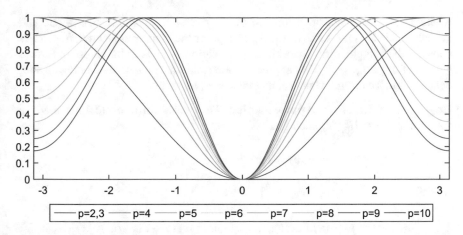

Fig. 3.4 Graph of f_p/M_{f_p} for $p = 2, \ldots, 10$

This reflects the fact that, as observed in Remark 8, $f_p(\theta)$ arises from the B-spline IgA collocation discretization of the second derivative $-\hat{u}''(\hat{x})$ on the parametric domain $[0, 1]$, which is a differential operator of order 2 (and it is nonnegative on $\{v \in C^2([0, 1]) : v(0) = v(1) = 0\}$); see also Remarks 2, 4 and 5.

Further properties of the functions $f_p(\theta)$, $g_p(\theta)$, $h_p(\theta)$ can be found in [22, Section 3]. In particular, it was proved therein that $f_p(\pi)/M_{f_p} \to 0$ exponentially as $p \to \infty$. Moreover, observing that $h_p(\theta)$ is defined by (3.125) for all degrees $p \geq 0$ (and we have $h_0(\theta) = h_1(\theta) = 1$ identically) provided that we use the standard convention that an empty sum like $\sum_{k=1}^0 \phi_{[1]}(1-k)\cos(k\theta)$ equals 0,[3] it was proved in [22] that, for all $p \geq 2$ and $\theta \in [-\pi, \pi]$,

$$f_p(\theta) = (2 - 2\cos\theta)h_{p-2}(\theta), \qquad \left(\frac{2}{\pi}\right)^{p-1} \leq h_{p-2}(\theta) \leq h_{p-2}(0) = 1. \qquad (3.141)$$

3.3.4.2 Galerkin B-Spline IgA Discretization of Convection-Diffusion-Reaction Equations

Consider the convection-diffusion-reaction problem

$$\begin{cases} -(a(x)u'(x))' + b(x)u'(x) + c(x)u(x) = f(x), & x \in \Omega, \\ u(x) = 0, & x \in \partial\Omega, \end{cases} \qquad (3.142)$$

---

[3]On the contrary, $f_p(\theta)$ and $g_p(\theta)$ are defined by (3.123) and (3.124) only for $p \geq 2$, because $\phi'_{[1]}(1)$ and $\phi''_{[1]}(1)$ do not exist.

where Ω is a bounded open interval of $\mathbb{R}$, $f \in L^2(\Omega)$ and $a, b, c \in L^\infty(\Omega)$. Problem (3.142) is the same as (3.81), except for the assumptions on a, b, c, f. We consider the isogeometric Galerkin approximation of (3.142) based on uniform B-splines of degree $p \geq 1$. This approximation technique is described below in some detail. For more on IgA Galerkin methods, see [18, 40].

Isogeometric Galerkin Approximation The weak form of (3.142) reads as follows: find $u \in H_0^1(\Omega)$ such that

$$\mathrm{a}(u, v) = \mathrm{f}(v), \qquad \forall v \in H_0^1(\Omega),$$

where

$$\mathrm{a}(u, v) = \int_\Omega \left(a(x)u'(x)v'(x) + b(x)u'(x)v(x) + c(x)u(x)v(x) \right) \mathrm{d}x,$$

$$\mathrm{f}(v) = \int_\Omega f(x)v(x)\mathrm{d}x.$$

In the standard Galerkin method, we look for an approximation $u_{\mathscr{W}}$ of u by choosing a finite dimensional vector space $\mathscr{W} \subset H_0^1(\Omega)$, the so-called approximation space, and by solving the following (Galerkin) problem: find $u_{\mathscr{W}} \in \mathscr{W}$ such that

$$\mathrm{a}(u_{\mathscr{W}}, v) = \mathrm{f}(v), \qquad \forall v \in \mathscr{W}.$$

If $\{\varphi_1, \ldots, \varphi_N\}$ is a basis of $\mathscr{W}$, then we can write $u_{\mathscr{W}} = \sum_{j=1}^N u_j \varphi_j$ for a unique vector $\mathbf{u} = (u_1, \ldots, u_N)^T$, and, by linearity, the computation of $u_{\mathscr{W}}$ (i.e., of $\mathbf{u}$) reduces to solving the linear system

$$A\mathbf{u} = \mathbf{f},$$

where $\mathbf{f} = \left[\mathrm{f}(\varphi_i) \right]_{i=1}^N$ and

$$A = \left[\mathrm{a}(\varphi_j, \varphi_i) \right]_{i,j=1}^N$$

$$= \left[\int_\Omega \left(a(x)\varphi_j'(x)\varphi_i'(x) + b(x)\varphi_j'(x)\varphi_i(x) + c(x)\varphi_j(x)\varphi_i(x) \right)\mathrm{d}x \right]_{i,j=1}^N$$

(3.143)

is the stiffness matrix.

Now, suppose that the physical domain Ω can be described by a global geometry function $G : [0, 1] \to \overline{\Omega}$, which is invertible and satisfies $G(\partial([0, 1])) = \partial\overline{\Omega}$. Let $\{\hat{\varphi}_1, \ldots, \hat{\varphi}_N\}$ be a set of basis functions defined on the parametric (or reference) domain $[0, 1]$ and vanishing on the boundary $\partial([0, 1])$. In the isogeometric Galerkin approach, we find an approximation $u_{\mathscr{W}}$ of u by using the standard Galerkin method, in which the approximation space is chosen as $\mathscr{W} = \mathrm{span}(\varphi_1, \ldots, \varphi_N)$, where

$$\varphi_i(x) = \hat{\varphi}_i(G^{-1}(x)) = \hat{\varphi}_i(\hat{x}), \qquad x = G(\hat{x}). \tag{3.144}$$

The resulting stiffness matrix A is given by (3.143), with the basis functions φ_i defined as in (3.144). Assuming that G and $\hat{\varphi}_i$, $i = 1, \ldots, N$, are sufficiently regular, we can apply standard differential calculus to obtain the following expression for A in terms of G and $\hat{\varphi}_i$, $i = 1, \ldots, N$:

$$A = \left[\int_{[0,1]} \left(a_G(\hat{x}) \hat{\varphi}_j'(\hat{x}) \hat{\varphi}_i'(\hat{x}) + \frac{b(G(\hat{x}))}{G'(\hat{x})} \hat{\varphi}_j'(\hat{x}) \hat{\varphi}_i(\hat{x}) \right. \right.$$
$$\left. \left. + c(G(\hat{x})) \hat{\varphi}_j(\hat{x}) \hat{\varphi}_i(\hat{x}) \right) |G'(\hat{x})| d\hat{x} \right]_{i,j=1}^{N}, \tag{3.145}$$

where $a_G(\hat{x})$ is the same as in (3.90),

$$a_G(\hat{x}) = \frac{a(G(\hat{x}))}{(G'(\hat{x}))^2}. \tag{3.146}$$

In the IgA framework, the functions $\hat{\varphi}_i$ are usually B-splines or NURBS. Here, the role of the $\hat{\varphi}_i$ will be played by B-splines over uniform knot sequences.

Galerkin B-Spline IgA Discretization Matrices As in the IgA collocation framework considered in Sect. 3.3.4.1, in the Galerkin B-spline IgA based on (uniform) B-splines, the functions $\hat{\varphi}_1, \ldots, \hat{\varphi}_N$ are chosen as the B-splines $N_{2,[p]}, \ldots, N_{n+p-1,[p]}$ defined in (3.96)–(3.98), i.e.,

$$\hat{\varphi}_i = N_{i+1,[p]}, \qquad i = 1, \ldots, n+p-2.$$

The boundary functions $N_{1,[p]}$ and $N_{n+p,[p]}$ are excluded because they do not vanish on $\partial([0, 1])$; see also Fig. 3.2. The stiffness matrix (3.145) resulting from this choice of the $\hat{\varphi}_i$ will be denoted by $A_{G,n}^{[p]}$:

$$A_{G,n}^{[p]} = \left[\int_{[0,1]} \left(a_G(\hat{x}) N_{j+1,[p]}'(\hat{x}) N_{i+1,[p]}'(\hat{x}) + \frac{b(G(\hat{x}))}{G'(\hat{x})} N_{j+1,[p]}'(\hat{x}) N_{i+1,[p]}(\hat{x}) \right. \right.$$
$$\left. \left. + c(G(\hat{x})) N_{j+1,[p]}(\hat{x}) N_{i+1,[p]}(\hat{x}) \right) |G'(\hat{x})| d\hat{x} \right]_{i,j=1}^{n+p-2}. \tag{3.147}$$

Note that $A_{G,n}^{[p]}$ can be decomposed as follows:

$$A_{G,n} = K_{G,n}^{[p]} + Z_{G,n}^{[p]}, \tag{3.148}$$

where

$$K_{G,n}^{[p]} = \left[\int_{[0,1]} a_G(\hat{x}) |G'(\hat{x})| N'_{j+1,[p]}(\hat{x}) N'_{i+1,[p]}(\hat{x}) \mathrm{d}\hat{x} \right]_{i,j=1}^{n+p-2} \qquad (3.149)$$

is the diffusion matrix, resulting from the discretization of the higher-order (diffusion) term in (3.142), and

$$Z_{G,n}^{[p]} = \left[\int_{[0,1]} \left(\frac{b(G(\hat{x}))}{G'(\hat{x})} N'_{j+1,[p]}(\hat{x}) N_{i+1,[p]}(\hat{x}) \right. \right.$$

$$\left. \left. + c(G(\hat{x})) N_{j+1,[p]}(\hat{x}) N_{i+1,[p]}(\hat{x}) \right) |G'(\hat{x})| \mathrm{d}\hat{x} \right]_{i,j=1}^{n+p-2} \qquad (3.150)$$

is the matrix resulting from the discretization of the lower-order (convection and reaction) terms. We will see that, as usual, the GLT analysis of a properly scaled version of the sequence $\{A_{G,n}^{[p]}\}_n$ reduces to the GLT analysis of its 'diffusion part' $\{K_{G,n}^{[p]}\}_n$, because $\|Z_{G,n}^{[p]}\|$ is negligible with respect to $\|K_{G,n}^{[p]}\|$ as $n \to \infty$.

Let

$$K_n^{[p]} = \left[\int_{[0,1]} N'_{j+1,[p]}(\hat{x}) N'_{i+1,[p]}(\hat{x}) \mathrm{d}\hat{x} \right]_{i,j=1}^{n+p-2}, \qquad (3.151)$$

$$H_n^{[p]} = \left[\int_{[0,1]} N'_{j+1,[p]}(\hat{x}) N_{i+1,[p]}(\hat{x}) \mathrm{d}\hat{x} \right]_{i,j=1}^{n+p-2}, \qquad (3.152)$$

$$M_n^{[p]} = \left[\int_{[0,1]} N_{j+1,[p]}(\hat{x}) N_{i+1,[p]}(\hat{x}) \mathrm{d}\hat{x} \right]_{i,j=1}^{n+p-2}. \qquad (3.153)$$

These matrices will play an important role in the GLT analysis of this section. In particular, it is necessary to understand their approximate structure. This is achieved by (approximately) construct them. We only construct their central submatrices

$$\left[(K_n^{[p]})_{ij} \right]_{i,j=p}^{n-1}, \qquad \left[(H_n^{[p]})_{ij} \right]_{i,j=p}^{n-1}, \qquad \left[(M_n^{[p]})_{ij} \right]_{i,j=p}^{n-1}, \qquad (3.154)$$

which are determined by the central basis functions in (3.111). For $i, j = p, \ldots, n-1$, noting that $[-i+p, n-i+p] \supseteq \mathrm{supp}(\phi_{[p]}) = [0, p+1]$ and using (3.109)–(3.110) and (3.112), we obtain

$$(K_n^{[p]})_{ij} = \int_{[0,1]} N'_{j+1,[p]}(\hat{x}) N'_{i+1,[p]}(\hat{x}) \mathrm{d}\hat{x}$$

$$= n^2 \int_{[0,1]} \phi'_{[p]}(n\hat{x} - j + p) \phi'_{[p]}(n\hat{x} - i + p) \mathrm{d}\hat{x}$$

$$= n \int_{[-i+p,\,n-i+p]} \phi'_{[p]}(t+i-j)\phi'_{[p]}(t)\mathrm{d}t$$

$$= n \int_{\mathbb{R}} \phi'_{[p]}(t+i-j)\phi'_{[p]}(t)\mathrm{d}t$$

$$= -n\,\phi''_{[2p+1]}(p+1+i-j) = -n\,\phi''_{[2p+1]}(p+1-i+j),$$

and similarly

$$(H_n^{[p]})_{ij} = \phi'_{[2p+1]}(p+1+i-j) = -\phi'_{[2p+1]}(p+1-i+j),$$

$$(M_n^{[p]})_{ij} = \frac{1}{n}\phi_{[2p+1]}(p+1+i-j) = \frac{1}{n}\phi_{[2p+1]}(p+1-i+j).$$

Since their entries depend only on the difference $i-j$, the submatrices (3.154) are Toeplitz matrices. More precisely,

$$\left[(K_n^{[p]})_{ij}\right]_{i,j=p}^{n-1} = n\left[-\phi''_{[2p+1]}(p+1-i+j)\right]_{i,j=p}^{n-1} = n\,T_{n-p}(f_p), \qquad (3.155)$$

$$\left[(H_n^{[p]})_{ij}\right]_{i,j=p}^{n-1} = \left[-\phi'_{[2p+1]}(p+1-i+j)\right]_{i,j=p}^{n-1} = \mathrm{i}\,T_{n-p}(g_p), \qquad (3.156)$$

$$\left[(M_n^{[p]})_{ij}\right]_{i,j=p}^{n-1} = \frac{1}{n}\left[\phi_{[2p+1]}(p+1-i+j)\right]_{i,j=p}^{n-1} = \frac{1}{n}T_{n-p}(h_p), \qquad (3.157)$$

where

$$f_p(\theta) = \sum_{k\in\mathbb{Z}} -\phi''_{[2p+1]}(p+1-k)\,\mathrm{e}^{\mathrm{i}k\theta}$$

$$= -\phi''_{[2p+1]}(p+1) - 2\sum_{k=1}^{p} \phi''_{[2p+1]}(p+1-k)\cos(k\theta), \qquad (3.158)$$

$$g_p(\theta) = -\mathrm{i}\sum_{k\in\mathbb{Z}} -\phi'_{[2p+1]}(p+1-k)\,\mathrm{e}^{\mathrm{i}k\theta}$$

$$= -2\sum_{k=1}^{p} \phi'_{[2p+1]}(p+1-k)\sin(k\theta), \qquad (3.159)$$

$$h_p(\theta) = \sum_{k\in\mathbb{Z}} \phi_{[2p+1]}(p+1-k)\,\mathrm{e}^{\mathrm{i}k\theta}$$

$$= \phi_{[2p+1]}(p+1) + 2\sum_{k=1}^{p} \phi_{[2p+1]}(p+1-k)\cos(k\theta); \qquad (3.160)$$

note that we used (3.108)–(3.109) to simplify the expressions of $f_p(\theta)$, $g_p(\theta)$, $h_p(\theta)$. It follows from (3.155) that $T_{n-p}(f_p)$ is the principal submatrix of both

$n^{-1} K_n^{[p]}$ and $T_{n+p-2}(f_p)$ corresponding to the set of indices $p, \ldots, n-1$. Similar results follow from (3.156)–(3.157), and so

$$n^{-1} K_n^{[p]} = T_{n+p-2}(f_p) + R_n^{[p]}, \qquad \text{rank}(R_n^{[p]}) \le 4(p-1), \qquad (3.161)$$

$$-i H_n^{[p]} = T_{n+p-2}(g_p) + S_n^{[p]}, \qquad \text{rank}(S_n^{[p]}) \le 4(p-1), \qquad (3.162)$$

$$n M_n^{[p]} = T_{n+p-2}(h_p) + V_n^{[p]}, \qquad \text{rank}(V_n^{[p]}) \le 4(p-1). \qquad (3.163)$$

Let us see two examples. In the case $p = 2$, the matrix $K_n^{[2]}$ is given by

$$K_n^{[2]} = \frac{n}{6}
\begin{bmatrix}
8 & -1 & -1 & & & & & & & \\
-1 & 6 & -2 & -1 & & & & & & \\
-1 & -2 & 6 & -2 & -1 & & & & & \\
 & -1 & -2 & 6 & -2 & -1 & & & & \\
 & & \ddots & \ddots & \ddots & \ddots & \ddots & & & \\
 & & & & -1 & -2 & 6 & -2 & -1 & \\
 & & & & & -1 & -2 & 6 & -2 & -1 \\
 & & & & & & -1 & -2 & 6 & -1 \\
 & & & & & & & -1 & -1 & 8
\end{bmatrix}.$$

The submatrix $T_{n-2}(f_2)$ appears in correspondence of the highlighted box and we have

$$f_2(\theta) = \frac{1}{6}(-e^{2i\theta} - 2e^{i\theta} + 6 - 2e^{-i\theta} - e^{-2i\theta}) = 1 - \frac{2}{3}\cos\theta - \frac{1}{3}\cos(2\theta),$$

as given by (3.158) for $p = 2$. In the case $p = 3$, the matrix $K_n^{[3]}$ is given by

$$K_n^{[3]} = \frac{n}{240}
\begin{bmatrix}
360 & 9 & -60 & -3 & & & & & & & & \\
9 & 162 & -8 & -47 & -2 & & & & & & & \\
-60 & -8 & 160 & -30 & -48 & -2 & & & & & & \\
-3 & -47 & -30 & 160 & -30 & -48 & -2 & & & & & \\
-2 & -48 & -30 & 160 & -30 & -48 & -2 & & & & & \\
 & -2 & -48 & -30 & 160 & -30 & -48 & -2 & & & & \\
 & & \ddots & \ddots & \ddots & \ddots & \ddots & \ddots & \ddots & & & \\
 & & & & -2 & -48 & -30 & 160 & -30 & -48 & -2 & \\
 & & & & & -2 & -48 & -30 & 160 & -30 & -48 & -2 \\
 & & & & & & -2 & -48 & -30 & 160 & -30 & -47 & -3 \\
 & & & & & & & -2 & -48 & -30 & 160 & -8 & -60 \\
 & & & & & & & & -2 & -47 & -8 & 162 & 9 \\
 & & & & & & & & & -3 & -60 & 9 & 360
\end{bmatrix}.$$

The submatrix $T_{n-3}(f_3)$ appears in correspondence of the highlighted box and we have

$$f_3(\theta) = \frac{1}{240}(-2e^{3i\theta} - 48e^{2i\theta} - 30e^{i\theta} + 160 - 30e^{-i\theta} - 48e^{-2i\theta} - 2e^{-3i\theta})$$

$$= \frac{2}{3} - \frac{1}{4}\cos\theta - \frac{2}{5}\cos(2\theta) - \frac{1}{60}\cos(3\theta),$$

as given by (3.158) for $p = 3$.

Remark 10 For every degree $q \geq 1$, the functions $f_q(\theta)$, $g_q(\theta)$, $h_q(\theta)$ defined by (3.158)–(3.160) for $p = q$ coincide with the functions $f_{2q+1}(\theta)$, $g_{2q+1}(\theta)$, $h_{2q+1}(\theta)$ defined by (3.123)–(3.125) for odd degree $p = 2q + 1$.

GLT Analysis of the Galerkin B-Spline IgA Discretization Matrices Assuming that the geometry map G is regular, i.e., $G \in C^1([0, 1])$ and $G'(\hat{x}) \neq 0$ for every $\hat{x} \in [0, 1]$, we show that, for any $p \geq 1$, $\{n^{-1}A^{[p]}_{G,n}\}_n$ is a GLT sequence whose symbol describes both its singular value and spectral distribution.

Theorem 13 *Let Ω be a bounded open interval of $\mathbb{R}$ and let $a, b, c \in L^\infty(\Omega)$. Let $p \geq 1$ and let $G : [0, 1] \to \overline{\Omega}$ be such that $G \in C^1([0, 1])$ and $G'(\hat{x}) \neq 0$ for all $\hat{x} \in [0, 1]$. Then*

$$\{n^{-1}A^{[p]}_{G,n}\}_n \sim_{\text{GLT}} f_{G,p} \tag{3.164}$$

and

$$\{n^{-1}A^{[p]}_{G,n}\}_n \sim_{\sigma, \lambda} f_{G,p}, \tag{3.165}$$

where

$$f_{G,p}(\hat{x}, \theta) = a_G(\hat{x})|G'(\hat{x})|f_p(\theta) = \frac{a(G(\hat{x}))}{|G'(\hat{x})|}f_p(\theta) \tag{3.166}$$

and $f_p(\theta)$ is defined in (3.158).

Proof We follow the same argument as in the proof of Theorem 10. Throughout the proof, the letter C will denote a generic constant independent of n.

Step 1 We show that

$$\|n^{-1}K^{[p]}_{G,n}\| \leq C \tag{3.167}$$

and

$$\|n^{-1}Z^{[p]}_{G,n}\| \leq C/n. \tag{3.168}$$

To prove (3.167), we note that $K_{G,n}^{[p]}$ is a banded matrix, with bandwidth at most equal to $2p + 1$. Indeed, due to the local support property (3.100), if $|i - j| > p$ then the supports of $N_{i+1,[p]}$ and $N_{j+1,[p]}$ intersect in at most one point, hence $(K_{G,n}^{[p]})_{ij} = 0$. Moreover, by (3.100) and (3.104), for all $i, j = 1, \ldots, n + p - 2$ we have

$$|(K_{G,n}^{[p]})_{ij}| = \left| \int_{[0,1]} a_G(\hat{x}) |G'(\hat{x})| N'_{j+1,[p]}(\hat{x}) \, N'_{i+1,[p]}(\hat{x}) \mathrm{d}\hat{x} \right|$$

$$= \left| \int_{[t_{i+1}, t_{i+p+2}]} \frac{a(G(\hat{x}))}{|G'(\hat{x})|} N'_{j+1,[p]}(\hat{x}) \, N'_{i+1,[p]}(\hat{x}) \mathrm{d}\hat{x} \right|$$

$$\leq \frac{4p^2 n^2 \|a\|_{L^\infty}}{\min_{\hat{x} \in [0,1]} |G'(\hat{x})|} \int_{[t_{i+1}, t_{i+p+2}]} \mathrm{d}\hat{x} \leq \frac{4p^2 (p + 1) n \|a\|_{L^\infty}}{\min_{\hat{x} \in [0,1]} |G'(\hat{x})|},$$

where in the last inequality we used the fact that $t_{k+p+1} - t_k \leq (p+1)/n$ for all $k = 1, \ldots, n + p$; see (3.94)–(3.95). In conclusion, the components of the banded matrix $n^{-1} K_{G,n}^{[p]}$ are bounded (in modulus) by a constant independent of n, and (3.167) follows from (3.1).

To prove (3.168), we follow the same argument as for the proof of (3.167). Due to the local support property (3.100), $Z_{G,n}^{[p]}$ is banded and, precisely, $(Z_{G,n}^{[p]})_{ij} = 0$ whenever $|i - j| > p$. Moreover, by (3.100) and (3.102)–(3.104), for all $i, j = 1, \ldots, n + p - 2$ we have

$$|(Z_{G,n}^{[p]})_{ij}| = \left| \int_{[t_{i+1}, t_{i+p+2}]} \left(\frac{b(G(\hat{x}))}{G'(\hat{x})} N'_{j+1,[p]}(\hat{x}) \, N_{i+1,[p]}(\hat{x}) \right. \right.$$

$$\left. \left. + c(G(\hat{x})) N_{j+1,[p]}(\hat{x}) \, N_{i+1,[p]}(\hat{x}) \right) |G'(\hat{x})| \mathrm{d}\hat{x} \right|$$

$$\leq 2p(p + 1) \|b\|_{L^\infty} + \frac{(p + 1) \|c\|_{L^\infty} \|G'\|_\infty}{n},$$

and (3.168) follows from (3.1).

Step 2 Consider the linear operator $K_n^{[p]}(\cdot) : L^1([0,1]) \to \mathbb{R}^{(n+p-2) \times (n+p-2)}$,

$$K_n^{[p]}(g) = \left[\int_{[0,1]} g(\hat{x}) N'_{j+1,[p]}(\hat{x}) N'_{i+1,[p]}(\hat{x}) \mathrm{d}\hat{x} \right]_{i,j=1}^{n+p-2}.$$

By (3.149), we have $K_{G,n}^{[p]} = K_n^{[p]}(a_G|G'|)$. The next three steps are devoted to show that

$$\{n^{-1} K_n^{[p]}(g)\}_n \sim_{\mathrm{GLT}} g(\hat{x}) f_p(\theta), \qquad \forall g \in L^1([0,1]). \tag{3.169}$$

Once this is done, the theorem is proved. Indeed, by applying (3.169) with $g = a_G|G'|$ we immediately obtain the relation $\{n^{-1}K_{G,n}^{[p]}\}_n \sim_{GLT} f_{G,p}(\hat{x}, \theta)$. Since $\{n^{-1}Z_{G,n}^{[p]}\}_n$ is zero-distributed by Step 1, (3.164) follows from the decomposition

$$n^{-1}A_{G,n}^{[p]} = n^{-1}K_{G,n}^{[p]} + n^{-1}Z_{G,n}^{[p]} \tag{3.170}$$

and from **GLT 3–GLT 4**; and the singular value distribution in (3.165) follows from **GLT 1**. If $b(x) = 0$ identically, then $n^{-1}A_{G,n}^{[p]}$ is symmetric and also the spectral distribution in (3.165) follows from **GLT 1**. If $b(x)$ is not identically 0, the spectral distribution in (3.165) follows from **GLT 2** applied to the decomposition (3.170), taking into account what we have seen in Step 1.

Step 3 We first prove (3.169) in the constant-coefficient case $g(\hat{x}) = 1$. In this case, we note that $K_n^{[p]}(1) = K_n^{[p]}$. Hence, the desired GLT relation $\{n^{-1}K_n^{[p]}(1)\}_n \sim_{GLT} f_p(\theta)$ follows from (3.161) and **GLT 3–GLT 4**, taking into account that $\{R_n^{[p]}\}_n$ is zero-distributed by **Z 1**.

Step 4 Now we prove (3.169) in the case where $g \in C([0,1])$. As in Step 4 of Sect. 3.3.3.1, the proof is based on the fact that the B-spline basis functions $N_{2,[p]}, \ldots, N_{n+p-1,[p]}$ are 'locally supported'. Indeed, the width of the support of the ith basis function $N_{i+1,[p]}$ is bounded by $(p+1)/n$ and goes to 0 as $n \to \infty$. Moreover, the support itself is located near the point $\frac{i}{n+p-2}$, because

$$\max_{\hat{x} \in [t_{i+1}, t_{i+p+2}]} \left| \hat{x} - \frac{i}{n+p-2} \right| \le \frac{C_p}{n} \tag{3.171}$$

for all $i = 2, \ldots, n + p - 1$ and for some constant C_p depending only on p. By (3.104) and (3.171), for all $i, j = 1, \ldots, n + p - 2$ we have

$$\left| (K_n^{[p]}(g))_{ij} - (D_{n+p-2}(g)K_n^{[p]}(1))_{ij} \right|$$

$$= \left| \int_{[0,1]} \left[g(\hat{x}) - g\left(\frac{i}{n+p-2} \right) \right] N'_{j+1,[p]}(\hat{x}) N'_{i+1,[p]}(\hat{x}) d\hat{x} \right|$$

$$\le 4p^2 n^2 \int_{[t_{i+1}, t_{i+p+2}]} \left| g(\hat{x}) - g\left(\frac{i}{n+p-2} \right) \right| d\hat{x} \le 4p^2(p+1)n\, \omega_g\left(\frac{C_p}{n} \right).$$

It follows that each entry of $Z_n = n^{-1}K_n^{[p]}(g) - n^{-1}D_{n+p-2}(g)K_n^{[p]}(1)$ is bounded in modulus by $C\omega_g(1/n)$. Moreover, Z_n is banded with bandwidth at most $2p + 1$, due to the local support property of the B-spline basis functions $N_{i,[p]}$. By (3.1) we conclude that $\|Z_n\| \le C\omega_g(1/n) \to 0$ as $n \to \infty$. Thus, $\{Z_n\}_n \sim_\sigma 0$, which implies (3.169) by Step 3 and **GLT 3–GLT 4**.

Step 5 Finally, we prove (3.169) in the general case where $g \in L^1([0,1])$. By the density of $C([0,1])$ in $L^1([0,1])$, there exist continuous functions $g_m \in C([0,1])$

such that $g_m \to g$ in $L^1([0, 1])$. By Step 4,

$$\{n^{-1} K_n^{[p]}(g_m)\}_n \sim_{\mathrm{GLT}} g_m(\hat{x}) f_p(\theta).$$

Moreover,

$$g_m(\hat{x}) f_p(\theta) \to g(\hat{x}) f_p(\theta) \quad \text{in measure}.$$

We show that

$$\{n^{-1} K_n^{[p]}(g_m)\}_n \xrightarrow{\text{a.c.s.}} \{n^{-1} K_n^{[p]}(g)\}_n.$$

Using (3.104) and (3.4), we obtain

$$\|K_n^{[p]}(g) - K_n^{[p]}(g_m)\|_1 \leq \sum_{i,j=1}^{n+p-2} \left| (K_n^{[p]}(g))_{ij} - (K_n^{[p]}(g_m))_{ij} \right|$$

$$= \sum_{i,j=1}^{n+p-2} \left| \int_{[0,1]} [g(\hat{x}) - g_m(\hat{x})] N'_{j+1,[p]}(\hat{x}) N'_{i+1,[p]}(\hat{x}) \mathrm{d}\hat{x} \right|$$

$$\leq \int_{[0,1]} |g(\hat{x}) - g_m(\hat{x})| \sum_{i,j=1}^{n+p-2} |N'_{j+1,[p]}(\hat{x})| \, |N'_{i+1,[p]}(\hat{x})| \mathrm{d}\hat{x}$$

$$\leq 4p^2 n^2 \int_{[0,1]} |g(\hat{x}) - g_m(\hat{x})| \mathrm{d}\hat{x}$$

and

$$\|n^{-1} K_n^{[p]}(g) - n^{-1} K_n^{[p]}(g_m)\|_1 \leq 4p^2 n \|g - g_m\|_{L^1}.$$

Thus, $\{n^{-1} K_n^{[p]}(g_m)\}_n \xrightarrow{\text{a.c.s.}} \{n^{-1} K_n^{[p]}(g)\}_n$ by **ACS 3**. The relation (3.169) now follows from **GLT 7**.   □

Remark 11 (Formal Structure of the Symbol) Problem (3.142) can be formally rewritten as in (3.82). If, for any $u : \overline{\Omega} \to \mathbb{R}$, we define $\hat{u} : [0, 1] \to \mathbb{R}$ as in (3.88), then u satisfies (3.82) if and only if $\hat{u}$ satisfies the corresponding transformed problem (3.89), in which the higher-order operator takes the form $-a_G(\hat{x})\hat{u}''(\hat{x})$. It is then clear that, similarly to the collocation case (see Remark 8), even in the Galerkin case the symbol $f_{G,p}(\hat{x}, \theta) = a_G(\hat{x})|G'(\hat{x})| f_p(\theta)$ preserves the formal structure of the higher-order operator associated with the transformed problem (3.89). However, in this Galerkin context we notice the appearance of the factor $|G'(\hat{x})|$, which is not present in the collocation setting; cf. (3.166) with (3.132).

Remark 12 (The Case $p = 1$) For $p = 1$, the symbol $f_p(\theta)$ in (3.158) is given by $f_1(\theta) = 2 - 2\cos\theta$. This should not come as a surprise, because the Galerkin B-spline IgA approximation with $p = 1$ (and G equal to the identity map over $[0, 1]$) coincides precisely with the linear FE approximation considered in Sect. 3.3.3.1; the only (unessential) difference is that the discretization step in Sect. 3.3.3.1 was chosen as $h = \frac{1}{n+1}$, while in this section we have $h = \frac{1}{n}$. In particular, the B-spline basis functions of degree 1, namely $N_{2,[1]}, \ldots, N_{n,[1]}$, are the hat-functions; cf. (3.98) (with $p = 1$) and (3.61).

Remark 13 The matrix $A_{G,n}^{[p]}$ in (3.147), which we decomposed as in (3.148), can also be decomposed as follows, according to the diffusion, convection and reaction terms:

$$A_{G,n}^{[p]} = K_{G,n}^{[p]} + H_{G,n}^{[p]} + M_{G,n}^{[p]},$$

where the diffusion, convection and reaction matrices are given by

$$K_{G,n}^{[p]} = \left[\int_{[0,1]} \frac{a(G(\hat{x}))}{|G'(\hat{x})|} N'_{j+1,[p]}(\hat{x}) N'_{i+1,[p]}(\hat{x}) d\hat{x} \right]_{i,j=1}^{n+p-2}, \tag{3.172}$$

$$H_{G,n}^{[p]} = \left[\int_{[0,1]} \frac{b(G(\hat{x}))|G'(\hat{x})|}{G'(\hat{x})} N'_{j+1,[p]}(\hat{x}) N_{i+1,[p]}(\hat{x}) d\hat{x} \right]_{i,j=1}^{n+p-2}, \tag{3.173}$$

$$M_{G,n}^{[p]} = \left[\int_{[0,1]} c(G(\hat{x}))|G'(\hat{x})| N_{j+1,[p]}(\hat{x}) N_{i+1,[p]}(\hat{x}) d\hat{x} \right]_{i,j=1}^{n+p-2}; \tag{3.174}$$

note that the diffusion matrix is the same as in (3.149). Let Ω be a bounded open interval of $\mathbb{R}$ and let $p \geq 1$. Then, the following results hold.

(a) Suppose

$$\frac{a(G(\hat{x}))}{|G'(\hat{x})|} \in L^1([0, 1]);$$

then $\{n^{-1} K_{G,n}^{[p]}\}_n \sim_{\text{GLT}} f_{G,p}$ and $\{n^{-1} K_{G,n}^{[p]}\}_n \sim_{\sigma,\lambda} f_{G,p}$, where

$$f_{G,p}(\hat{x}, \theta) = \frac{a(G(\hat{x}))}{|G'(\hat{x})|} f_p(\theta) \tag{3.175}$$

and $f_p(\theta)$ is defined in (3.158); note that $f_{G,p}(\hat{x}, \theta)$ is the same as in (3.166).

(b) Suppose

$$\frac{b(G(\hat{x}))|G'(\hat{x})|}{G'(\hat{x})} \in C([0, 1]);$$

then $\{-\mathrm{i}\, H_{G,n}^{[p]}\}_n \sim_{\mathrm{GLT}} g_{G,p}$ and $\{-\mathrm{i}\, H_{G,n}^{[p]}\}_n \sim_{\sigma,\lambda} g_{G,p}$, where

$$g_{G,p}(\hat{x}, \theta) = \frac{b(G(\hat{x}))|G'(\hat{x})|}{G'(\hat{x})} g_p(\theta) \qquad (3.176)$$

and $g_p(\theta)$ is defined in (3.159).

(c) Suppose

$$c(G(\hat{x}))|G'(\hat{x})| \in L^1([0, 1]);$$

then $\{n M_{G,n}^{[p]}\}_n \sim_{\mathrm{GLT}} h_{G,p}$ and $\{n M_{G,n}^{[p]}\}_n \sim_{\sigma,\lambda} h_{G,p}$, where

$$h_{G,p}(\hat{x}, \theta) = c(G(\hat{x}))|G'(\hat{x})| h_p(\theta) \qquad (3.177)$$

and $h_p(\theta)$ is defined in (3.160).

While the proof of (b) requires some work, the proofs of (a) and (c) can be done by following the same argument as in the proof of Theorem 13. The proofs of (a)–(c) can be found in [29, solution to Exercise 10.5].

3.3.4.3 Galerkin B-Spline IgA Discretization of Second-Order Eigenvalue Problems

Let $\mathbb{R}^+$ be the set of positive real numbers. Consider the following second-order eigenvalue problem: find eigenvalues $\lambda_j \in \mathbb{R}^+$ and eigenfunctions u_j, for $j = 1, 2, \ldots, \infty$, such that

$$\begin{cases} -(a(x)u_j'(x))' = \lambda_j c(x)u_j(x), & x \in \Omega, \\ u_j(x) = 0, & x \in \partial\Omega, \end{cases} \qquad (3.178)$$

where Ω is a bounded open interval of $\mathbb{R}$ and we assume $a, c \in L^1(\Omega)$ and $a, c > 0$ a.e. in Ω. It can be shown that the eigenvalues λ_j must necessarily be real and positive. This can be formally seen by multiplying (3.178) by $u_j(x)$ and integrating over Ω:

$$\lambda_j = \frac{-\int_\Omega (a(x)u_j'(x))' u_j(x)\,\mathrm{d}x}{\int_\Omega c(x)(u_j(x))^2\,\mathrm{d}x} = \frac{\int_\Omega a(x)(u_j'(x))^2\,\mathrm{d}x}{\int_\Omega c(x)(u_j(x))^2\,\mathrm{d}x} > 0.$$

Isogeometric Galerkin Approximation The weak form of (3.178) reads as follows: find eigenvalues $\lambda_j \in \mathbb{R}^+$ and eigenfunctions $u_j \in H_0^1(\Omega)$, for $j = 1, 2, \ldots, \infty$, such that

$$\mathrm{a}(u_j, w) = \lambda_j(c\, u_j, w), \qquad \forall w \in H_0^1(\Omega),$$

where

$$a(u_j, w) = \int_\Omega a(x) u'_j(x) w'(x) dx,$$

$$(c\, u_j, w) = \int_\Omega c(x) u_j(x) w(x) dx.$$

In the standard Galerkin method, we choose a finite dimensional vector space $\mathscr{W} \subset H_0^1(\Omega)$, the so-called approximation space, we let $N = \dim \mathscr{W}$ and we look for approximations of the eigenpairs (λ_j, u_j), $j = 1, 2, \ldots, \infty$, by solving the following discrete (Galerkin) problem: find $\lambda_{j,\mathscr{W}} \in \mathbb{R}^+$ and $u_{j,\mathscr{W}} \in \mathscr{W}$, for $j = 1, \ldots, N$, such that

$$a(u_{j,\mathscr{W}}, w) = \lambda_{j,\mathscr{W}}(c\, u_{j,\mathscr{W}}, w), \qquad \forall w \in \mathscr{W}. \qquad (3.179)$$

Assuming that both the exact and numerical eigenvalues are arranged in non-decreasing order, the pair $(\lambda_{j,\mathscr{W}}, u_{j,\mathscr{W}})$ is taken as an approximation to the pair (λ_j, u_j) for all $j = 1, 2, \ldots, N$. The numbers $\lambda_{j,\mathscr{W}}/\lambda_j - 1$, $j = 1, \ldots, N$, are referred to as the (relative) eigenvalue errors. If $\{\varphi_1, \ldots, \varphi_N\}$ is a basis of $\mathscr{W}$, we can identify each $w \in \mathscr{W}$ with its coefficient vector relative to this basis. With this identification in mind, solving the discrete problem (3.179) is equivalent to solving the generalized eigenvalue problem

$$K \mathbf{u}_{j,\mathscr{W}} = \lambda_{j,\mathscr{W}} M \mathbf{u}_{j,\mathscr{W}}, \qquad (3.180)$$

where $\mathbf{u}_{j,\mathscr{W}}$ is the coefficient vector of $u_{j,\mathscr{W}}$ with respect to $\{\varphi_1, \ldots, \varphi_N\}$ and

$$K = \left[\int_\Omega a(x) \varphi'_j(x) \varphi'_i(x) dx \right]_{i,j=1}^N, \qquad (3.181)$$

$$M = \left[\int_\Omega c(x) \varphi_j(x) \varphi_i(x) dx \right]_{i,j=1}^N. \qquad (3.182)$$

The matrices K and M are referred to as the stiffness and mass matrix, respectively. Due to our assumption that $a, c > 0$ a.e., both K and M are symmetric positive definite, regardless of the chosen basis functions $\varphi_1, \ldots, \varphi_N$. Moreover, it is clear from (3.180) that the numerical eigenvalues $\lambda_{j,\mathscr{W}}$, $j = 1, \ldots, N$, are just the eigenvalues of the matrix

$$L = M^{-1} K. \qquad (3.183)$$

In the isogeometric Galerkin method, we assume that the physical domain Ω is described by a global geometry function $G : [0, 1] \rightarrow \overline{\Omega}$, which is invertible and satisfies $G(\partial([0, 1])) = \partial\overline{\Omega}$. We fix a set of basis functions $\{\hat{\varphi}_1, \ldots, \hat{\varphi}_N\}$ defined on the reference (parametric) domain $[0, 1]$ and vanishing on the boundary $\partial([0, 1])$, and we find approximations to the exact eigenpairs (λ_j, u_j), $j = 1, 2, \ldots, \infty$, by using the standard Galerkin method described above, in which the approximation space is chosen as $\mathscr{W} = \text{span}(\varphi_1, \ldots, \varphi_N)$, where

$$\varphi_i(x) = \hat{\varphi}_i(G^{-1}(x)) = \hat{\varphi}_i(\hat{x}), \qquad x = G(\hat{x}). \tag{3.184}$$

The resulting stiffness and mass matrices K and M are given by (3.181)–(3.182), with the basis functions φ_i defined as in (3.184). If we assume that G and $\hat{\varphi}_i$, $i = 1, \ldots, N$, are sufficiently regular, we can apply standard differential calculus to obtain for K and M the following expressions:

$$K = \left[\int_{[0,1]} \frac{a(G(\hat{x}))}{|G'(\hat{x})|} \hat{\varphi}'_j(\hat{x}) \hat{\varphi}'_i(\hat{x}) \mathrm{d}\hat{x} \right]_{i,j=1}^{N}, \tag{3.185}$$

$$M = \left[\int_{[0,1]} c(G(\hat{x}))|G'(\hat{x})|\hat{\varphi}_j(\hat{x}) \hat{\varphi}_i(\hat{x}) \mathrm{d}\hat{x} \right]_{i,j=1}^{N}. \tag{3.186}$$

GLT Analysis of the Galerkin B-Spline IgA Discretization Matrices Following the approach of Sects. 3.3.4.1–3.3.4.2, we choose the basis functions $\hat{\varphi}_i$, $i = 1, \ldots, N$, as the B-splines $N_{i+1,[p]}$, $i = 1, \ldots, n + p - 2$. The resulting stiffness and mass matrices (3.185)–(3.186) are given by

$$K_{G,n}^{[p]} = \left[\int_{[0,1]} \frac{a(G(\hat{x}))}{|G'(\hat{x})|} N'_{j+1,[p]}(\hat{x}) N'_{i+1,[p]}(\hat{x}) \mathrm{d}\hat{x} \right]_{i,j=1}^{n+p-2},$$

$$M_{G,n}^{[p]} = \left[\int_{[0,1]} c(G(\hat{x}))|G'(\hat{x})| N_{j+1,[p]}(\hat{x}) N_{i+1,[p]}(\hat{x}) \mathrm{d}\hat{x} \right]_{i,j=1}^{n+p-2},$$

and it is immediately seen that they are the same as the diffusion and reaction matrices in (3.172) and (3.174). The numerical eigenvalues will be henceforth denoted by $\lambda_{j,n}$, $j = 1, \ldots, n + p - 2$; as noted above, they are simply the eigenvalues of the matrix

$$L_{G,n}^{[p]} = (M_{G,n}^{[p]})^{-1} K_{G,n}^{[p]}.$$

Theorem 14 *Let Ω be a bounded open interval of $\mathbb{R}$ and let $a, c \in L^1(\Omega)$ with $a, c > 0$ a.e. Let $p \geq 1$ and let $G : [0, 1] \to \overline{\Omega}$ be such that*

$$\frac{a(G(\hat{x}))}{|G'(\hat{x})|} \in L^1([0, 1]).$$

Then

$$\{n^{-2} L_{G,n}^{[p]}\}_n \sim_{\text{GLT}} e_{G,p}(\hat{x}, \theta) \tag{3.187}$$

and

$$\{n^{-2}L_{G,n}^{[p]}\}_n \sim_{\sigma,\lambda} e_{G,p}(\hat{x}, \theta), \tag{3.188}$$

where

$$e_{G,p}(\hat{x}, \theta) = (h_{G,p}(\theta))^{-1} f_{G,p}(\theta) = \frac{a(G(\hat{x}))}{c(G(\hat{x}))(G'(\hat{x}))^2} e_p(\theta), \tag{3.189}$$

$$e_p(\theta) = (h_p(\theta))^{-1} f_p(\theta), \tag{3.190}$$

and $f_p(\theta)$, $h_p(\theta)$, $f_{G,p}(\hat{x}, \theta)$, $h_{G,p}(\hat{x}, \theta)$ are given by (3.158), (3.160), (3.175), (3.177), respectively.

Proof We have $a(G(\hat{x}))/|G'(\hat{x})| \in L^1([0, 1])$ by assumption and $c(G(\hat{x}))|G'(\hat{x})| \in L^1([0, 1])$ because $c \in L^1(\Omega)$ by assumption and

$$\int_{[0,1]} c(G(\hat{x}))|G'(\hat{x})|\mathrm{d}\hat{x} = \int_{\Omega} c(x)\mathrm{d}x.$$

Hence, by Remark 13,

$$\{n^{-1}K_{G,n}^{[p]}\}_n \sim_{\text{GLT}} f_{G,p}, \qquad \{nM_{G,n}^{[p]}\}_n \sim_{\text{GLT}} h_{G,p},$$

and the relations (3.187)–(3.188) follow from Theorem 1, taking into account that $h_{G,p}(\hat{x}, \theta) \neq 0$ a.e. by our assumption that $c(x) > 0$ a.e. and by the positivity of $h_p(\theta)$; see (3.141) and Remark 10.   □

For $p = 1, 2, 3, 4$, Eq. (3.190) gives

$$e_1(\theta) = \frac{6(1 - \cos\theta)}{2 + \cos\theta},$$

$$e_2(\theta) = \frac{20(3 - 2\cos\theta - \cos(2\theta))}{33 + 26\cos\theta + \cos(2\theta)},$$

$$e_3(\theta) = \frac{42(40 - 15\cos\theta - 24\cos(2\theta) - \cos(3\theta))}{1208 + 1191\cos\theta + 120\cos(2\theta) + \cos(3\theta)},$$

$$e_4(\theta) = \frac{72(1225 - 154\cos\theta - 952\cos(2\theta) - 118\cos(3\theta) - \cos(4\theta))}{78095 + 88234\cos\theta + 14608\cos(2\theta) + 502\cos(3\theta) + \cos(4\theta)}.$$

These equations are the analogs of formulas (117), (130), (135), (140) obtained by engineers in [43]; see also formulas (32), (33) in [17], formulas (23), (56) in [41], and formulas (23), (24) in [46]. We may therefore conclude that (3.189) is a generalization of these formulas to any degree $p \geq 1$ and also to the variable-coefficient case with nontrivial geometry map, because it should be noted that the engineering papers [17, 41, 43, 46] only addressed the constant-coefficient case with identity geometry map (that is, the case in which $a(x) = c(x) = 1$ identically and G is the identity map on $\overline{\Omega} = [0, 1]$).

Remark 14 Contrary to the B-spline IgA discretizations investigated herein and in [43], the authors of [17, 41, 46] considered NURBS IgA discretizations. However, the same formulas are obtained in both cases. This can be easily explained in view of the results of [34], where it is shown that the symbols f_p, g_p, h_p in (3.158)–(3.160) are exactly the same in the B-spline and NURBS IgA frameworks. Note also that paper [34] addresses the general variable-coefficient case with nontrivial geometry map.

Remark 15 (A GLT Program for the Future) An extension of the results obtained in this section can be found in [35]. In particular, it is numerically shown in [35] and formally proved in [27] that

$$e_p(\theta) \to e_\infty(\theta) = \theta^2, \qquad \theta \in [0, \pi].$$

By Theorem 14, $e_p(\theta)$ is the symbol of the matrix-sequence $\{n^{-2}L_{G,n}^{[p]}\}_n$ obtained in the constant-coefficient case $a(x) = c(x) = 1$ with identity geometry map $G(x) = x : [0, 1] \to [0, 1]$. Note that in this case the eigenvalue problem (3.178) simplifies to

$$\begin{cases} -u_j''(x) = \lambda_j u_j(x), & x \in (0, 1), \\ u_j(0) = u_j(1) = 0, \end{cases}$$

and the corresponding eigenvalues are given by $\lambda_j = j^2\pi^2$ for $j = 1, 2, \ldots, \infty$. In particular, the first n eigenvalues are obtained as $n^{-2}\lambda_j = e_\infty(\frac{j\pi}{n})$ for $j = 1, 2, \ldots, n$. Based on these observations and on further insights arising from [35], we here outline a general research program for the future, which highlights once again the potential impact of the theory of GLT sequences.

1. Consider a *general* well-posed eigenvalue problem

$$\mathcal{K}u_j = \lambda_j \mathcal{M}u_j, \qquad j = 1, 2, \ldots, \infty. \tag{3.191}$$

   This could be, for example, a variable-coefficient eigenvalue problem involving low regularity coefficients, such as (3.178). It could also be a more complicated problem, defined over a multidimensional non-rectangular domain; in this case, the following step 4 will require the multidimensional version of the theory of GLT sequences [30].
2. Let $\mathscr{F}_n^{[\nu]}$ be a family of numerical methods for the discretization of (3.191). Here, n is the mesh fineness parameter, while ν is related to the approximation order of the method: the larger is ν, the higher is the precision of the method. In the IgA framework, ν could be for example the degree p.

3. For any n and v, let

$$K_n^{[v]} \mathbf{u}_j = \lambda_{j,n}^{[v]} M_n^{[v]} \mathbf{u}_j, \qquad j = 1, \ldots, N_n^{[v]}, \tag{3.192}$$

be the discrete eigenvalue problem arising from the approximation of the continuous eigenvalue problem (3.191) through the numerical method $\mathscr{F}_n^{[v]}$, and set

$$L_n^{[v]} = (M_n^{[v]})^{-1} K_n^{[v]}.$$

Note that the discrete eigenvalues $\lambda_{j,n}^{[v]}$ are just the eigenvalues of $L_n^{[v]}$.

4. Compute the symbol e_v of a properly normalized version of $\{L_n^{[v]}\}_n$. As illustrated throughout this work and especially in this section, this step can be efficiently performed through the theory of GLT sequences.

5. Suppose the following conditions are met.

   - For each v there is $\epsilon = \epsilon_v > 0$ such that $\lambda_{j,n}^{[v]} \to \lambda_j$ as $n \to \infty$ for $1 \le j \le \epsilon N_n^{[v]}$. In other words, we are assuming that, for every v, a nonzero portion of the discrete eigenvalues converge to the corresponding exact eigenvalues.
   - e_v converges to e_∞ as $v \to \infty$.

Then we expect that the limit symbol e_∞ will provide a description of the eigenvalue distribution of the continuous eigenvalue problem (3.191).

The program outlined in the above five steps could represent a general recipe for determining the distribution of the eigenvalues of a *general* eigenvalue problem of the form (3.191).

Acknowledgements Carlo Garoni is a Marie-Curie fellow of the Italian INdAM under grant agreement PCOFUND-GA-2012-600198. This work has been partially supported by INdAM-GNCS.

References

1. F. Auricchio, L. Beirão da Veiga, T.J.R. Hughes, A. Reali, G. Sangalli, Isogeometric collocation methods. Math. Models Methods Appl. Sci. **20**, 2075–2107 (2010)
2. F. Avram, On bilinear forms in Gaussian random variables and Toeplitz matrices. Probab. Theory Relat. Fields **79**, 37–45 (1988)
3. G. Barbarino, Equivalence between GLT sequences and measurable functions. Linear Algebra Appl. **529**, 397–412 (2017)
4. B. Beckermann, A.B.J. Kuijlaars, Superlinear convergence of conjugate gradients. SIAM J. Numer. Anal. **39**, 300–329 (2001)
5. B. Beckermann, S. Serra-Capizzano, On the asymptotic spectrum of finite element matrix sequences. SIAM J. Numer. Anal. **45**, 746–769 (2007)
6. M. Benzi, G.H. Golub, J. Liesen, Numerical solution of saddle point problems. Acta Numer. **14**, 1–137 (2005)

7. R. Bhatia, *Matrix Analysis* (Springer, New York, 1997)
8. D.A. Bini, M. Capovani, O. Menchi, *Metodi Numerici per l'Algebra Lineare* (Zanichelli, Bologna, 1988)
9. A. Böttcher, C. Garoni, S. Serra-Capizzano, Exploration of Toeplitz-like matrices with unbounded symbols is not a purely academic journey. Sb. Math. **208**, 1602–1627 (2017)
10. A. Böttcher, S.M. Grudsky, *Toeplitz Matrices, Asymptotic Linear Algebra, and Functional Analysis* (Birkhäuser Verlag, Basel, 2000)
11. A. Böttcher, S.M. Grudsky, *Spectral Properties of Banded Toeplitz Matrices* (SIAM, Philadelphia, 2005)
12. A. Böttcher, B. Silbermann, *Introduction to Large Truncated Toeplitz Matrices* (Springer, New York, 1999)
13. A. Böttcher, B. Silbermann, *Analysis of Toeplitz Operators*, 2nd edn. (Springer, Berlin, 2006)
14. H. Brezis, *Functional Analysis, Sobolev Spaces and Partial Differential Equations* (Springer, New York, 2011)
15. C.K. Chui, *An Introduction to Wavelets* (Academic Press, San Diego, 1992)
16. A. Cicone, C. Garoni, S. Serra-Capizzano, Spectral and convergence analysis of the Discrete ALIF method (submitted)
17. J.A. Cottrell, A. Reali, Y. Bazilevs, T.J.R. Hughes, Isogeometric analysis of structural vibrations. Comput. Methods Appl. Mech. Eng. **195**, 5257–5296 (2006)
18. J.A. Cottrell, T.J.R. Hughes, Y. Bazilevs, *Isogeometric Analysis: Toward Integration of CAD and FEA* (Wiley, Chichester, 2009)
19. C. De Boor, *A Practical Guide to Splines*, revised edn. (Springer, New York, 2001)
20. M. Donatelli, C. Garoni, C. Manni, S. Serra-Capizzano, H. Speleers, Robust and optimal multi-iterative techniques for IgA Galerkin linear systems. Comput. Methods Appl. Mech. Eng. **284**, 230–264 (2015)
21. M. Donatelli, C. Garoni, C. Manni, S. Serra-Capizzano, H. Speleers, Robust and optimal multi-iterative techniques for IgA collocation linear systems. Comput. Methods Appl. Mech. Eng. **284**, 1120–1146 (2015)
22. M. Donatelli, C. Garoni, C. Manni, S. Serra-Capizzano, H. Speleers, Spectral analysis and spectral symbol of matrices in isogeometric collocation methods. Math. Comput. **85**, 1639–1680 (2016)
23. M. Donatelli, M. Mazza, S. Serra-Capizzano, Spectral analysis and structure preserving preconditioners for fractional diffusion equations. J. Comput. Phys. **307**, 262–279 (2016)
24. M. Donatelli, C. Garoni, C. Manni, S. Serra-Capizzano, H. Speleers, Symbol-based multigrid methods for Galerkin B-spline isogeometric analysis. SIAM J. Numer. Anal. **55**, 31–62 (2017)
25. A. Dorostkar, M. Neytcheva, S. Serra-Capizzano, Spectral analysis of coupled PDEs and of their Schur complements via the notion of generalized locally Toeplitz sequences. Technical Report 2015-008 (2015), Department of Information Technology, Uppsala University. http://www.it.uu.se/research/publications/reports/2015-008/
26. A. Dorostkar, M. Neytcheva, S. Serra-Capizzano, Spectral analysis of coupled PDEs and of their Schur complements via generalized locally Toeplitz sequences in 2D. Comput. Methods Appl. Mech. Eng. **309**, 74–105 (2016)
27. S.E. Ekström, I. Furci, C. Garoni, C. Manni, S. Serra-Capizzano, H. Speleers, Are the eigenvalues of the B-spline isogeometric analysis approximation of $-\Delta u = \lambda u$ known in almost closed form? Numer. Linear Algebra Appl. https://doi.org/10.1002/nla.2198
28. C. Garoni, Spectral distribution of PDE discretization matrices from isogeometric analysis: the case of L^1 coefficients and non-regular geometry. J. Spectral Theory **8**, 297–313 (2018)
29. C. Garoni, S. Serra-Capizzano, *Generalized Locally Toeplitz Sequences: Theory and Applications, Volume I* (Springer, Cham, 2017)
30. C. Garoni, S. Serra-Capizzano, *Generalized Locally Toeplitz Sequences: Theory and Applications, Volume II* (Springer, to appear)
31. C. Garoni, C. Manni, F. Pelosi, S. Serra-Capizzano, H. Speleers, On the spectrum of stiffness matrices arising from isogeometric analysis. Numer. Math. **127**, 751–799 (2014)

32. C. Garoni, C. Manni, S. Serra-Capizzano, D. Sesana, H. Speleers, Spectral analysis and spectral symbol of matrices in isogeometric Galerkin methods. Math. Comput. **86**, 1343–1373 (2017)
33. C. Garoni, C. Manni, S. Serra-Capizzano, D. Sesana, H. Speleers, Lusin theorem, GLT sequences and matrix computations: an application to the spectral analysis of PDE discretization matrices. J. Math. Anal. Appl. **446**, 365–382 (2017)
34. C. Garoni, C. Manni, S. Serra-Capizzano, H. Speleers, NURBS versus B-splines in isogeometric discretization methods: a spectral analysis (submitted)
35. C. Garoni, H. Speleers, S.-E. Ekström, A. Reali, S. Serra-Capizzano, T.J.R. Hughes, Symbol-based analysis of finite element and isogeometric B-spline discretizations of eigenvalue problems: exposition and review. ICES Report 18-16, Institute for Computational Engineering and Sciences, The University of Texas at Austin (2018). https://www.ices.utexas.edu/media/reports/2018/1816.pdf
36. G.H. Golub, C.F. Van Loan, *Matrix Computations*, 4th edn. (The Johns Hopkins University Press, Baltimore, 2013)
37. U. Grenander, G. Szegő, *Toeplitz Forms and Their Applications*, 2nd edn. (AMS Chelsea Publishing, New York, 1984)
38. N.J. Higham, *Functions of Matrices: Theory and Computation* (SIAM, Philadelphia, 2008)
39. L. Hörmander, Pseudo-differential operators and non-elliptic boundary problems. Ann. Math. **83**, 129–209 (1966)
40. T.J.R. Hughes, J.A. Cottrell, Y. Bazilevs, Isogeometric analysis: CAD, finite elements, NURBS, exact geometry and mesh refinement. Comput. Methods Appl. Mech. Eng. **194**, 4135–4195 (2005)
41. T.J.R. Hughes, A. Reali, G. Sangalli, Duality and unified analysis of discrete approximations in structural dynamics and wave propagation: comparison of p-method finite elements with k-method NURBS. Comput. Methods Appl. Mech. Eng. **197**, 4104–4124 (2008)
42. T.J.R. Hughes, A. Reali, G. Sangalli, Efficient quadrature for NURBS-based isogeometric analysis. Comput. Methods Appl. Mech. Eng. **199**, 301–313 (2010)
43. T.J.R. Hughes, J.A. Evans, A. Reali, Finite element and NURBS approximations of eigenvalue, boundary-value, and initial-value problems. Comput. Methods Appl. Mech. Eng. **272**, 290–320 (2014)
44. S.V. Parter, On the distribution of the singular values of Toeplitz matrices. Linear Algebra Appl. **80**, 115–130 (1986)
45. A. Quarteroni, *Numerical Models for Differential Problems*, 2nd edn. (Springer Italia, Milan, 2014)
46. A. Reali, An isogeometric analysis approach for the study of structural vibrations. J. Earthquake Eng. **10**, 1–30 (2006)
47. A. Reali, T.J.R. Hughes, An introduction to isogeometric collocation methods, in *Isogeometric Methods for Numerical Simulation*, ed. by G. Beer, S. Bordas, Chap. 4 (CISM, Udine, 2015)
48. F. Roman, C. Manni, H. Speleers, Spectral analysis of matrices in Galerkin methods based on generalized B-splines with high smoothness. Numer. Math. **135**, 169–216 (2017)
49. D. Schillinger, J.A. Evans, A. Reali, M.A. Scott, T.J.R. Hughes, Isogeometric collocation: cost comparison with Galerkin methods and extension to adaptive hierarchical NURBS discretizations. Comput. Methods Appl. Mech. Eng. **267**, 170–232 (2013)
50. S. Serra-Capizzano, Generalized locally Toeplitz sequences: spectral analysis and applications to discretized partial differential equations. Linear Algebra Appl. **366**, 371–402 (2003)
51. S. Serra-Capizzano, The GLT class as a generalized Fourier analysis and applications. Linear Algebra Appl. **419**, 180–233 (2006)
52. S. Serra-Capizzano, C. Tablino-Possio, Analysis of preconditioning strategies for collocation linear systems. Linear Algebra Appl. **369**, 41–75 (2003)
53. G.D. Smith, *Numerical Solution of Partial Differential Equations: Finite Difference Methods*, 3rd edn. (Oxford University Press, New York, 1985)
54. H. Speleers, Inner products of box splines and their derivatives. BIT Numer. Math. **55**, 559–567 (2015)

55. P. Tilli, A note on the spectral distribution of Toeplitz matrices. Linear Multilinear Algebra **45**, 147–159 (1998)
56. P. Tilli, Locally Toeplitz sequences: spectral properties and applications. Linear Algebra Appl. **278**, 91–120 (1998)
57. P. Tilli, Some results on complex Toeplitz eigenvalues. J. Math. Anal. Appl. **239**, 390–401 (1999)
58. E.E. Tyrtyshnikov, A unifying approach to some old and new theorems on distribution and clustering. Linear Algebra Appl. **232**, 1–43 (1996)
59. E.E. Tyrtyshnikov, N.L. Zamarashkin, Spectra of multilevel Toeplitz matrices: advanced theory via simple matrix relationships. Linear Algebra Appl. **270**, 15–27 (1998)
60. N.L. Zamarashkin, E.E. Tyrtyshnikov, Distribution of eigenvalues and singular values of Toeplitz matrices under weakened conditions on the generating function. Sb. Math. **188**, 1191–1201 (1997)

Chapter 4
Isogeometric Analysis: Mathematical and Implementational Aspects, with Applications

Thomas J. R. Hughes, Giancarlo Sangalli, and Mattia Tani

Abstract Isogeometric analysis (IGA) is a recent and successful extension of classical finite element analysis. IGA adopts smooth splines, NURBS and generalizations to approximate problem unknowns, in order to simplify the interaction with computer aided geometric design (CAGD). The same functions are used to parametrize the geometry of interest. Important features emerge from the use of smooth approximations of the unknown fields. When a careful implementation is adopted, which exploit its full potential, IGA is a powerful and efficient high-order discretization method for the numerical solution of PDEs. We present an overview of the mathematical properties of IGA, discuss computationally efficient isogeometric algorithms, and present some significant applications.

4.1 Introduction

Isogeometric Analysis (IGA) was proposed in the seminal paper [70], with a fundamental motivation: to improve the interoperability between computer aided geometric design (CAGD) and the analysis, i.e., numerical simulation. In IGA, the functions that are used in CAGD geometry description (these are splines, NURBS, etcetera) are used also for the representation of the unknowns of Partial Differential Equations (PDEs) that model physical phenomena of interest.

In the last decade Isogeometric methods have been successfully used on a variety of engineering problems. The use of splines and NURBS in the representation of

T. J. R. Hughes
Institute for Computational Engineering and Sciences, The University of Texas at Austin,
Austin, TX, USA
e-mail: hughes@ices.utexas.edu

G. Sangalli (✉) · M. Tani
Dipartimento di Matematica, Università di Pavia and Istituto di Matematica Applicata
e Tecnologie Informatiche "E. Magenes" (CNR), Pavia, Italy
e-mail: Giancarlo.sangalli@unipv.it; mattia.tani@unipv.it

© Springer Nature Switzerland AG 2018
T. Lyche et al. (eds.), *Splines and PDEs: From Approximation Theory to Numerical
Linear Algebra*, Lecture Notes in Mathematics 2219,
https://doi.org/10.1007/978-3-319-94911-6_4

unknown fields yields important features, with respect to standard finite element methods. This is due to the spline smoothness which not only allows direct approximation of PDEs of order higher than two,[1] but also increases accuracy per degree of freedom (comparing to standard C^0 finite elements) and the spectral accuracy,[2] and moreover facilitates construction of spaces that can be used in schemes that preserve specific fundamental properties of the PDE of interest (for example, smooth divergence-free isogeometric spaces, see [37, 38, 57] and [58]). Spline smoothness is the key ingredient of isogeometric collocation methods, see [7] and [100].

The mathematics of isogeometric methods is based on the classical spline theory (see, e.g., [51, 102]), but also poses new challenges. The study of h-refinement of tensor-product isogeometric spaces is addressed in [15] and [22]. The study of k-refinement, that is, the use of splines and NURBS of high order and smoothness (C^{p-1} continuity for p-degree splines) is developed in [19, 40, 59, 115]. With a suitable code design, k-refinement boosts both accuracy and computational efficiency, see [97]. Stability of mixed isogeometric methods with a saddle-point form is the aim of the works [6, 21, 32, 33, 37, 56–58, 118].

Recent overview of IGA and its mathematical properties are [25] and [69].

We present in the following sections an introduction of the construction of isogeometric scalar and vector spaces, their approximation and spectral properties, of the computationally-efficient algorithms that can be used to construct and solve isogeometric linear systems, and finally report (from the literature) some significant isogeometric analyses of benchmark applications.

4.2 Splines and NURBS: Definition and Properties

This section contains a quick introduction to B-splines and NURBS and their use in geometric modeling and CAGD. Reference books on this topic are [25, 44, 51, 91, 93, 102].

---

[1]IGA of Cahn-Hilliard 4th-order model of phase separation is studied in [62, 63]; Kirchhoff-Love 4th-order model of plates and shells in [21, 26, 78, 79]; IGA of crack propagation is studied in [126], with 4th- and 6th-order gradient-enhanced theories of damage [127]; 4th-order phase-field fracture models are considered in [30] and [29], where higher-order convergence rates to sharp-interface limit solutions are numerically demonstrated.

[2]The effect of regularity on the spectral behavior of isogeometric discretizations has been studied in [49, 71, 73].

4.2.1 Univariate Splines

Given two positive integers p and n, we say that $\varXi := \{\xi_1, \ldots, \xi_{n+p+1}\}$ is a *p-open knot vector* if

$$\xi_1 = \ldots = \xi_{p+1} < \xi_{p+2} \leq \ldots \leq \xi_n < \xi_{n+1} = \ldots = \xi_{n+p+1},$$

where repeated knots are allowed, and $\xi_1 = 0$ and $\xi_{n+p+1} = 1$. The vector $Z = \{\zeta_1, \ldots, \zeta_N\}$ contains the *breakpoints*, that is the knots without repetitions, where m_j is the multiplicity of the breakpoint ζ_j, such that

$$\varXi = \{\underbrace{\zeta_1, \ldots, \zeta_1}_{m_1 \text{ times}}, \underbrace{\zeta_2, \ldots, \zeta_2}_{m_2 \text{ times}}, \ldots, \underbrace{\zeta_N, \ldots, \zeta_N}_{m_N \text{ times}}\}, \tag{4.1}$$

with $\sum_{i=1}^{N} m_i = n + p + 1$. We assume $m_j \leq p + 1$ for all internal knots. The points in Z form a mesh, and the local mesh size of the element $I_i = (\zeta_i, \zeta_{i+1})$ is denoted $h_i = \zeta_{i+1} - \zeta_i$, for $i = 1, \ldots, N - 1$.

B-spline functions of degree p are defined by the well-known Cox-DeBoor recursion:

$$\widehat{B}_{i,0}(\zeta) = \begin{cases} 1 \text{ if } \xi_i \leq \zeta < \xi_{i+1}, \\ 0 \text{ otherwise,} \end{cases} \tag{4.2}$$

and

$$\widehat{B}_{i,p}(\zeta) = \frac{\zeta - \xi_i}{\xi_{i+p} - \xi_i} \widehat{B}_{i,p-1}(\zeta) + \frac{\xi_{i+p+1} - \zeta}{\xi_{i+p+1} - \xi_{i+1}} \widehat{B}_{i+1,p-1}(\zeta), \tag{4.3}$$

where $0/0 = 0$. This gives a set of n B-splines that are non-negative, form a partition of unity, have local support, are linear independent.

The $\{\widehat{B}_{i,p}\}$ form a basis for the space of univariate *splines*, that is, piecewise polynomials of degree p with $k_j := p - m_j$ continuous derivatives at the points ζ_j, for $j = 1, \ldots, N$:

$$S^p(\varXi) = \text{span}\{\widehat{B}_{i,p}, \ i = 1, \ldots, n\}. \tag{4.4}$$

Remark 1 The notation S_r^p will be adopted to refer to the space $S^p(\varXi)$ when the multiplicity m_j of all internal knots is $p - r$. Then, S_r^p is a spline space with continuity C^r.

The maximum multiplicity allowed, $m_j = p + 1$, gives $k_j = -1$, which represents a discontinuity at ζ_j. The *regularity vector* $\mathbf{k} = \{k_1, \ldots, k_N\}$ will collect the regularity of the basis functions at the internal knots, with $k_1 = k_N = -1$ for the boundary knots. An example of B-splines is given in Fig. 4.1. B-splines are interpolatory at knots ζ_j if and only if the multiplicity $m_j \geq p$, that is where the B-spline is at most C^0.

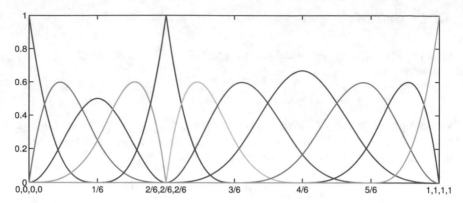

Fig. 4.1 Cubic B-splines and the corresponding knot vector with repetitions

Each B-spline $\widehat{B}_{i,p}$ depends only on $p+2$ knots, which are collected in the *local knot vector*

$$\Xi_{i,p} := \{\xi_i, \ldots, \xi_{i+p+1}\}.$$

When needed, we will stress this fact by adopting the notation

$$\widehat{B}_{i,p}(\zeta) = \widehat{B}[\Xi_{i,p}](\zeta). \tag{4.5}$$

The support of each basis function is exactly $\mathrm{supp}(\widehat{B}_{i,p}) = [\xi_i, \xi_{i+p+1}]$.

A *spline curve* in $\mathbb{R}^d$, $d = 2, 3$ is a curve parametrized by a linear combination of B-splines and control points as follows:

$$\mathbf{C}(\zeta) = \sum_{i=1}^{n} \mathbf{c}_i \, \widehat{B}_{i,p}(\zeta) \qquad \mathbf{c}_i \in \mathbb{R}^d, \tag{4.6}$$

where $\{\mathbf{c}_i\}_{i=1}^{n}$ are called control points. Given a spline curve $\mathbf{C}(\zeta)$, its control polygon $\mathbf{C}_P(\zeta)$ is the piecewise linear interpolant of the control points $\{\mathbf{c}_i\}_{i=1}^{n}$ (see Fig. 4.2).

In general, conic sections cannot be parametrized by polynomials but can be parametrized with rational polynomials, see [91, Sect. 1.4]. This has motivated the introduction of *Non-Uniform Rational B-Splines* (NURBS). In order to define NURBS, we set the *weight function* $W(\zeta) = \sum_{\ell=1}^{n} w_\ell \widehat{B}_{\ell,p}(\zeta)$ where the positive coefficients $w_\ell > 0$ for $\ell = 1, \ldots, n$ are called *weights*. We define the NURBS basis functions as

$$\widehat{N}_{i,p}(\zeta) = \frac{w_i \widehat{B}_{i,p}(\zeta)}{\sum_{\ell=1}^{n} w_\ell \widehat{B}_{\ell,p}(\zeta)} = \frac{w_i \widehat{B}_{i,p}(\zeta)}{W(\zeta)}, \qquad i = 1, \ldots, n, \tag{4.7}$$

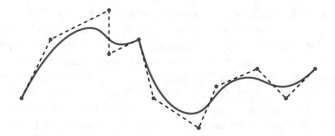

Fig. 4.2 Spline curve (solid line), control polygon (dashed line) and control points (red dots)

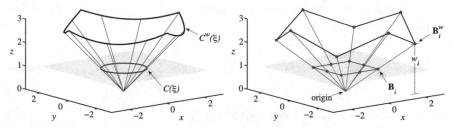

Fig. 4.3 On the left, the NURBS function $\xi \mapsto C(\xi)$ parametrizes the red circumference of a circle, given as the projection of the non-rational black spline curve, parametrized by the spline $\xi \mapsto C^w(\xi)$. The NURBS and spline control points are denoted $\mathbf{B}_i$ and $\mathbf{B}_i^w$, respectively, in the right plot

which are rational B-splines. NURBS (4.7) inherit the main properties of B-splines mentioned above, that is they are non-negative, form a partition of unity, and have local support. We denote the NURBS space they span by

$$N^p(\varXi, W) = \text{span}\{\widehat{N}_{i,p}, \ i = 1, \ldots, n\}. \tag{4.8}$$

Similarly to splines, a NURBS curve is defined by associating one control point to each basis function, in the form:

$$\mathbf{C}(\zeta) = \sum_{i=1}^{n} \mathbf{c}_i \, \widehat{N}_{i,p}(\zeta) \qquad \mathbf{c}_i \in \mathbb{R}^d. \tag{4.9}$$

Actually, the NURBS curve is a projection into $\mathbb{R}^d$ of a non-rational B-spline curve in the space $\mathbb{R}^{d+1}$, which is defined by

$$\mathbf{C}^w(\zeta) = \sum_{i=1}^{n} \mathbf{c}_i^w \, \widehat{B}_{i,p}(\zeta),$$

where $\mathbf{c}_i^w = [\mathbf{c}_i, w_i] \in \mathbb{R}^{d+1}$ (see Fig. 4.3).

For splines and NURBS curves, refinement is performed by *knot insertion* and *degree elevation*. In IGA, these two algorithms generate two kinds of refinement (see [70]): *h-refinement* which corresponds to mesh refinement and is obtained by insertion of new knots, and *p-refinement* which corresponds to degree elevation while maintaining interelement regularity, that is, by increasing the multiplicity of all knots. Furthermore, in IGA literature *k-refinement* denotes degree elevation, with increasing interelement regularity. This is not refinement in the sense of nested spaces, since the sequence of spaces generated by *k*-refinement has increasing global smoothness.

Having defined the spline space $S^p(\Xi)$, the next step is to introduce suitable projectors onto it. We focus on so called *quasi-interpolants* A common way to define them is by giving a dual basis, i.e.,

$$\Pi_{p,\Xi} : C^\infty([0,1]) \to S^p(\Xi), \qquad \Pi_{p,\Xi}(f) = \sum_{j=1}^{n} \lambda_{j,p}(f)\widehat{B}_{j,p}, \qquad (4.10)$$

where $\lambda_{j,p}$ are a set of dual functionals satisfying

$$\lambda_{j,p}(\widehat{B}_{k,p}) = \delta_{jk}, \qquad (4.11)$$

δ_{jk} being the Kronecker symbol. The quasi-interpolant $\Pi_{p,\Xi}$ preserves splines, that is

$$\Pi_{p,\Xi}(f) = f, \quad \forall f \in S^p(\Xi). \qquad (4.12)$$

From now on we assume local quasi-uniformity of the knot vector $\zeta_1, \zeta_2, \ldots, \zeta_N$, that is, there exists a constant $\theta \geq 1$ such that the mesh sizes $h_i = \zeta_{i+1} - \zeta_i$ satisfy the relation $\theta^{-1} \leq h_i/h_{i+1} \leq \theta$, for $i = 1, \ldots, N-2$. Among possible choices for the dual basis $\{\lambda_{j,p}\}$, $j = 1, \ldots, n$, a classical one is given in [102, Sect. 4.6], yielding to the following stability property (see [15, 25, 102]).

Proposition 1 *For any non-empty knot span $I_i = (\zeta_i, \zeta_{i+1})$,*

$$\|\Pi_{p,\Xi}(f)\|_{L^2(I_i)} \leq C\|f\|_{L^2(\widetilde{I}_i)}, \qquad (4.13)$$

where the constant C depends only upon the degree p, and $\widetilde{I}_i$ is the support extension, i.e., the interior of the union of the supports of basis functions whose support intersects I_i

4.2.2 *Multivariate Splines and NURBS*

Multivariate B-splines are defined from univariate B-splines by tensorization. Let d be the space dimensions (in practical cases, $d = 2, 3$). Assume $n_\ell \in \mathbb{N}$, the degree $p_\ell \in \mathbb{N}$ and the p_ℓ-open knot vector $\Xi_\ell = \{\xi_{\ell,1}, \dots, \xi_{\ell,n_\ell+p_\ell+1}\}$ are given, for $\ell = 1, \dots, d$. We define a polynomial degree vector $\mathbf{p} = (p_1, \dots, p_d)$ and $\Xi = \Xi_1 \times \dots \times \Xi_d$. The corresponding knot values without repetitions are given for each direction ℓ by $Z_\ell = \{\zeta_{\ell,1}, \dots, \zeta_{\ell,N_\ell}\}$. The knots Z_ℓ form a Cartesian grid in the *parametric domain* $\widehat{\Omega} = (0, 1)^d$, giving the *Bézier mesh*, which is denoted by $\widehat{\mathscr{M}}$:

$$\widehat{\mathscr{M}} = \{Q_{\mathbf{j}} = I_{1,j_1} \times \dots \times I_{d,j_d} \text{ such that } I_{\ell,j_\ell} = (\zeta_{\ell,j_\ell}, \zeta_{\ell,j_\ell+1}) \text{ for } 1 \le j_\ell \le n_{\text{EL},\ell} - 1\}. \tag{4.14}$$

For a generic Bézier element $Q_{\mathbf{j}} \in \widehat{\mathscr{M}}$, we also define its *support extension* $\widetilde{Q}_{\mathbf{j}} = \widetilde{I}_{1,j_1} \times \dots \times \widetilde{I}_{d,j_d}$, with $\widetilde{I}_{\ell,j_\ell}$ the univariate support extension as defined in Proposition 1. We make use of the set of multi-indices $\mathbf{I} = \{\mathbf{i} = (i_1, \dots, i_d) : 1 \le i_\ell \le n_\ell\}$, and for each multi-index $\mathbf{i} = (i_1, \dots, i_d)$, we define the local knot vector $\Xi_{\mathbf{i},\mathbf{p}} = \Xi_{i_1,p_1} \times \dots \times \Xi_{i_d,p_d}$. Then we introduce the set of multivariate B-splines

$$\left\{ \widehat{B}_{\mathbf{i},\mathbf{p}}(\zeta) = \widehat{B}[\Xi_{i_1,p_1}](\zeta_1) \dots \widehat{B}[\Xi_{i_d,p_d}](\zeta_d), \; \forall \mathbf{i} \in \mathbf{I} \right\}. \tag{4.15}$$

The spline space in the parametric domain $\widehat{\Omega}$ is then

$$S^{\mathbf{p}}(\Xi) = \text{span}\{\widehat{B}_{\mathbf{i},\mathbf{p}}(\zeta), \; \mathbf{i} \in \mathbf{I}\}, \tag{4.16}$$

which is the space of piecewise polynomials of degree $\mathbf{p}$ with the regularity across Bézier elements given by the knot multiplicities.

Multivariate NURBS are defined as rational tensor product B-splines. Given a set of *weights* $\{w_{\mathbf{i}}, \; \mathbf{i} \in \mathbf{I}\}$, and the weight function $W(\zeta) = \sum_{\mathbf{j} \in \mathbf{I}} w_{\mathbf{j}} \widehat{B}_{\mathbf{j},\mathbf{p}}(\zeta)$, we define the NURBS basis functions

$$\widehat{N}_{\mathbf{i},\mathbf{p}}(\zeta) = \frac{w_{\mathbf{i}} \widehat{B}_{\mathbf{i},\mathbf{p}}(\zeta)}{\sum_{\mathbf{j} \in \mathbf{I}} w_{\mathbf{j}} \widehat{B}_{\mathbf{j},\mathbf{p}}(\zeta)} = \frac{w_{\mathbf{i}} \widehat{B}_{\mathbf{i},\mathbf{p}}(\zeta)}{W(\zeta)}.$$

The NURBS space in the parametric domain $\widehat{\Omega}$ is then

$$N^{\mathbf{p}}(\Xi, W) = \text{span}\{\widehat{N}_{\mathbf{i},\mathbf{p}}(\zeta), \; \mathbf{i} \in \mathbf{I}\}.$$

As in the case of NURBS curves, the choice of the weights depends on the geometry to parametrize, and in IGA applications it is preserved by refinement.

Tensor-product B-splines and NURBS (4.7) are non-negative, form a partition of unity and have local support. As for curves, we define spline (NURBS, respectively) parametrizations of multivariate geometries in $\mathbb{R}^m$, $m = 2, 3$. A spline (NURBS, respectively) parametrization is then any linear combination of B-spline (NURBS, respectively) basis functions via control points $c_i \in \mathbb{R}^m$

$$\mathbf{F}(\zeta) = \sum_{i \in I} \mathbf{c_i} \widehat{B}_{\mathbf{i},\mathbf{p}}(\zeta), \qquad \text{with } \zeta \in \widehat{\Omega}. \tag{4.17}$$

Depending on the values of d and m, the map (4.17) can define a planar surface in $\mathbb{R}^2$ ($d = 2, m = 2$), a manifold in $\mathbb{R}^3$ ($d = 2, m = 3$), or a volume in $\mathbb{R}^3$ ($d = 3, m = 3$).

The definition of the control polygon is generalized for multivariate splines and NURBS to a control mesh, the mesh connecting the control points c_i. Since B-splines and NURBS are not interpolatory, the control mesh is not a mesh on the domain Ω. Instead, the image of the Bézier mesh in the parametric domain through $\mathbf{F}$ gives the physical Bézier mesh in Ω, simply denoted the Bézier mesh if there is no risk of confusion (see Fig. 4.4).

The interpolation and quasi-interpolation projectors can be also extended to the multi-dimensional case by a tensor product construction. Let, for $i = 1, \ldots, d$, the notation $\Pi^i_{p_i}$ denote the univariate projector $\Pi_{p,\Xi}$ onto the space $S^{p_i}(\Xi_i)$, then define

$$\Pi_{\mathbf{p}}(f) = (\Pi^1_{p_1} \otimes \ldots \otimes \Pi^d_{p_d})(f). \tag{4.18}$$

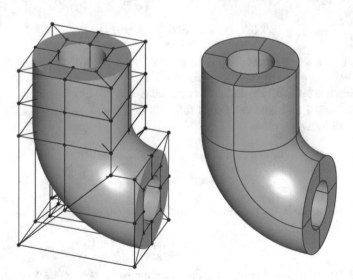

Fig. 4.4 The control mesh (left) and the physical Bézier mesh (right) for a pipe elbow is represented

Analogously, the multivariate quasi-interpolant is also defined from a dual basis (see
[51, Chapter XVII]). Indeed, we have

$$\Pi_{\mathbf{p},\Xi}(f) = \sum_{\mathbf{i}\in\mathbf{I}} \lambda_{\mathbf{i},\mathbf{p}}(f)\widehat{B}_{\mathbf{i},\mathbf{p}},$$

where each dual functional is defined from the univariate dual bases as $\lambda_{\mathbf{i},\mathbf{p}} = \lambda_{i_1,p_1} \otimes \ldots \otimes \lambda_{i_d,p_d}$.

4.2.3   Splines Spaces with Local Tensor-Product Structure

A well developed research area concerns extensions of splines spaces beyond
the tensor product structure, and allow local mesh refinement: for example T-
splines, Locally-refinable (LR) splines, and hierarchical splines. T-splines have been
proposed in [107] and have been adopted for isogeometric methods since [16].
They have been applied to shell problems [68], fluid-structure interaction problems
[17] and contact mechanics simulation [52]. The algorithm for local refinement
has evolved since its introduction in [108] (see, e.g., [104]), in order to overcome
some initial limitations (see, e.g.,[55]). Other possibilities are LR-splines [53] and
hierarchical splines [36, 128].

We summarize here the definition of a T-spline and its main properties, following
[25]. A T-mesh is a mesh that allows T-junctions. See Fig. 4.5 (left) for an example.
A T-spline set

$$\left\{ \widehat{B}_{\mathbf{A},\mathbf{p}}, \ \mathbf{A} \in \mathscr{A} \right\}, \tag{4.19}$$

is a generalization of the tensor-product set of multivariate splines (4.15). Indeed
the functions in (4.19) have the structure

$$\widehat{B}_{\mathbf{A},\mathbf{p}}(\boldsymbol{\zeta}) = \widehat{B}[\Xi_{\mathbf{A},1,p_1}](\zeta_1)\ldots\widehat{B}[\Xi_{\mathbf{A},d,p_d}](\zeta_d) \tag{4.20}$$

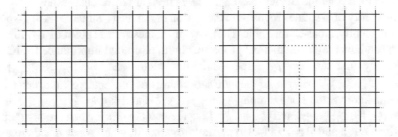

Fig. 4.5 A T-mesh with two T-junctions (on the left) and the same T-mesh with the T-junction
extensions (on the right). The degree for this example is cubic and the T-mesh is analysis suitable
since the extensions, one horizontal and the other vertical, do not intersect

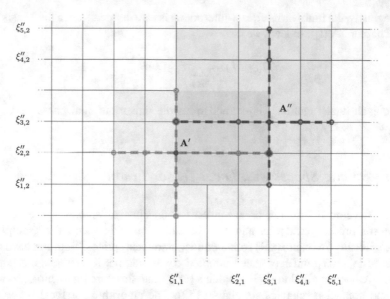

Fig. 4.6 Two bi-cubic T-spline anchors $\mathbf{A}'$ and $\mathbf{A}''$ and related local knot vectors. In particular, the local knot vectors for $\mathbf{A}''$ are $\varXi_{\mathbf{A}'',d,3} = \{\xi_{1,d}'', \xi_{2,d}'', \xi_{3,d}'', \xi_{4,d}'', \xi_{5,d}'', \}, d = 1, 2$. In this example, the two T-splines $\widehat{B}_{\mathbf{A}',\mathbf{p}}$ and $\widehat{B}_{\mathbf{A}'',\mathbf{p}}$ partially overlap (the overlapping holds in the horizontal direction)

where the set of indices, usually referred to as *anchors*, $\mathscr{A}$ and the associated local knot vectors $\varXi_{\mathbf{A},\ell,p_\ell}$, for all $\mathbf{A} \in \mathscr{A}$ are obtained from the T-mesh. If the polynomial degree is odd (in all directions) the anchors are associated with the vertices of the T-mesh, if the polynomial degree is even (in all directions) the anchors are associated with the elements. Different polynomial degrees in different directions are possible. The local knot vectors are obtained from the anchors by moving along one direction and recording the knots corresponding to the intersections with the mesh. See the example in Fig. 4.6.

On the parametric domain $\widehat{\varOmega}$ we can define a Bézier mesh $\widehat{\mathscr{M}}$ as the collection of the maximal open sets $Q \subset \widehat{\varOmega}$ where the T-splines of (4.19) are polynomials in Q. We remark that the Bézier mesh and the T-mesh are different meshes.

The theory of T-splines focuses on the notion of Analysis-Suitable (AS) T-splines or, equivalently, Dual-Compatible (DC) T-splines: these are a subset of T-splines for which fundamental mathematical properties hold, of crucial importance for IGA.

We say that the two p-degree local knot vectors $\varXi' = \{\xi_1', \dots \xi_{p+2}'\}$ and $\varXi'' = \{\xi_1'', \dots \xi_{p+2}''\}$ *overlap* if they are sub-vectors of consecutive knots taken from the same knot vector. For example $\{\xi_1, \xi_2, \xi_3, \xi_5, \xi_7\}$ and $\{\xi_3, \xi_5, \xi_7, \xi_8, \xi_9\}$ overlap, while $\{\xi_1, \xi_2, \xi_3, \xi_5, \xi_7\}$ and $\{\xi_3, \xi_4, \xi_5, \xi_6, \xi_8\}$ do not overlap. Then we say that two T-splines $\widehat{B}_{\mathbf{A}',\mathbf{p}}$ and $\widehat{B}_{\mathbf{A}'',\mathbf{p}}$ in (4.19) partially overlap if, when $\mathbf{A}' \neq \mathbf{A}''$, there exists a direction ℓ such that the local knot vectors $\varXi_{\mathbf{A}',\ell,p_\ell}$ and $\varXi_{\mathbf{A}'',\ell,p_\ell}$ are different and overlap. This is the case of Fig. 4.6. Finally, the set (4.19) is a Dual-Compatible

(DC) set of T-splines if each pair of T-splines in it partially overlaps. Its span

$$S^{\mathbf{p}}(\mathscr{A}) = \operatorname{span}\left\{\widehat{B}_{\mathbf{A},\mathbf{p}}, \ \mathbf{A} \in \mathscr{A}\right\}, \qquad (4.21)$$

is denoted a Dual-Compatible (DC) T-spline space. The definition of a DC set of T-splines simplifies in two dimension ([23]): when $d = 2$, a T-spline space is a DC set of splines if and only if each pair of T-splines in it have overlapping local knot vector in at least one direction.

A full tensor-product space (see Sect. 4.2.2) is a particular case of a DC spline space. In general, partial overlap is sufficient for the construction of a dual basis, as in the full tensor-product case. We only need, indeed, a univariate dual basis (e.g., the one in [102]), and denote by $\lambda[\varXi_{\mathbf{A},\ell,p_\ell}]$ the univariate functional as in (4.11), depending on the local knot vector $\varXi_{\mathbf{A},\ell,p_\ell}$ and dual to each univariate B-spline with overlapping knot vector.

Proposition 2 *Assume that* (4.19) *is a DC set, and consider an associated set of functionals*

$$\left\{\lambda_{\mathbf{A},\mathbf{p}}, \ \mathbf{A} \in \mathscr{A}\right\}, \qquad (4.22)$$

$$\lambda_{\mathbf{A},\mathbf{p}} = \lambda[\varXi_{\mathbf{A},1,p_1}] \otimes \ldots \otimes \lambda[\varXi_{\mathbf{A},d,p_d}]. \qquad (4.23)$$

Then (4.22) *is a dual basis for* (4.19).

Above, we assume that the local knot vectors in (4.23) are the same as in (4.19), (4.20). The proof of Proposition 2 can be found in [25]. The existence of dual functionals implies important properties for a DC set (4.19) and the related space $S^{\mathbf{p}}(\mathscr{A})$ in (4.21), as stated in the following theorem.

Theorem 1 *The T-splines in a DC set* (4.19) *are linearly independent. If the constant function belongs to* $S^{\mathbf{p}}(\mathscr{A})$, *they form a partition of unity. If the space of global polynomials of multi-degree* **p** *is contained in* $S^{\mathbf{p}}(\mathscr{A})$, *then the DC T-splines are locally linearly independent, that is, given* $Q \in \mathscr{M}$, *then the non-null T-splines restricted to the element* Q *are linearly independent.*

An important consequence of Proposition 2 is that we can build a projection operator $\varPi_{\mathbf{p}} : L^2(\widehat{\varOmega}) \to S^{\mathbf{p}}(\mathscr{A})$ by

$$\varPi_{\mathbf{p}}(f)(\zeta) = \sum_{\mathbf{A} \in \mathscr{A}} \lambda_{\mathbf{A},\mathbf{p}}(f)\widehat{B}_{\mathbf{A},\mathbf{p}}(\zeta) \qquad \forall f \in L^2(\widehat{\varOmega}), \ \forall \zeta \in \widehat{\varOmega}. \qquad (4.24)$$

This is the main tool (as mentioned in Sect. 4.4.1) to prove optimal approximation properties of T-splines.

In general, for DC T-splines, and in particular for tensor-product B-splines, we can define so called Greville abscissae. Each Greville abscissa

$$\boldsymbol{\gamma}_{\mathbf{A}} = \left(\gamma[\varXi_{\mathbf{A},1,p_1}], \ldots, \gamma[\varXi_{\mathbf{A},d,p_d}]\right)$$

is a point in the parametric domain $\widehat{\Omega}$ and its d-component $\gamma[\varXi_{\mathbf{A},\ell,p_\ell}]$ is the average of the p_ℓ internal knots of $\varXi_{\mathbf{A},\ell,p_\ell}$. They are the coefficients of the identity function in the T-spline expansion. Indeed, assuming that linear polynomials belong to the space $S^{\mathbf{p}}(\mathscr{A})$, we have that

$$\zeta_\ell = \sum_{\mathbf{A}\in\mathscr{A}} \gamma[\varXi_{\mathbf{A},\ell,p_\ell}]\widehat{B}_{\mathbf{A},\mathbf{p}}(\zeta), \quad \forall \zeta \in \widehat{\Omega}, \quad 1 \leq \ell \leq d. \tag{4.25}$$

Greville abscissae are used as interpolation points (see [51]) and therefore for collocation based IGA [7, 8, 100].

A useful result, proved in [20, 23], is that a T-spline set is DC if and only if (under minor technical assumptions) it comes from a T-mesh that is Analysis-Suitable. The latter is a topological condition for the T-mesh [16] and it refers to dimension $d = 2$. A horizontal T-junction *extension* is a horizontal line that extend the T-mesh from a T-junctions of kind $\vdash$ and $\dashv$ in the direction of the missing edge for a length of $\lceil p_1/2 \rceil$ elements, and in the opposite direction for $\lfloor p_1/2 \rfloor$ elements; analogously a vertical T-junction *extension* is a vertical line that extend the T-mesh from a T-junctions of kind $\perp$ or $\top$ in the direction of the missing edge for a length of $\lceil p_2/2 \rceil$ elements, and in the opposite direction for $\lfloor p_2/2 \rfloor$ elements, see Fig. 4.6 (right). Then, a T-mesh is *Analysis-Suitable* (AS) if horizontal T-junction extensions do not intersect vertical T-junction extensions.

4.2.4 Beyond Tensor-Product Structure

Multivariate unstructured spline spaces are spanned by basis functions that are not, in general, tensor products. Non-tensor-product basis functions appear around so-called *extraordinary points*. Subdivision schemes, but also multipatch or T-splines spaces in the most general setting, are unstructured spaces. The construction and mathematical study of these spaces is important especially for IGA and is one of the most important recent research activities, see [35, 98]. We will further address this topic in Sects. 4.3.3 and 4.4.3.

4.3 Isogeometric Spaces: Definition

In this section, following [25], we give the definition of isogeometric spaces. We consider a single patch domain, i.e., the physical domain Ω is the image of the unit square, or the unit cube (the parametric domain $\widehat{\Omega}$) by a single NURBS parametrization. Then, for a given degree vector $\mathbf{p}^0$, knot vectors $\varXi^0$ and a weight function $W \in S_{\mathbf{p}^0}(\varXi^0)$, a map $\mathbf{F} \in (N^{\mathbf{p}^0}(\varXi^0, W))^d$ is given such that $\Omega = \mathbf{F}(\widehat{\Omega})$, as in Fig. 4.7.

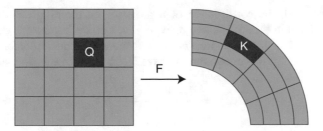

Fig. 4.7 Mesh $\widehat{\mathcal{M}}$ in the parametric domain, and its image $\mathcal{M}$ in the physical domain

After having introduced the parametric Bézier mesh $\widehat{\mathcal{M}}$ in (4.14), as the mesh associated to the knot vectors $\boldsymbol{\Xi}$, we now define the physical Bézier mesh (or simply Bézier mesh) as the image of the elements in $\widehat{\mathcal{M}}$ through $\mathbf{F}$:

$$\mathcal{M} := \{K \subset \Omega : K = \mathbf{F}(Q),\, Q \in \widehat{\mathcal{M}}\}, \tag{4.26}$$

see Fig. 4.7. The meshes for the coarsest knot vector $\boldsymbol{\Xi}^0$ will be denoted by $\widehat{\mathcal{M}_0}$ and $\mathcal{M}_0$. For any element $K = \mathbf{F}(Q) \in \mathcal{M}$, we define its support extension as $\widetilde{K} = \mathbf{F}(\widetilde{Q})$, with $\widetilde{Q}$ the support extension of Q. We denote the element size of any element $Q \in \widehat{\mathcal{M}}$ by $h_Q = \mathrm{diam}(Q)$, and the global mesh size by $h = \max\{h_Q : Q \in \widehat{\mathcal{M}}\}$. Analogously, we define the element sizes $h_K = \mathrm{diam}(K)$ and $h_{\widetilde{K}} = \mathrm{diam}(\widetilde{K})$.

For the sake of simplicity, we assume that the parametrization $\mathbf{F}$ is regular, that is, the inverse parametrization $\mathbf{F}^{-1}$ is well defined, and piecewise differentiable of any order with bounded derivatives. Assuming $\mathbf{F}$ is regular ensures that $h_Q \simeq h_K$. The case a of singular parametrization, that is, non-regular parametrization, will be discussed in Sect. 4.4.4.

4.3.1 Isoparametric Spaces

Isogeometric spaces are constructed as push-forward through $\mathbf{F}$ of (refined) splines or NURBS spaces. In detail, let $\widehat{V}_h = N^{\mathbf{p}}(\boldsymbol{\Xi}, W)$ be a refinement of $N^{\mathbf{p}^0}(\boldsymbol{\Xi}^0, W)$, we define the scalar isogeometric space as:

$$V_h = \{f \circ \mathbf{F}^{-1} : f \in \widehat{V}_h\}. \tag{4.27}$$

Analogously,

$$V_h = \mathrm{span}\{N_{\mathbf{i},\mathbf{p}}(\mathbf{x}) := \widehat{N}_{\mathbf{i},\mathbf{p}} \circ \mathbf{F}^{-1}(\mathbf{x}),\, \mathbf{i} \in \mathbf{I}\}, \tag{4.28}$$

that is, the functions $N_{\mathbf{i},\mathbf{p}}$ form a basis of the space V_h. Isogeometric spaces with boundary conditions are defined straightforwardly.

Following [15], the construction of a projector on the NURBS isogeometric space V_h (defined in (4.27)) is based on a pull-back on the parametric domain, on a decomposition of the function into a numerator and weight denominator, and finally a spline projection of the numerator. We have $\Pi_{V_h} : \mathscr{V}(\Omega) \to V_h$ defined as

$$\Pi_{V_h} f := \frac{\Pi_{\mathbf{p}}(W(f \circ \mathbf{F}))}{W} \circ \mathbf{F}^{-1}, \tag{4.29}$$

where $\Pi_{\mathbf{p}}$ is the spline projector (4.18) and $\mathscr{V}(\Omega)$ is a suitable function space. The approximation properties of Π_{V_h} will be discussed in Sect. 4.4.

The isogeometric vector space, as introduced in [70], is just $(V_h)^d$, that is a space of vector-valued functions whose components are in V_h. In parametric coordinates a spline isogeometric vector field of this kind reads

$$\mathbf{u}(\boldsymbol{\zeta}) = \sum_{i \in I} \mathbf{u_i} \widehat{B}_{\mathbf{i},\mathbf{p}}(\boldsymbol{\zeta}), \qquad \text{with } \boldsymbol{\zeta} \in \widehat{\Omega}, \tag{4.30}$$

where $\mathbf{u_i}$ are the degrees-of-freedom, also referred as *control variables* since they play the role of the control points of the geometry parametrization (4.17). This is an isoparametric construction.

4.3.2 De Rham Compatible Spaces

The following diagram

$$\mathbb{R} \longrightarrow H^1(\Omega) \xrightarrow{\ \mathbf{grad}\ } \mathbf{H}(\mathbf{curl};\Omega) \xrightarrow{\ \mathbf{curl}\ } \mathbf{H}(\mathrm{div};\Omega) \xrightarrow{\ \mathrm{div}\ } L^2(\Omega) \longrightarrow 0 \tag{4.31}$$

is the De Rham cochain complex. The Sobolev spaces involved are the two standard scalar-valued, $H^1(\Omega)$ and $L^2(\Omega)$, and the two vector valued

$$\mathbf{H}(\mathbf{curl};\ \Omega) = \{\mathbf{u} \in L^2(\Omega)^3\ :\ \mathbf{curl}\,\mathbf{u} \in L^2(\Omega)^3\}$$

$$\mathbf{H}(\mathrm{div};\ \Omega) = \{\mathbf{u} \in L^2(\Omega)^3\ :\ \mathrm{div}\ \mathbf{u} \in L^2(\Omega)\}.$$

Furthermore, as in general for complexes, the image of a differential operator in (4.31) is subset of the kernel of the next: for example, constants have null **grad**, gradients are **curl**-free fields, and so on. De Rham cochain complexes are related to the well-posedness of PDEs of key importance, for example in electromagnetic or fluid applications. This is why it is important to discretize (4.31) while preserving its structure. This is a well developed area of research for classical finite elements, called Finite Element Exterior Calculus (see the reviews [4, 5]) and likewise a successful development of IGA.

For the sake of simplicity, again, we restrict to a single patch domain and we do not include boundary conditions in the spaces. The dimension here is $d = 3$. The construction of isogeometric De Rham compatible spaces involves two stages.

The first stage is the definition of spaces on the parametric domain $\widehat{\Omega}$. These are tensor-product spline spaces, as (4.16), with a specific choice for the degree and regularity in each direction. For that, we use the expanded notation $S^{p_1,p_2,p_3}(\Xi_1, \Xi_2, \Xi_3)$ for $S^{\mathbf{p}}(\Xi)$. Given degrees p_1, p_2, p_3 and knot vectors Ξ_1, Ξ_2, Ξ_3 we then define on $\widehat{\Omega}$ the spaces:

$$
\begin{aligned}
\widehat{X}_h^0 &= S^{p_1,p_2,p_3}(\Xi_1, \Xi_2, \Xi_3), \\
\widehat{X}_h^1 &= S^{p_1-1,p_2,p_3}(\Xi_1', \Xi_2, \Xi_3) \times S^{p_1,p_2-1,p_3}(\Xi_1, \Xi_2', \Xi_3) \\
&\quad \times S^{p_1,p_2,p_3-1}(\Xi_1, \Xi_2, \Xi_3'), \\
\widehat{X}_h^2 &= S^{p_1,p_2-1,p_3-1}(\Xi_1, \Xi_2', \Xi_3') \times S^{p_1-1,p_2,p_3-1}(\Xi_1', \Xi_2, \Xi_3') \\
&\quad \times S^{p_1-1,p_2-1,p_3}(\Xi_1', \Xi_2', \Xi_3), \\
\widehat{X}_h^3 &= S^{p_1-1,p_2-1,p_3-1}(\Xi_1', \Xi_2', \Xi_3'),
\end{aligned}
\tag{4.32}
$$

where, given $\Xi_\ell = \{\xi_{\ell,1}, \ldots, \xi_{\ell,n_\ell+p_\ell+1}\}$, Ξ_ℓ' is defined as the knot vector $\{\xi_{\ell,2}, \ldots, \xi_{\ell,n_\ell+p_\ell}\}$, and we assume the knot multiplicities $1 \leq m_{\ell,i} \leq p_\ell$, for $i = 2, \ldots, N_\ell - 1$ and $\ell = 1, 2, 3$. With this choice, the functions in $\widehat{X}_h^0$ are at least continuous. Then, $\widehat{\mathbf{grad}}\,(\widehat{X}_h^0) \subset \widehat{X}_h^1$, and analogously, from the definition of the curl and the divergence operators we get $\widehat{\mathbf{curl}}(\widehat{X}_h^1) \subset \widehat{X}_h^2$, and $\widehat{\mathrm{div}}\,(\widehat{X}_h^2) \subset \widehat{X}_h^3$. This follows easily from the action of the derivative operator on tensor-product splines, for example:

$$
\frac{\partial}{\partial \zeta_1} : S^{p_1,p_2,p_3}(\Xi_1, \Xi_2, \Xi_3) \to S^{p_1-1,p_2,p_3}(\Xi_1', \Xi_2, \Xi_3)
$$

It is also proved in [38] that the kernel of each operator is exactly the image of the preceding one. In other words, these spaces form an exact sequence:

$$
\mathbb{R} \longrightarrow \widehat{X}_h^0 \xrightarrow{\widehat{\mathbf{grad}}} \widehat{X}_h^1 \xrightarrow{\widehat{\mathbf{curl}}} \widehat{X}_h^2 \xrightarrow{\widehat{\mathrm{div}}} \widehat{X}_h^3 \longrightarrow 0
\tag{4.33}
$$

This is consistent with (4.31).

The second stage is the push forward of the isogeometric De Rham compatible spaces from the parametric domain $\widehat{\Omega}$ onto Ω. The classical isoparametric transformation on all spaces does not preserve the structure of the De Rham cochain complex. We need to use the transformations:

$$
\begin{aligned}
\iota^0(f) &:= f \circ \mathbf{F}, \quad f \in H^1(\Omega), \\
\iota^1(\mathbf{f}) &:= (D\mathbf{F})^T (\mathbf{f} \circ \mathbf{F}), \quad \mathbf{f} \in \mathbf{H}(\mathbf{curl}; \Omega), \\
\iota^2(\mathbf{f}) &:= \det(D\mathbf{F})(D\mathbf{F})^{-1}(\mathbf{f} \circ \mathbf{F}), \quad \mathbf{f} \in \mathbf{H}(\mathrm{div}; \Omega), \\
\iota^3(f) &:= \det(D\mathbf{F})(f \circ \mathbf{F}), \quad f \in L^2(\Omega),
\end{aligned}
\tag{4.34}
$$

where DF is the Jacobian matrix of the mapping $\mathbf{F} : \widehat{\Omega} \to \Omega$. The transformation above preserve the structure of the De Rham cochain complex, in the sense of the following commuting diagram (see [66, Sect. 2.2] and [86, Sect. 3.9]):

$$
\begin{array}{ccccccccc}
\mathbb{R} & \longrightarrow & \widehat{X}_h^0 & \xrightarrow{\ \widehat{\mathrm{grad}}\ } & \widehat{X}_h^1 & \xrightarrow{\ \widehat{\mathrm{curl}}\ } & \widehat{X}_h^2 & \xrightarrow{\ \widehat{\mathrm{div}}\ } & \widehat{X}_h^3 & \longrightarrow & 0 \\
 & & \ \ \uparrow \iota^0 & & \ \ \uparrow \iota^1 & & \ \ \uparrow \iota^2 & & \ \ \uparrow \iota^3 & & \\
\mathbb{R} & \longrightarrow & X_h^0 & \xrightarrow{\ \mathrm{grad}\ } & X_h^1 & \xrightarrow{\ \mathrm{curl}\ } & X_h^2 & \xrightarrow{\ \mathrm{div}\ } & X_h^3 & \longrightarrow & 0
\end{array}
\qquad (4.35)
$$

Note that the diagram above implicitly defines the isogeometric De Rham compatible spaces on Ω, that is X_h^0, X_h^1, X_h^2 and X_h^3; for example:

$$
X_h^2 = \left\{ \mathbf{f} : \Omega \to \mathbb{R}^3 \text{ such that } \det(DF)(DF)^{-1}(\mathbf{f} \circ \mathbf{F}) \in \widehat{X}_h^2 \right\}. \qquad (4.36)
$$

In this setting, the geometry parametrization $\mathbf{F}$ can be either a spline in $(\widehat{X}_h^0)^3$ or a NURBS.

In fact, thanks to the smoothness of splines, isogeometric De Rham compatible spaces enjoy a wider applicability than their finite element counterpart. For example, assuming $m_{\ell,i} \le p_\ell - 1$, for $i = 2, \ldots, N_\ell - 1$ and $\ell = 1, 2, 3$, then the space X_h^2 is subset of $(H^1(\Omega))^3$. Furthermore there exists a subset $K_h \subset X_h^2$ of divergence-free isogeometric vector fields, i.e.,

$$
K_h = \left\{ \mathbf{f} \in X_h^2 \text{ such that } \operatorname{div} \mathbf{f} = 0 \right\}, \qquad (4.37)
$$

that can be characterized as

$$
\mathbf{f} \in K_h \quad \Leftrightarrow \quad \int_\Omega (\operatorname{div} \mathbf{f}) v = 0, \ \forall v \in X_h^3, \qquad (4.38)
$$

as well as

$$
\mathbf{f} \in K_h \quad \Leftrightarrow \quad \exists \mathbf{v} \in X_h^1 \text{ such that } \mathbf{curl}\, \mathbf{v} = \mathbf{f}. \qquad (4.39)
$$

Both K_h and X_h^2 play an important role in the IGA of incompressible fluids, allowing exact point-wise divergence-free solutions that are difficult to achieve by finite element methods, or in linear small-deformation elasticity for incompressible materials, allowing point-wise preservation of the linearized volume under deformation. We refer to [37, 56–58, 123] and the numerical tests of Sect. 4.7. We should also mention that for large deformation elasticity the volume preservation constraint becomes $\det \mathbf{f} = 1$, $\mathbf{f}$ denoting the deformation gradient, and the construction of isogeometric spaces that allow its exact preservation is an open and very challenging problem.

4.3.3 Extensions

Isogeometric spaces can be constructed from non-tensor-product or unstructured spline spaces, as the ones listed in Sect. 4.2.3.

Unstructured multipatch isogeometric spaces may have C^0 continuity at patch interfaces, of higher continuity. The implementation of C^0-continuity over multi-patch domains is well understood (see e.g. [81, 106] for strong and [34] for weak imposition of C^0 conditions). Some papers have tackled the problem of constructing isogeometric spaces of higher order smoothness, such as [35, 48, 76, 89, 98]. The difficulty is how to construct analysis-suitable unstructured isogeometric spaces with global C^1 or higher continuity. The main question concerns the approximation properties of these spaces, see Sect. 4.4.3.

An important operation, derived from CAGD, and applied to isogeometric spaces is trimming, see [85] Indeed trimming is very common in geometry representation, since it is the natural outcome of Boolean operations (union, intersection, subtraction of domains). One possibility is to approximate (up to some prescribed tolerance) the trimmed domain by an untrimmed multipatch or T-spline parametrized domain, see [109]. Another possibility is to use directly the trimmed geometry and deal with the two major difficulties that arise: efficient quadrature and imposition of boundary conditions, see [94, 95, 99].

4.4 Isogeometric Spaces: Approximation Properties

4.4.1 h-Refinement

The purpose of this section is to summarize the approximation properties of the isogeometric space V_h defined in (4.27). We focus on the convergence analysis under h-refinement, presenting results first obtained in [15] and [22]. To express the error bounds, we will make use of Sobolev spaces on a domain D, that can be either Ω or $\widehat{\Omega}$ or subsets such as Q, $\widetilde{Q}$, K or $\widetilde{K}$. For example, $H^s(D)$, $s \in \mathbb{N}$ is the space of square integrable functions $f \in L^2(\Omega)$ such that its derivatives up to order s are square integrable. However, conventional Sobolev spaces are not enough. Indeed, since the mapping $\mathbf{F}$ is not arbitrarily regular across mesh lines, even if a scalar function f in physical space satisfies $f \in H^s(\Omega)$, its pull-back $\widehat{f} = f \circ \mathbf{F}$ is not in general in $H^s(\widehat{\Omega})$. As a consequence, the natural function space in parametric space, in order to study the approximation properties of mapped NURBS, is not the standard Sobolev space H^s but rather a "bent" version that allows for less regularity across mesh lines. In the following, as usual, C will denote a constant, possibly different at each occurrence, but independent of the mesh-size h. Note that, unless noted otherwise, C depends on the polynomial degree p and regularity.

Let $d = 1$ first. We recall that $I_i = (\zeta_i, \zeta_{i+1})$ are the intervals of the partition of $I = (0, 1)$ given by the knot vector. We define for any $q \in \mathbb{N}$ the piecewise polynomial space

$$\mathscr{P}_q(\varXi) = \{v \in L^2(I) \text{ such that } v|_{I_i} \text{ is a } q \text{ -degree polynomial}, \forall i = 1, \dots, N-1\}.$$

Given $s \in \mathbb{N}$ and any sub-interval $E \subset I$, we indicate by $H^s(E)$ the usual Sobolev space endowed with norm $\| \cdot \|_{H^s(E)}$ and semi-norm $| \cdot |_{H^s(E)}$. We define the *bent Sobolev space* (see [15]) on I as

$$\mathscr{H}^s(I) = \left\{ \begin{array}{l} f \in L^2(I) \text{ such that } f|_{I_i} \in H^s(I_i) \; \forall \, i = 1, \dots, N-1, \text{ and} \\[4pt] D_-^k f(\zeta_i) = D_+^k f(\zeta_i), \; \forall k = 0, \dots, \min\{s-1, k_i\}, \forall i = 2, \dots, N-1, \end{array} \right\}$$

(4.40)

where $D_\pm^k$ denote the kth-order left and right derivative (or left and right limit for $k = 0$), and k_i is the number of continuous derivatives at the break point ζ_i. We endow the above space with the broken norm and semi-norms

$$\|f\|_{\mathscr{H}^s(I)}^2 = \sum_{j=0}^{s} |f|_{\mathscr{H}^j(I)}^2, \quad |f|_{\mathscr{H}^j(I)}^2 = \sum_{i=1}^{N-1} |f|_{H^j(I_i)}^2 \quad \forall j = 0, 1, \dots, s,$$

where $| \cdot |_{H^0(I_i)} = \| \cdot \|_{L^2(I_i)}$.

In higher dimensions, the tensor product bent Sobolev spaces are defined as follows. Let $\mathbf{s} = (s_1, s_2, \dots, s_d)$ in $\mathbb{N}^d$. By a tensor product construction starting from (4.40), we define the tensor product bent Sobolev spaces in the parametric domain $\widehat{\Omega} := (0, 1)^d$

$$\mathscr{H}^{\mathbf{s}}(\widehat{\Omega}) := \mathscr{H}^{s_1}(0, 1) \otimes \mathscr{H}^{s_2}(0, 1) \otimes \dots \otimes \mathscr{H}^{s_d}(0, 1),$$

endowed with the tensor-product norm and seminorms. The above definition clearly extends immediately to the case of any hyper-rectangle $E \subset \widehat{\Omega}$ that is a union of elements in $\mathscr{M}$.

We restrict, for simplicity of exposition, to the two-dimensional case. As in the one-dimensional case, we assume local quasi-uniformity of the mesh in each direction. Let $\Pi_{p_i, \varXi_i} : L^2(I) \to S^{p_i}(\varXi_i)$, for $i = 1, 2$, indicate the univariate quasi-interpolant associated to the knot vector $\varXi_i$ and polynomial degree p_i. Let moreover $\Pi_{\mathbf{p}, \varXi} = \Pi_{p_1, \varXi_1} \otimes \Pi_{p_2, \varXi_2}$ from $L^2(\Omega)$ to $S^{\mathbf{p}}(\varXi)$ denote the tensor product quasi-interpolant built using the $\Pi_{p_i, \varXi_i}$ defined in (4.18) for $d = 2$. In what follows, given any sufficiently regular function $f : \widehat{\Omega} \to \mathbb{R}$, we will indicate the partial derivative operators with the symbol

$$\widehat{D}^{\mathbf{r}} f = \frac{\partial^{r_1} \partial^{r_2} f}{\partial \zeta_1^{r_1} \partial \zeta_2^{r_2}} \qquad \mathbf{r} = (r_1, r_2) \in \mathbb{N}^2.$$

(4.41)

Let $E \subset \widehat{\Omega}$ be any union of elements $Q \in \mathcal{M}$ of the spline mesh. We will adopt the notation

$$\|f\|^2_{L^2_h(E)} := \sum_{Q \in \mathcal{M} \text{ s.t. } Q \subset E} \|f\|^2_{L^2(Q)}.$$

The element size of a generic element $Q_{\mathbf{i}} = I_{1,i_1} \times \ldots \times I_{d,i_d} \in \mathcal{M}$ will be denoted by $h_{Q_{\mathbf{i}}} = \mathrm{diam}(Q_{\mathbf{i}})$. We will indicate the length of the edges of $Q_{\mathbf{i}}$ by $h_{1,i_1}, h_{2,i_2},$. Because of the local quasi-uniformity of the mesh in each direction, the length of the two edges of the extended patch $\widetilde{Q}_{\mathbf{i}}$ are bounded from above by h_{1,i_1} and h_{2,i_2}, up to a multiplicative factor. The quasi-uniformity constant is denoted θ. We have the following result (see [22, 25] for its proof), that can be established for spaces with boundary conditions as well.

Proposition 3 *Given integers* $0 \leq r_1 \leq s_1 \leq p_1 + 1$ *and* $0 \leq r_2 \leq s_2 \leq p_2 + 1$, *there exists a constant* C *depending only on* $\mathbf{p}, \theta$ *such that for all elements* $Q_{\mathbf{i}} \in \mathcal{M}$,

$$\|\widehat{D}^{(r_1,r_2)}(f - \Pi_{\mathbf{p},\varXi} f)\|_{L^2(Q_{\mathbf{i}})}$$

$$\leq C\left((h_{1,i_1})^{s_1-r_1}\|\widehat{D}^{(s_1,r_2)}f\|_{L^2_h(\widetilde{Q}_{\mathbf{i}})} + (h_{2,i_2})^{s_2-r_2}\|\widehat{D}^{(r_1,s_2)}f\|_{L^2_h(\widetilde{Q}_{\mathbf{i}})}\right)$$

for all f *in* $\mathscr{H}^{(s_1,r_2)}(\widehat{\Omega}) \cap \mathscr{H}^{(r_1,s_2)}(\widehat{\Omega})$.

We can state the approximation estimate for the projection operator on the isogeometric space V_h, that is $\Pi_{V_h} : L^2(\Omega) \rightarrow V_h$, defined in (4.29). In the physical domain $\Omega = \mathbf{F}(\widehat{\Omega})$, we introduce the coordinate system naturally induced by the geometrical map $\mathbf{F}$, referred to as the $\mathbf{F}$-coordinate system, that associates to a point $\mathbf{x} \in \Omega$ the Cartesian coordinates in $\widehat{\Omega}$ of its counter-image $\mathbf{F}^{-1}(\mathbf{x})$. At each $\mathbf{x} \in K \in \mathcal{M}_0$ (more generally, at each $\mathbf{x}$ where $\mathbf{F}$ is differentiable) the tangent *base vectors* $\mathbf{g}_1$ and $\mathbf{g}_2$ of the $\mathbf{F}$-coordinate system can be defined as

$$\mathbf{g}_i = \mathbf{g}_i(\mathbf{x}) = \frac{\partial \mathbf{F}}{\partial \zeta_i}(\mathbf{F}^{-1}(\mathbf{x})), \quad i = 1, 2; \tag{4.42}$$

these are the images of the canonical base vectors $\widehat{\mathbf{e}}_i$ in $\widehat{\Omega}$, and represent the axis directions of the $\mathbf{F}$-coordinate system (see Fig. 4.8).

Analogously to the derivatives in the parametric domain (4.41), the derivatives of $f : \Omega \rightarrow \mathbb{R}$ in Cartesian coordinates are denoted by

$$D^{\mathbf{r}}f = \frac{\partial^{r_1}\partial^{r_2}f}{\partial x_1^{r_1}\partial x_2^{r_2}} \quad \mathbf{r} = (r_1, r_2) \in \mathbb{N}^2.$$

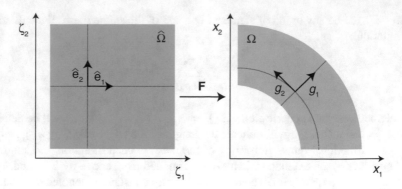

Fig. 4.8 Illustration of the **F**-coordinate system in the physical domain

We also consider the derivatives of $f : \Omega \rightarrow \mathbb{R}$ with respect to the **F**-coordinates. These are just the directional derivatives: for the first order we have

$$\frac{\partial f}{\partial \mathbf{g}_i}(\mathbf{x}) = \nabla f(\mathbf{x}) \cdot \mathbf{g}_i(\mathbf{x}) = \lim_{t \to 0} \frac{f(\mathbf{x} + t\mathbf{g}_i(\mathbf{x})) - f(\mathbf{x})}{t}, \tag{4.43}$$

which is well defined for any $\mathbf{x}$ in the (open) elements of the coarse triangulation $\mathcal{M}_0$, as already noted. Higher order derivatives are defined by recursion

$$\frac{\partial^{r_i} f}{\partial \mathbf{g}_i^{r_i}} = \frac{\partial}{\partial \mathbf{g}_i}\left(\frac{\partial^{r_i-1} f}{\partial \mathbf{g}_i^{r_i-1}}\right) = \left(\frac{\partial}{\partial \mathbf{g}_i}\left(\cdots\left(\frac{\partial}{\partial \mathbf{g}_i}\left(\frac{\partial f}{\partial \mathbf{g}_i}\right)\right)\right)\right);$$

more generally, we adopt the notation

$$D_{\mathbf{F}}^{\mathbf{r}} f = \frac{\partial^{r_1}}{\partial \mathbf{g}_1^{r_1}} \frac{\partial^{r_2} f}{\partial \mathbf{g}_2^{r_2}} \qquad \mathbf{r} = (r_1, r_2) \in \mathbb{N}^2. \tag{4.44}$$

Derivatives with respect to the **F**-coordinates are directly related to derivatives in the parametric domain, by

$$D_{\mathbf{F}}^{\mathbf{r}} f = \left(\widehat{D}^{\mathbf{r}}(f \circ \mathbf{F})\right) \circ \mathbf{F}^{-1}. \tag{4.45}$$

Let E be a union of elements $K \in \mathcal{M}$. We introduce the broken norms and seminorms

$$\|f\|_{\mathscr{H}_{\mathbf{F}}^{(s_1,s_2)}(E)}^2 = \sum_{r_1=0}^{s_1} \sum_{r_2=0}^{s_2} |f|_{\mathscr{H}_{\mathbf{F}}^{(r_1,r_2)}(E)}^2, \tag{4.46}$$

$$|f|_{\mathscr{H}_{\mathbf{F}}^{(s_1,s_2)}(E)}^2 = \sum_{K \in \mathcal{M} \text{ s.t. } K \subset E} |f|_{H_{\mathbf{F}}^{(s_1,s_2)}(K)}^2,$$

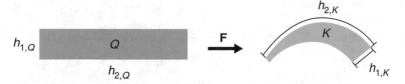

Fig. 4.9 Q is mapped by the geometrical map $\mathbf{F}$ to K

where

$$|f|_{H_{\mathbf{F}}^{(s_1,s_2)}(K)} = \left\| D_{\mathbf{F}}^{(s_1,s_2)} f \right\|_{L^2(K)}.$$

We also introduce the following space

$$H_{\mathbf{F}}^{(s_1,s_2)}(\Omega) = \text{closure of } C^\infty(\Omega) \text{ with respect to the norm } \| \cdot \|_{\mathscr{H}_{\mathbf{F}}^{(s_1,s_2)}(\Omega)}.$$

The following theorem from [22] states the main estimate for the approximation error of $\Pi_{V_h} f$ and, making use of derivatives in the $\mathbf{F}$-coordinate system, it is suitable for anisotropic meshes. For a generic element $K_{\mathbf{i}} = \mathbf{F}(Q_{\mathbf{i}}) \in \mathcal{M}$, the notation $\widetilde{K}_{\mathbf{i}} = \mathbf{F}(\widetilde{Q}_{\mathbf{i}})$ indicates its support extension (Fig. 4.9).

Theorem 2 *Given integers* r_i, s_i, *such that* $0 \leq r_i \leq s_i \leq p_i + 1$, $i = 1, 2$, *there exists a constant C depending only on* $\mathbf{p}, \theta, \mathbf{F}, W$ *such that for all elements* $K_{\mathbf{i}} = \mathbf{F}(Q_{\mathbf{i}}) \in \mathcal{M}$,

$$|f - \Pi_{V_h} f|_{\mathscr{H}_{\mathbf{F}}^{(r_1,r_2)}(K_{\mathbf{i}})} \leq C \Big((h_{1,i_1})^{s_1-r_1} \|f\|_{\mathscr{H}_{\mathbf{F}}^{(s_1,r_2)}(\widetilde{K}_{\mathbf{i}})} + (h_{2,i_2})^{s_2-r_2} \|f\|_{\mathscr{H}_{\mathbf{F}}^{(r_1,s_2)}(\widetilde{K}_{\mathbf{i}})} \Big)$$
$$(4.47)$$

for all f in $H_{\mathbf{F}}^{(s_1,r_2)}(\Omega) \cap H_{\mathbf{F}}^{(r_1,s_2)}(\Omega)$.

We have the following corollary of Theorem 2, similar to [15, Theorem 3.1], or [25, Theorem 4.24] (the case with boundary conditions is handled similarly).

Corollary 1 *Given integers r, s, such that* $0 \leq r \leq s \leq \min(p_1, \ldots, p_d)+1$, *there exists a constant C depending only on* $\mathbf{p}, \theta, \mathbf{F}, W$ *such that*

$$\|f - \Pi_{V_h} f\|_{H^r(K_{\mathbf{i}})} \leq C(h_{K_{\mathbf{i}}})^{s-r} \|f\|_{H^s(\widetilde{K}_{\mathbf{i}})} \quad \forall K_{\mathbf{i}} \in \mathcal{M},$$
$$\|f - \Pi_{V_h} f\|_{H^r(\Omega)} \leq Ch^{s-r} \|f\|_{H^s(\Omega)},$$
$$(4.48)$$

for all f in $H^s(\Omega)$.

The error bound above straightforwardly covers isogeometric/isoparametric vector fields. The error theory is possible also for isogeometric De Rham compatible vector fields. In this framework there exists commuting projectors, i.e., projectors that make the diagram

$$
\begin{array}{ccccccccc}
\mathbb{R} & \longrightarrow & H^1(\Omega) & \xrightarrow{\ \mathbf{grad}\ } & \mathbf{H}(\mathbf{curl};\Omega) & \xrightarrow{\ \mathbf{curl}\ } & \mathbf{H}(\mathrm{div};\Omega) & \xrightarrow{\ \mathrm{div}\ } & L^2(\Omega) & \longrightarrow & 0 \\
 & & \Big\downarrow{\scriptstyle \Pi^0} & & \Big\downarrow{\scriptstyle \Pi^1} & & \Big\downarrow{\scriptstyle \Pi^2} & & \Big\downarrow{\scriptstyle \Pi^3} & & \\
\mathbb{R} & \longrightarrow & X_h^0 & \xrightarrow{\ \mathbf{grad}\ } & X_h^1 & \xrightarrow{\ \mathbf{curl}\ } & X_h^2 & \xrightarrow{\ \mathrm{div}\ } & X_h^3 & \longrightarrow & 0.
\end{array}
$$

$$(4.49)$$

commutative. These projectors not only are important for stating approximation estimates, but also play a fundamental role in the stability of isogeometric schemes; see [4, 38].

4.4.2 *p-Refinement and k-Refinement*

Approximation estimates in Sobolev norms have the general form

$$
\inf_{f_h \in V_h} \| f - f_h \|_{H^r(\Omega)} \le C(h, p, k; r, s) \| f \|_{H^s(\Omega)} \tag{4.50}
$$

where the optimal constant is therefore

$$
C(h, p, k; r, s) = \sup_{f \in B^s(\Omega)} \inf_{f_h \in V_h} \| f - f_h \|_{H^r(\Omega)} \tag{4.51}
$$

where $B^s(\Omega) = \{ f \in H^s(\Omega)$ such that $\| f \|_{H^s(\Omega)} \le 1 \}$ is the unit ball in $H^s(\Omega)$. The study in Sect. 4.4.1 covers the approximation under h-refinement, giving an asymptotic bound to (4.51) with respect to h which is sharp, for $s \le p + 1$,

$$
C(h, p, k; r, s) \approx C(p, k; r, s) h^{s-r}, \quad \text{for } h \to 0. \tag{4.52}
$$

This is the fundamental and most standard analysis, but it does not explain the benefits of k-refinement, a unique feature of IGA. High-degree, high-continuity splines and NURBS are superior to standard high-order finite elements when considering accuracy per degree-of-freedom. The study of k-refinement is still incomplete even though some important results are available in the literature. In particular, [19] contains h, p, k-explicit approximation bounds for spline spaces of degree $2q + 1$ and up to C^q global continuity, while the recent work [115] contains the error estimate

$$
\inf_{f_h \in V_h} | f - f_h |_{H^r(\Omega)} \le (\sqrt{2}h)^{q-r} | f |_{H^q(\Omega)}
$$

for univariate C^{p-1}, p-degree splines, with $0 \le r \le q \le p + 1$ on uniform knot vectors.

An innovative approach, and alternative to standard error analysis, is developed in [59]. There a theoretical/numerical investigation provides clear evidence of the importance of k-refinement. The space of smooth splines is shown to be very close to a best approximation space in the Sobolev metric. The approach is as follows: given the isogeometric space V_h, with $N = \dim V_h$ together with (4.51), we consider the *Kolmogorov N-width*:

$$d_N(B^s(\Omega), H^r(\Omega)) = \inf_{\substack{W_h \subset H^s(\Omega) \\ \dim W_h = N}} \sup_{f \in B^s(\Omega)} \inf_{f_h \in W_h} \| f - f_h \|_{H^r(\Omega)}. \qquad (4.53)$$

Then the optimality ratio is defined as

$$\Lambda(B^s(\Omega), V_h, H^r(\Omega)) = \frac{C(h, p, k; r, s)}{d_N(B^s(\Omega), H^r(\Omega))}. \qquad (4.54)$$

In general, the quantity $\Lambda(B^s(\Omega), V_h, H^r(\Omega))$ is hard to compute analytically but can be accurately approximated numerically, by solving suitable generalized eigenvalue problems (see [59]). In Fig. 4.10 we compare smooth C^3 quartic splines and standard quartic finite elements (that is, C^0 splines) under h-refinement. An interesting result is that smooth splines asymptotically achieve optimal approximation in the context considered, that is, they tend to be an optimal approximation space given the number of degrees-of-freedom, since $\Lambda(B^5(0, 1), S_3^4, L^2(0, 1)) \to 1$. This is not surprising as it is known that uniform periodic spline spaces are optimal

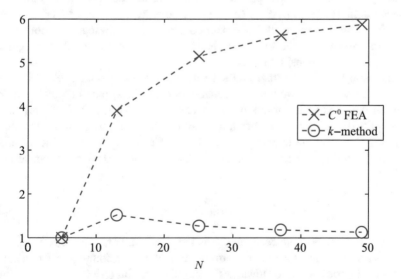

Fig. 4.10 Optimality ratios: comparison between quartic C^3 splines (i.e., $\Lambda(B^5(0, 1), S_3^4, L^2(0, 1))$, blue line with circles) and C^0 finite elements (i.e., $\Lambda(B^5(0, 1), S_0^4, L^2(0, 1))$, red line with crosses) on the unit interval for different mesh-sizes h (the total number of degrees-of-freedom N is the abscissa)

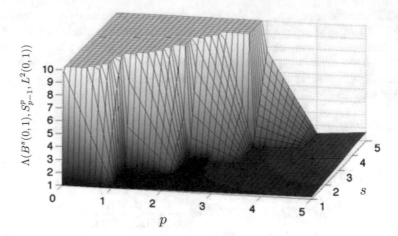

Fig. 4.11 Optimality ratios: for different Sobolev regularity s and for different spline degree p with maximal smoothness. The number of degrees-of-freedom is $N = 30$. The surface plot is capped at 10 for purposes of visualization. Note that if $p \geq s - 1$ the optimality ratio is near 1. Even for low regularity (i.e., low s), smooth splines (i.e., high p) produce optimality ratios near 1. This supports the claim that "smooth splines are always good"

in the periodic setting. On the contrary, C^0 finite elements are far from optimal. In Fig. 4.11 we plot the optimality ratios for the L^2 error for different Sobolev regularity s and for smooth splines with different degrees p. There is numerical evidence that $\Lambda(B^s(0, 1), S_{p-1}^p, L^2(0, 1))$ is bounded and close to 1 for all $p \geq s - 1$. It is a surprising result, but in fact confirms that high-degree smooth splines are accurate even when the solution to be approximated has low Sobolev regularity (see [59] for further considerations).

This issue has been further studied in [40], for the special case of solutions that are piecewise analytic with a localized singularity, which is typical of elliptic PDEs on domains with corners or sharp edges. The work [40] focus instead on the simplified one-dimensional problem, and consider a model singular solution $f(\zeta) = \zeta^\alpha - \zeta$ on the interval $[0, 1]$, with $0 < \alpha < 1$. From the theory of hp-FEMs (i.e., hp finite elements; see [103]) it is known that exponential convergence is achieved, precisely

$$|f - f_h|_{H^1(0,1)} \leq C e^{-b\sqrt{N}} \tag{4.55}$$

where C and b are positive constants, N is the total number of degrees-of-freedom, and f_h is a suitable finite element approximation of f. The bound (4.55) holds if the mesh is geometrically refined towards the singularity point $\zeta = 0$ and with a suitable selection of the polynomial degree, growing from left (the singularity) to the right of the interval $[0, 1]$. The seminal paper [10] gives the reference hp-FEM convergence rate which is reported in Fig. 4.12. Likewise, exponential convergence occurs with C^{p-1}, p-degree spline approximation on a geometrically graded knot span, as

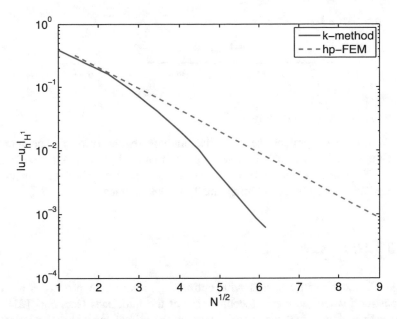

Fig. 4.12 Energy norm error versus the (square root of) number of degrees-of-freedom N for the approximation of the solution $u(x) = x^{0.7} - x$ of the problem $-u'' = f$ with homogeneous Dirichlet boundary conditions. The mesh is geometrically graded (with ratio $q = 0.35$ for IGA) and the spline degree is proportional to the number of elements for IGA, and the smoothness is maximal, that is the spline space is C^{p-1} globally continuous. Mesh-size and degrees are optimally selected for hp-FEM, according to the criteria of [10]. Exponential convergence $|u - u_h|_{H^1} \leq C \exp(-b\sqrt{N})$ is evidenced in both cases, with larger b for IGA

reported in the same figure. Remarkably, convergence is faster (with the constant b in (4.55) that appears to be higher) for smooth splines, even though for splines the degree p is the same for all mesh elements, and grows proportionally with the total number of elements, whereas for hp-FEM a locally varying polynomial degree is utilized on an element-by-element basis.

Exponential convergence for splines is proved in the main theorem of [40], reported below.

Theorem 3 *Assume that $f \in H_0^1(0, 1)$ and*

$$\left\| \zeta^{\beta+k-2} \frac{\partial^k f}{\partial^k \zeta} \right\|_{L^2(0,1)} \leq C_u d_u^{k-2}(k-2)!, \qquad k = 2, 3, \ldots \qquad (4.56)$$

for some $0 < \beta \leq 1$ and $C_u, d_u > 0$. Then there exist $b > 0$ and $C > 0$ such that for any $q > 1$, for any σ with $0 < \sigma < 1$ and $1 > \sigma > (1 + 2/d_u)^{-1}$,

$$\inf_{f_h \in S^p(\Xi)} \| f - f_h \|_{H^1(\Omega)} \leq C e^{-b(\sigma,\beta)\sqrt{N}}, \qquad (4.57)$$

where $p = 2q + 1$,

$$\Xi = \{\underbrace{0, \ldots, 0}_{p \text{ times}}, \underbrace{\sigma^{p-1}, \ldots, \sigma^{p-1}}_{q \text{ times}}, \underbrace{\sigma^{p-2}, \ldots, \sigma^{p-2}}_{q \text{ times}}, \ldots, \underbrace{\sigma, \ldots, \sigma}_{q \text{ times}}, \underbrace{1, \ldots, 1}_{p \text{ times}}\},$$

$$(4.58)$$

and N is the dimension of $S^p(\Xi)$.

Condition (4.56) expresses the piecewise analytic regularity of f. Theorem 3 is based on [19], and as such it covers approximation by $2q + 1$ degree splines having C^q global continuity. However, as is apparent from Fig. 4.12, exponential convergence is also observed for maximally smooth splines.

4.4.3 Multipatch

While C^0 isogeometric spaces with optimal approximation properties are easy to construct, when the mesh is conforming at the interfaces (see, e.g., [25]), the construction of smooth isogeometric spaces with optimal approximation properties on unstructured geometries is a challenging problem and still open in its full generality. The problem is related to one of accurate representation (fitting) of smooth surfaces having complex topology, which is a fundamental area of research in the community of CAGD.

There are mainly two strategies for constructing smooth multipatch geometries and corresponding isogeometric spaces. One strategy is to adopt a geometry parametrization which is globally smooth almost everywhere, with the exception of a neighborhood of the extraordinary points (or edges in 3D), see Fig. 4.13 (left). The other strategy is to use geometry parametrizations that are only C^0 at patch interfaces; see Fig. 4.13 (right). The first option includes subdivision surfaces [43] and the T-spline construction in [105] and, while possessing attractive features, typically lacks optimal approximation properties [76, 89]. One exception is the recent works [120], where a specific construction is shown to achieve optimal

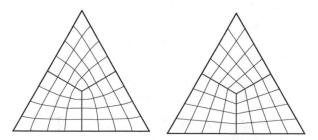

Fig. 4.13 Two possible parametrization schemes: C^1 away from the extraordinary point (left) and C^0 at patch interfaces (right)

order in h-refinement. On the other hand, some optimal constructions have been recently obtained also following the second strategy, pictured in Fig. 4.13 (right) (see [27, 48, 77, 87]). We summarize here the main concepts and results from [48], referring to the paper itself for a complete presentation.

Consider a planar ($d = 2$) spline multipatch domain of interest

$$\Omega = \Omega^{(1)} \cup \ldots \cup \Omega^{(N)} \subset \mathbb{R}^2, \tag{4.59}$$

where the closed sets $\Omega^{(i)}$ form a regular partition without hanging nodes. Assume each $\Omega^{(i)}$ is a non-singular spline patch, with at least C^1 continuity within each patch, and that there exist parametrizations

$$\mathbf{F}^{(i)} : [0, 1] \times [0, 1] = \widehat{\Omega} \to \Omega^{(i)}, \tag{4.60}$$

where

$$\mathbf{F}^{(i)} \in \mathscr{S}^p(\boldsymbol{\Xi}) \times \mathscr{S}^p(\boldsymbol{\Xi}) \subset C^1(\widehat{\Omega}); \tag{4.61}$$

Furthermore, assume global continuity of the patch parametrizations. This means the following. Let us fix $\Gamma = \Gamma^{(i,j)} = \Omega^{(i)} \cap \Omega^{(j)}$. Let $\mathbf{F}^{(L)}, \mathbf{F}^{(R)}$ be given such that

$$
\begin{aligned}
\mathbf{F}^{(L)} &: [-1, 0] \times [0, 1] = \widehat{\Omega}^{(L)} \to \Omega^{(L)} = \Omega^{(i)}, \\
\mathbf{F}^{(R)} &: [0, 1] \times [0, 1] = \widehat{\Omega}^{(R)} \to \Omega^{(R)} = \Omega^{(j)},
\end{aligned}
\tag{4.62}
$$

where $(\mathbf{F}^{(L)})^{-1} \circ \mathbf{F}^{(i)}$ and $(\mathbf{F}^{(R)})^{-1} \circ \mathbf{F}^{(j)}$ are linear transformations. The set $[-1, 1] \times [0, 1]$ plays the role of a combined parametric domain. The coordinates in $[-1, 1] \times [0, 1]$ are denoted u and v. The global continuity condition states that the parametrizations agree at $u = 0$, i.e., there is an $\mathbf{F}_0 : [0, 1] \to \mathbb{R}^2$ with

$$\Gamma = \{\mathbf{F}_0(v) = \mathbf{F}^{(L)}(0, v) = \mathbf{F}^{(R)}(0, v), v \in [0, 1]\}. \tag{4.63}$$

For the sake of simplicity we assume that the knot vectors of all patches and in each direction coincide, are open and uniform. An example is depicted in Fig. 4.14.

The multipatch isogeometric space is given as

$$\mathscr{V} = \left\{ \phi : \Omega \to \mathbb{R} \text{ such that } \phi \circ \mathbf{F}^{(i)} \in \mathscr{S}_r^p(\widehat{\Omega}), i = 1, \ldots, N \right\}; \tag{4.64}$$

the space of continuous isogeometric functions is

$$\mathscr{V}^0 = \mathscr{V} \cap C^0(\Omega), \tag{4.65}$$

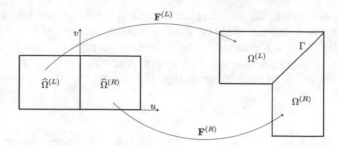

Fig. 4.14 Example of the setting of (4.62)–(4.63)

and the space of C^1 isogeometric functions is

$$\mathscr{V}^1 = \mathscr{V} \cap C^1(\Omega). \qquad (4.66)$$

The graph $\Sigma \subset \Omega \times \mathbb{R}$ of an isogeometric function $\phi : \Omega \to \mathbb{R}$ splits into patches Σ^i having the parametrization

$$\begin{bmatrix} \mathbf{F}^{(i)} \\ g^{(i)} \end{bmatrix} : [0, 1] \times [0, 1] = \widehat{\Omega} \to \Sigma^{(i)} \qquad (4.67)$$

where $g^{(i)} = \phi \circ \mathbf{F}^{(i)}$. As in (4.62), we can select a patch interface $\Gamma = \Gamma^{(i,j)} = \Omega^{(i)} \cap \Omega^{(j)}$, define $g^{(L)}$, $g^{(R)}$ such that

$$\begin{aligned} \begin{bmatrix} \mathbf{F}^{(L)} \\ g^{(L)} \end{bmatrix} &: [-1, 0] \times [0, 1] = \widehat{\Omega}^{(L)} \to \Sigma^{(i)} = \Sigma^{(L)}, \\[2ex] \begin{bmatrix} \mathbf{F}^{(R)} \\ g^{(R)} \end{bmatrix} &: [0, 1] \times [0, 1] = \widehat{\Omega}^{(R)} \to \Sigma^{(j)} = \Sigma^{(R)}, \end{aligned} \qquad (4.68)$$

see Fig. 4.15. Continuity of ϕ is implied by the continuity of the graph parametrization, then we set

$$g_0(v) = g^{(L)}(0, v) = g^{(R)}(0, v), \qquad (4.69)$$

for all $v \in [0, 1]$, analogous to (4.63).

Under suitable conditions, smoothness of a function is equivalent to the smoothness of the graph, considered as a geometric entity. In particular, for an isogeometric function that is C^1 within each patch and globally continuous, the global C^1 continuity is then equivalent to the *geometric continuity of order* 1 (in short G^1) of its graph parametrization. Geometric continuity of the graph parametrization means

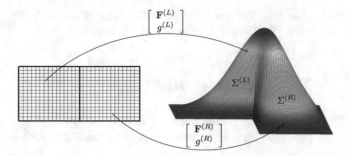

Fig. 4.15 Example of the general setting of (4.68)

that, on each patch interface, with notation (4.68), the tangent vectors

$$\begin{bmatrix} D_u \mathbf{F}^{(L)}(0, v) \\ D_u g^{(L)}(0, v) \end{bmatrix}, \begin{bmatrix} D_v \mathbf{F}_0(v) \\ D_v g_0(v) \end{bmatrix} \text{ and } \begin{bmatrix} D_u \mathbf{F}^{(R)}(0, v) \\ D_u g^{(R)}(0, v) \end{bmatrix},$$

are co-planar, i.e., linearly dependent. In the CAGD literature, G^1 continuity is commonly stated as below (see, e.g., [18, 82, 90]).

Definition 1 (G^1-**Continuity at** $\Sigma^{(i)} \cap \Sigma^{(j)}$)   Given the parametrizations $\mathbf{F}^{(L)}$, $\mathbf{F}^{(R)}$, $g^{(L)}$, $g^{(R)}$ as in (4.62), (4.68), fulfilling (4.61) and (4.69), we say that the graph parametrization is G^1 at the interface $\Sigma^{(i)} \cap \Sigma^{(j)}$ if there exist $\alpha^{(L)} : [0, 1] \to \mathbb{R}$, $\alpha^{(R)} : [0, 1] \to \mathbb{R}$ and $\beta : [0, 1] \to \mathbb{R}$ such that for all $v \in [0, 1]$,

$$\alpha^{(L)}(v)\alpha^{(R)}(v) > 0 \tag{4.70}$$

and

$$\alpha^{(R)}(v) \begin{bmatrix} D_u \mathbf{F}^{(L)}(0, v) \\ D_u g^{(L)}(0, v) \end{bmatrix} - \alpha^{(L)}(v) \begin{bmatrix} D_u \mathbf{F}^{(R)}(0, v) \\ D_u g^{(R)}(0, v) \end{bmatrix} + \beta(v) \begin{bmatrix} D_v \mathbf{F}_0(v) \\ D_v g_0(v) \end{bmatrix} = \mathbf{0}.$$

$$\tag{4.71}$$

Since the first two equations of (4.71) are linearly independent, $\alpha^{(L)}$, $\alpha^{(R)}$ and β are uniquely determined, up to a common multiplicative factor, by $\mathbf{F}^{(L)}$ and $\mathbf{F}^{(R)}$, i.e. from the equation

$$\alpha^{(R)}(v) D_u \mathbf{F}^{(L)}(0, v) - \alpha^{(L)}(v) D_u \mathbf{F}^{(R)}(0, v) + \beta(v) D_v \mathbf{F}_0(v) = \mathbf{0}. \tag{4.72}$$

We have indeed the following proposition (see [48] and [90]).

Proposition 4 *Given any* $\mathbf{F}^{(L)}$, $\mathbf{F}^{(R)}$ *then* (4.72) *holds if and only if* $\alpha^{(S)}(v) = \gamma(v)\bar{\alpha}^{(S)}(v)$, *for* $S \in \{L, R\}$, *and* $\beta(v) = \gamma(v)\bar{\beta}(v)$, *where*

$$\bar{\alpha}^{(S)}(v) = \det \left[D_u \mathbf{F}^{(S)}(0, v) \ D_v \mathbf{F}_0(v) \right], \tag{4.73}$$

$$\bar{\beta}(v) = \det \left[D_u \mathbf{F}^{(L)}(0, v) \ D_u \mathbf{F}^{(R)}(0, v) \right], \tag{4.74}$$

and $\gamma : [0, 1] \to \mathbb{R}$ *is any scalar function. In addition,* $\gamma(v) \neq 0$ *if and only if* (4.70) *holds. Moreover, there exist functions* $\beta^{(S)}(v)$, *for* $S \in \{L, R\}$, *such that*

$$\beta(v) = \alpha^{(L)}(v)\beta^{(R)}(v) - \alpha^{(R)}(v)\beta^{(L)}(v). \tag{4.75}$$

In the context of isogeometric methods we consider Ω and its parametrization given. Then for each interface $\alpha^{(L)}$, $\alpha^{(R)}$ and β are determined from (4.72) as stated in Proposition 4. It should be observed that for planar domains, there always exist $\alpha^{(L)}$, $\alpha^{(R)}$ and β fulfilling (4.72) (this is not the case for surfaces, see [48]). Then, the C^1 continuity of isogeometric functions is equivalent to the last equation in (4.71), that is

$$\alpha^{(R)}(v) D_u g^{(L)}(0, v) - \alpha^{(L)}(v) D_u g^{(R)}(0, v) + \beta(v) D_v g_0(v) = 0 \tag{4.76}$$

for all $v \in [0, 1]$. Optimal approximation properties of the isogeometric space on Ω holds under restrictions on $\alpha^{(L)}$, $\alpha^{(R)}$ and β, i.e. on the geometry parametrization. This leads to the definition below ([48]).

Definition 2 (Analysis-Suitable G^1-Continuity) $\mathbf{F}^{(L)}$ *and* $\mathbf{F}^{(R)}$ *are analysis-suitable* G^1-*continuous at the interface* Γ *(in short, AS* G^1*) if there exist* $\alpha^{(L)}, \alpha^{(R)}, \beta^{(L)}, \beta^{(R)} \in \mathscr{P}^1([0, 1])$ *such that* (4.72) *and* (4.75) *hold.*

The class of planar AS G^1 parametrizations contains all the bilinear ones and more, see Fig. 4.16.

In [48], the structure of C^1 isogeometric spaces over AS G^1 geometries is studied, providing an explanation of the optimal convergence of the space of p-degree isogeometric functions, having up to C^{p-2} continuity within the patches (and global C^1 continuity). On the other hand, no convergence under h-refinement occurs for C^{p-1} continuity within the patches. This phenomenon is referred to as

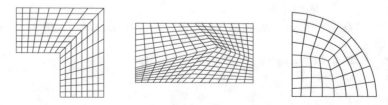

Fig. 4.16 Examples of planar domain having an AS G^1 parametrization

C^1 *locking.* Moreover, it is shown that AS G^1 geometries are needed to guarantee optimal convergence, in general.

4.4.4 Singular Parametrizations

The theory of isogeometric spaces we have reviewed in previous sections assumes that the geometry parametrization is regular. However, singular parametrizations are used in IGA, as they allow more flexibility in the geometry representation. Figure 4.17 shows two examples of this kind, for a single-patch parametrization of the circle. Typically, a singularity appears when some of the control points near the boundary coincide or are collinear.

Isogeometric spaces with singular mapping have been studied in the papers [112–115]. The paper [113] addresses a class of singular geometries that includes the two circles of Fig. 4.17. It is shown that in these cases the standard isogeometric spaces, as they are constructed in the non-singular case, are not in $H^1(\Omega)$. However, [113] identifies the subspace of H^1 isogeometric functions, and constructs a basis. The study is generalized to H^2 smoothness in [114]. In [112], function spaces of higher-order smoothness C^k are explicitly constructed on polar parametrizations that are obtained by linear transformation and degree elevation from a triangular Bézier patch. See also [119]. For general parametrizations, [116] gives a representation of the derivatives of isogeometric functions.

Singular parametrizations can be used to design smooth isogeometric spaces on unstructured multipatch domains. A different C^1 constructions is proposed in [88]. In both cases, the singular mapping is employed at the extraordinary vertices.

From the practical point of view, isogeometric methods are surprisingly robust with respect to singular parametrizations. Even if some of the integrals appearing in the linear system matrix are divergent, the use of Gaussian quadrature hides the trouble and the Galerkin variational formulation returns the correct approximation. However, it is advisable to use the correct subspace basis, given in [113] and [114], to avoid ill-conditioning of the isogeometric formulation.

In [12], the authors use isogeometric analysis on the sphere with a polar parametrization (the extension of Fig. 4.17a), and benchmark the h-convergence in H^2 and H^3 norms, for solution of 4th and 6th order differential equations, respectively. It is shown that enforcing C^0 continuity at the poles yields optimal convergence, that is, the higher-order smoothness of the isogeometric solution at the poles is naturally enforced by the variational formulation.

(a)

(b)

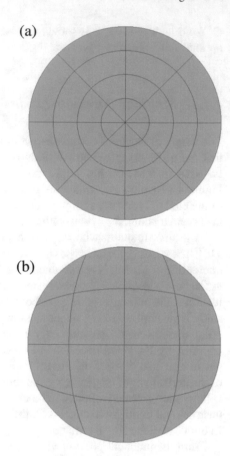

4.5   Isogeometric Spaces: Spectral Properties

We are interested in the Galerkin approximation of the eigenvalues and eigenfunction of the Laplacian differential operator, as a model problem. We will consider
mainly the univariate case. As we will see in this section, the use of C^{p-1}-
continuous splines yields advantages when compared to standard C^0 FEM. The
results shown here are taken from [71, 73]; we refer to that works for more
details. Contrary to the previous Sect. 4.4, the error analysis considered here is *not*
asymptotic, rather it may be characterized as a *global analysis* approach.

The asymptotic approach is more commonly found in the literature. Classical
functional analysis results state that, given an eigenvalue of the differential operator,
for a small enough mesh size this eigenvalue is well approximated in the discrete
problem. However, for a given mesh size, this kind of analysis offers no information
about which discrete modes are a good approximation of the exact modes, and which
ones are not.

What happens in practice is that only the lowest discrete modes are accurate. In general, a large portion of the eigenvalue/eigenfunction spectrum, the so-called "higher modes," are not approximations of their exact counterparts in any meaningful sense. It is well-known in the structural engineering discipline that the higher modes are grossly inaccurate, but the precise point in the spectrum where the eigenvalues and eigenfunctions cease to approximate their corresponding exact counterparts is never known in realistic engineering situations.

First, we focus on the approximations of eigenvalues from a global perspective, that is, we study the approximation errors of the full spectrum. This is done for the simplest possible case, that is the second derivative operator. Based on Fourier/von Neumann analysis, we show that, per degree-of-freedom and for the same polynomial degree p, C^{p-1} splines (i.e., k-method) are more accurate than C^0 splines (p-method), i.e., finite elements.

Then, we study the accuracy of k-method and p-method approximations to the eigenfunctions of the elliptic eigenvalue problem. The inaccuracy of p-method higher modal eigenvalues has been known for quite some time. We show that there are large error spikes in the L^2-norms of the eigenfunction errors centered about the transitions between branches of the p-method eigenvalue spectrum. The k-method errors are better behaved in every respect. The L^2-norms of the eigenfunction errors are indistinguishable from the L^2 best approximation errors of the eigenfunctions. As shown in [73], when solving an elliptic boundary-value problem, or a parabolic or an hyperbolic problem, the error can be expressed entirely in terms of the eigenfunction and eigenvalue errors. This is an important result but the situation is potentially very different for elliptic boundary-value problems and for parabolic and hyperbolic problems. In these cases, all modes may participate in the solution to some extent and inaccurate higher modes may not always be simply ignored. The different mathematical structures of these cases lead to different conclusions. The inaccuracy of the higher p-method modes becomes a significant concern primarily for the hyperbolic initial-value problem, while the k-method produces accurate results in the same circumstances.

4.5.1 Spectrum and Dispersion Analysis

We consider as a model problem for the eigenvalue study the one of free vibrations of a linear (∞-dimensional) structural system, without damping and force terms:

$$\mathcal{M}\frac{d^2\mathbf{u}}{dt^2} + \mathcal{K}\mathbf{u} = \mathbf{0}, \tag{4.77}$$

where $\mathcal{M}$ and $\mathcal{K}$ are, respectively, the mass and stiffness operators, and $\mathbf{u} = \mathbf{u}(t, \mathbf{x})$ is the displacement. The nth normal mode $\boldsymbol{\phi}_n$ and its frequency ω_n are obtained from the eigenvalue problem $\mathcal{K}\boldsymbol{\phi}_n = \omega_n^2\mathcal{M}\boldsymbol{\phi}_n$. Separating the variables as $\mathbf{u}(t, \mathbf{x}) =$

$\sum_n \widehat{u}_n(t)\phi_n(\mathbf{x})$, and, using Eq. (4.77), we obtain

$$\frac{d^2\widehat{u}_n(t)}{dt^2} + \omega_n^2 \widehat{u}_n(t) = 0;$$

Then $\widehat{u}_n(t) = C_- e^{-\iota\omega_n t} + C_+ e^{\iota\omega_n t}$, that is each modal coefficient $\widehat{u}_n$ oscillates at a frequency ω_n. After discretization, the following discrete equations of motion are obtained

$$\mathbf{M}\frac{d^2\mathbf{u}^h}{dt^2} + \mathbf{K}\mathbf{u}^h = \mathbf{0}, \tag{4.78}$$

where $\mathbf{M}$ and $\mathbf{K}$ are, respectively, the finite-dimensional consistent mass and stiffness matrices, and $\mathbf{u}^h = \mathbf{u}^h(t, \mathbf{x})$ is the discrete displacement vector. Analogously to the continuum case, the discrete normal modes $\boldsymbol{\phi}_n^h$ and the frequencies ω_n^h are obtained from the eigenproblem

$$\mathbf{K}\boldsymbol{\phi}_n^h = (\omega_n^h)^2 \mathbf{M}\boldsymbol{\phi}_n^h, \tag{4.79}$$

and separating the variables as $\mathbf{u}^h(t, \mathbf{x}) = \sum_n \widehat{u}_n^h(t)\boldsymbol{\phi}_n^h(\mathbf{x})$, we end up with $\widehat{u}_n^h$ oscillating at a frequency ω_n^h, that is: $\widehat{u}_n^h = C_- e^{-\iota\omega_n^h t} + C_+ e^{\iota\omega_n^h t}$. The nth discrete normal mode $\boldsymbol{\phi}_n^h$ is in general different from the nth exact normal mode $\boldsymbol{\phi}_n$ (Fig. 4.18), for $n = 1, \ldots, N$, N being the total number of degrees-of-freedom. The corresponding discrete and exact frequencies will be different The target of the frequency analysis is to evaluate how well the discrete spectrum approximates the exact spectrum.

We begin dealing with the eigenproblem (4.79) associated to a linear ($p = 1$) approximation on the one-dimensional domain $(0, L)$. We employ a uniform mesh $0 = \zeta_0 < \zeta_1 < \ldots < \zeta_A < \ldots < \zeta_{N+1} = L$, where the number of elements is $n_{el} = N + 1$ and the mesh-size is $h = L/n_{el}$. Considering homogeneous Dirichlet (fixed-fixed) boundary conditions, the eigenproblem (4.79) can be written as

$$\frac{1}{h}(\phi_{A-1} - 2\phi_A + \phi_{A+1}) + \frac{h(\omega^h)^2}{6}(\phi_{A-1} + 4\phi_A + \phi_{A+1}) = 0, \quad A = 1, \ldots, N, \tag{4.80}$$

$$\phi_0 = \phi_{N+1} = 0, \tag{4.81}$$

where N is the total number of degrees-of-freedom, and $\phi_A = \phi^h(\zeta_A)$ is the nodal value of the discrete normal mode at node ζ_A. Equation (4.80) solutions are linear combinations of exponential functions $\phi_A = (\rho_1)^A$ and $\phi_A = (\rho_2)^A$, where ρ_1 and ρ_2 are the distinct roots of the characteristic polynomial

$$(1 - 2\rho + \rho^2) + \frac{(\omega^h h)^2}{6}(1 + 4\rho + \rho^2) = 0. \tag{4.82}$$

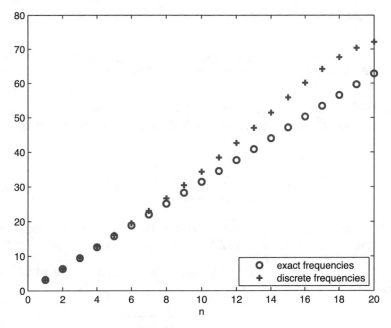

Fig. 4.18 Exact and discrete natural frequencies for the one-dimensional model problem of free vibration of an elastic rod with homogeneous Dirichlet boundary conditions. The discrete method is based on linear finite elements

Actually, (4.82) admits distinct roots when $\omega^h h \neq 0, \sqrt{12}$; for $\omega^h h = 0$, (4.82) admits the double root $\rho = 1$ (in this case, solutions of (4.80) are combinations of $\phi_A \equiv 1$ and $\phi_A = A$, that is, the affine functions), while for $\omega^h h = \sqrt{12}$ there is a double root $\rho = -1$ (and solutions of (4.80) are combinations of $\phi_A = (-1)^A$ and $\phi_A = A(-1)^A$). Observe that, in general, $\rho_2 = \rho_1^{-1}$. For the purpose of spectrum analysis, we are interested in $0 < \omega^h h < \sqrt{12}$, which we assume for the remainder of this section. In this case, $\rho_{1,2}$ are complex conjugate (we assume $Im(\rho_1) \geq 0$) and of unit modulus. Moreover, in order to compare the discrete spectrum to the exact spectrum, it is useful to represent the solutions of (4.80) as linear combinations of $e^{\pm iA\omega h}$ (that is, $\phi_A = C_- e^{-iA\omega h} + C_+ e^{iA\omega h}$), by introducing ω such that $e^{i\omega h} = \rho_1$. With this hypothesis, ω is real and, because of periodicity, we restrict to $0 \leq \omega h \leq \pi$. Using this representation in (4.82) and using the identity $2\cos(\alpha) = e^{i\alpha} + e^{-i\alpha}$, after simple computations the relation between ωh and $\omega^h h$ is obtained:

$$\frac{(\omega^h h)^2}{6}(2 + \cos(\omega h)) - (1 - \cos(\omega h)) = 0. \qquad (4.83)$$

Solving for $\omega^h h \geq 0$, we get

$$\omega^h h = \sqrt{6\frac{1 - \cos(\omega h)}{2 + \cos(\omega h)}}. \qquad (4.84)$$

Furthermore, taking into account the boundary conditions, (4.80)–(4.81) admit the non-null solution

$$\phi_A = C \frac{e^{+iAn\pi/(N+1)} - e^{-iAn\pi/(N+1)}}{2i} \equiv C \sin\left(\frac{An\pi}{N+1}\right) \tag{4.85}$$

for all $\omega = \pi/L, 2\pi/L, \ldots, N\pi/L$. Precisely, (4.85) is the nth discrete normal mode, associated to the corresponding nth discrete natural frequency ω^h, given by (4.84):

$$\omega^h = \frac{N+1}{L} \sqrt{6 \frac{1 - \cos(n\pi/(N+1))}{2 + \cos(n\pi/(N+1))}}. \tag{4.86}$$

The nth discrete mode $\phi_A = C \sin(An\pi/(N+1))$ is the nodal interpolant of the nth exact mode $\phi(x) = C \sin(n\pi x/L)$, whose natural frequency is $\omega = n\pi/L$. The quantity $\dfrac{\omega^h}{\omega} - 1 = \dfrac{\omega^h - \omega}{\omega}$ represents the relative error for the natural frequency. The plot of

$$\frac{\omega^h}{\omega} = \frac{1}{\omega h} \sqrt{6 \frac{1 - \cos(\omega h)}{2 + \cos(\omega h)}} \tag{4.87}$$

is shown in Fig. 4.19.

We now consider the quadratic p-method for the eigenproblem (4.79). Assuming to have the same mesh as in the linear case, there are $N = 2n_{el} - 1$ degrees-of-freedom. If we consider the usual Lagrange nodal basis, the corresponding stencil equation is different for element-endpoint degrees-of-freedom and bubble (internal to element) degrees-of-freedom: one has

$$\frac{1}{3h}(-\phi_{A-1} + 8\phi_{A-1/2} - 14\phi_A + 8\phi_{A+1/2} - \phi_{A+1})$$

$$+ (\omega^h)^2 \frac{h}{30}(-\phi_{A-1} + 2\phi_{A-1/2} + 8\phi_A + 2\phi_{A+1/2} - \phi_{A+1}) = 0, \quad A = 1, \ldots, N. \tag{4.88}$$

and

$$\frac{1}{3h}(8\phi_A - 16\phi_{A+1/2} + 8\phi_{A+1}) + (\omega^h)^2 \frac{h}{30}(2\phi_A + 16\phi_{A+1/2} + 2\phi_{A+1}) = 0, \tag{4.89}$$

$$A = 1, \ldots, N,$$

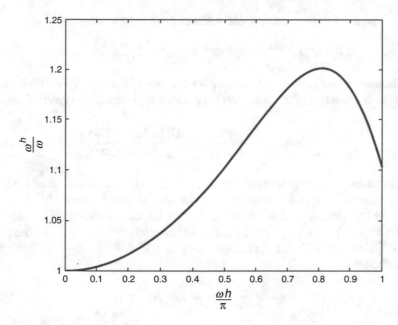

Fig. 4.19 Discrete-to-exact frequencies ratio for linear approximation

respectively. We also have the boundary conditions $\phi_0 = \phi_{N+1} = 0$. The bubble degrees-of-freedom can be calculated as

$$\phi_{A+1/2} = \frac{40 + (\omega^h h)^2}{8(10 - (\omega^h h)^2)}(\phi_A + \phi_{A+1}). \tag{4.90}$$

Eliminating them, we obtain a system of equations for the element-endpoints degrees of freedom:

$$
\begin{aligned}
\frac{1}{3h} &\left[\left(\frac{30 + 2(\omega^h h)^2}{10 - (\omega^h h)^2} \right) \phi_{A-1} + \left(\frac{-60 + 16(\omega^h h)^2}{10 - (\omega^h h)^2} \right) \phi_A \right. \\
&\left. + \left(\frac{30 + 2(\omega^h h)^2}{10 - (\omega^h h)^2} \right) \phi_{A+1} \right] \\
+ k^2 &\frac{h}{30} \left[\left(\frac{5(\omega^h h)^2}{40 - 4(\omega^h h)^2} \right) \phi_{A-1} + \left(\frac{200 - 15(\omega^h h)^2}{20 - 2(\omega^h h)^2} \right) \phi_A \right. \\
&\left. + \left(\frac{5(\omega^h h)^2}{40 - 4(\omega^h h)^2} \right) \phi_{A+1} \right] = 0.
\end{aligned}
\tag{4.91}
$$

for $A = 1, \ldots, N$. The bubble elimination is not possible when the bubble equation (4.89) is singular for $u_{A+1/2}$, that happens for $\omega^h h = \sqrt{10}$.

Normal modes at the element-endpoints nodes can be written as

$$\phi_A = C_- e^{-\imath \omega h A} + C_+ e^{\imath \omega h A}, \qquad A = 1, \ldots, N. \tag{4.92}$$

The boundary condition $\phi_0 = 0$ determines $C_- = -C_+$, while $\phi_{n_{el}} = 0$ determines $\frac{\omega L}{\pi} \in \mathbb{Z}$. Substituting (4.92) into (4.91), we obtain the relation between $\omega^h h$ and ωh:

$$\cos(\omega h) = \frac{3 (\omega^h h)^4 - 104 (\omega^h h)^2 + 240}{(\omega^h h)^4 + 16 (\omega^h h)^2 + 240}. \tag{4.93}$$

The natural frequencies are obtained solving (4.93) with respect to $\omega^h h$. Unlike the linear case, each real value of ωh is associated with two values of $\omega^h h$, on two different branches, termed *acoustical* and *optical*. It can be shown that a monotone $\omega^h h$ versus ωh relation is obtained representing the two branches in the range $\omega h \in [0, \pi]$ and $\omega h \in [\pi, 2\pi]$ respectively (see Figs. 4.20 and 4.21). Therefore, we associate to

$$\omega h = \frac{n\pi}{n_{el}}, \qquad n = 1, \ldots n_{el} - 1, \tag{4.94}$$

the smallest positive root of (4.93), obtaining the acoustical branch, and we associate to

$$\omega h = \frac{n\pi}{n_{el}}, \qquad n = n_{el} + 1, \ldots 2n_{el} - 1 \equiv N; \tag{4.95}$$

the highest root of (4.93), obtaining the optical branch. These roots are the natural frequencies that can be obtained by bubble elimination. The frequency $\omega^h h = \sqrt{10}$, which gives bubble resonance is associated with the normal mode

$$\begin{aligned}
\phi_A &= 0, & \forall A = 0, \ldots, n_{el}, \\
\phi_{A+1/2} &= C(-1)^A & \forall A = 0, \ldots, n_{el} - 1.
\end{aligned} \tag{4.96}$$

Since $\omega^h h = \sqrt{10}$ is located between the two branches, this frequency is associated with mode number $n = n_{el}$. Then, all normal modes at element endpoints are given by

$$\phi_A = C \sin\left(\frac{An\pi}{N+1}\right), \qquad A = 0, 1, \ldots n_{el}, \tag{4.97}$$

n being the mode number. Therefore, (4.97) is an interpolate of the exact modes (at element endpoint nodes).

The numerical error in the calculation of natural frequencies is visualized by the graph of ω^h / ω versus ωh, shown in Fig. 4.21.

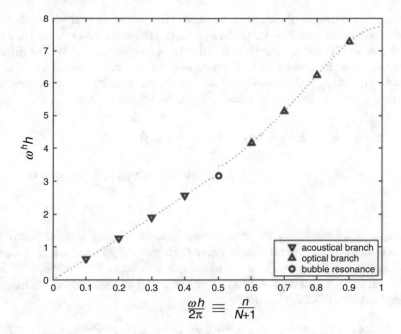

Fig. 4.20 Analytically computed (discrete) natural frequencies for the quadratic *p*-method (*N* = 9)

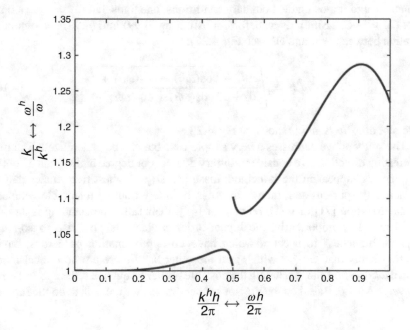

Fig. 4.21 Analytically computed (discrete) natural frequencies for the quadratic *p*-method

Finally, we discuss the quadratic k-method. A rigorous analysis of this case would be too technical; here we prefer to maintain the discussion informal and refer the reader to [71] for the technical details. The equations of (4.79) have different expression for the interior stencil points and for the stencil points close to the boundary (the first and last two equations). We also have for the boundary conditions $\phi_0 = \phi_{N+1} = 0$. In the interior stencil points, the equations read

$$\frac{1}{6h}(\phi_{A-2} + 2\phi_{A-1} - 6\phi_A + 2\phi_{A+1} + \phi_{A+2})$$

$$+ (\omega^h)^2 \frac{h}{120}(\phi_{A-2} + 26\phi_{A-1} + 66\phi_A + 26\phi_{A+1} + \phi_{A+2}) = 0, \quad (4.98)$$

$$\forall A = 3, \ldots, N-2.$$

A major difference from the cases considered previously is that (4.98) is a homogeneous recurrence relation of order 4. Because of its structure, its solutions can be written as linear combinations of the four solutions $e^{\pm \imath \omega h A}$ and $e^{\pm \imath \widetilde{\omega}h A}$. Here ω^h is real and positive while $\widetilde{\omega}^h$ has a nonzero imaginary part. More precisely, the general solution of (4.98) has the form

$$\phi_A = C_+ e^{\imath \omega h A} + C_- e^{-\imath \omega h A} + \widetilde{C}_+ e^{\imath \widetilde{\omega}h A} + \widetilde{C}_- e^{-\imath \widetilde{\omega}h A}, \quad (4.99)$$

for any constants $C_+, C_-, \widetilde{C}_+, \widetilde{C}_-$. Plugging this expression of ϕ_A into the boundary equations and imposing the boundary conditions, one finds that $\widetilde{C}_+ = \widetilde{C}_- = 0$ and that $C_+ = -C_-$. Similarly as before, substituting (4.99) into (4.98), we obtain the relation between $\omega^h h$ and ωh (see Fig. 4.22):

$$\omega^h h = \sqrt{\frac{20(2 - \cos(\omega h) - \cos(\omega h)^2)}{16 + 13\cos(\omega h) + \cos(\omega h)^2}}. \quad (4.100)$$

The plot of $\omega^h h$ vs. ωh is shown in Fig. 4.23.

The study above addresses a very simple case but can be generalized. The most interesting direction is to consider arbitrary degree. For degree higher than 2 "outlier frequencies" appear in the k-method: these are $O(p)$ highest frequencies that are numerically spurious and, though they can be filtered out by a suitable geometric parametrization [71] or mesh refinement [45], their full understanding is an open problem. Most importantly, the higher-order p-elements give rise to so-called "optical branches" to spectra, which have no approximation properties, having relative errors that *diverge* with p; on the other hand there are no optical modes with the k-method and, excluding the possible outlier frequencies, the spectral errors *converge* with p. Based on the previous observations, we are able to confidently use

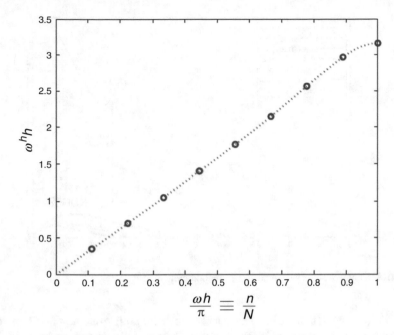

Fig. 4.22 Analytically computed (discrete) natural frequencies for the quadratic k-method ($N = 9$)

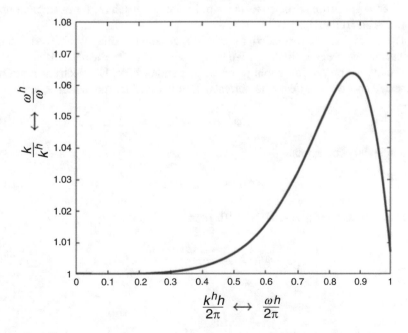

Fig. 4.23 Analytically computed (discrete) natural frequencies for the quadratic k-method

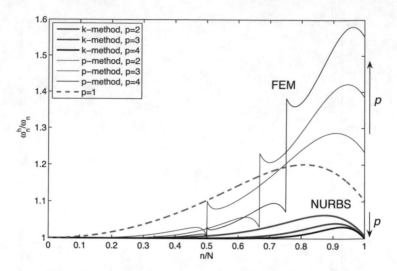

Fig. 4.24 Comparison of k-method and p-method numerical spectra

numerics to calculate invariant analytical spectra for both p-method and k-method. This comparison is reported in Fig. 4.24 and registers a significant advantage for the latter. These results may at least partially explain why classical higher-order finite elements have not been widely adopted in problems for which the upper part of the discrete spectrum participates in a significant way, such as, for example, impact problems and turbulence.

The study can be extended to multidimensional problems as well, mainly confirming the previous findings. We refer again to [71] for the details.

Finally, we present a simple problem that shows how the spectrum properties presented above may affect a numerical solution. Consider the model equation

$$\phi'' + k\phi = 0, \tag{4.101}$$

with boundary conditions

$$\phi(0) = 1, \quad \phi(1) = 0. \tag{4.102}$$

The solution to problem (4.101)–(4.102) can be written as

$$\phi(x, k) = \frac{\sin(k(1 - x))}{\sin(k)}. \tag{4.103}$$

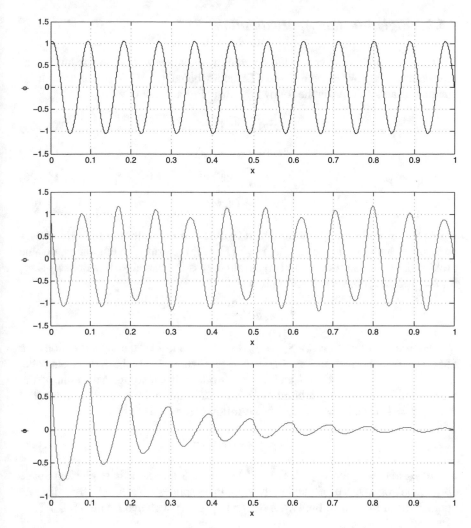

Fig. 4.25 Solutions of the boundary value problem (4.101)–(4.102) for $p = 3$ computed with $k = 71$: exact solution (top), k-method (31 degrees-of-freedom, center) and p-method (31 degrees-of-freedom, bottom)

We numerically solve (4.101)–(4.102) for $k = 71$, selecting $p = 3$ and 31 degrees-of-freedom for the k- and p-method. The results are reported in Fig. 4.25. The k-method is able to reproduce correctly the oscillations of the exact solutions (phase and amplitude are approximately correct). There are no stopping bands for the k-method. On the contrary, since $k = 71$ is within the 2nd stopping band of the p-method, a spurious attenuation is observed. We refer to [71] for the complete study.

4.5.2   Eigenfunction Approximation

Let Ω be a bounded and connected domain in $\mathbb{R}^d$, where $d \in \mathbb{Z}^+$ is the number of space dimensions. We assume Ω has a Lipschitz boundary $\partial\Omega$. We assume both are continuous and coercive in the following sense: For all $v, w \in \mathcal{V}$,

$$a(v, w) \leq \|v\|_E \|w\|_E \tag{4.104}$$

$$\|w\|_E^2 = a(w, w) \tag{4.105}$$

$$(v, w) \leq \|v\| \|w\| \tag{4.106}$$

$$\|w\|^2 = (v, w) \tag{4.107}$$

where $\|\cdot\|_E$ is the energy-norm which is assumed equivalent to the $(H^m(\Omega))^n$-norm on $\mathcal{V}$ and $\|\cdot\|$ is the $(L^2(\Omega))^n = (H^0(\Omega))^n$ norm. The elliptic eigenvalue problem is stated as follows: Find eigenvalues $\lambda_l \in \mathbb{R}^+$ and eigenfunctions $u_l \in \mathcal{V}$, for $l = 1, 2, \ldots, \infty$, such that, for all $w \in \mathcal{V}$,

$$\lambda_l(w, u_l) = a(w, u_l) \tag{4.108}$$

It is well-known that $0 < \lambda_1 \leq \lambda_2 \leq \lambda_3 \leq \ldots$, and that the eigenfunctions are $(L^2(\Omega))^n$-orthonormal, that is, $(u_k, u_l) = \delta_{kl}$ where δ_{kl} is the Kronecker delta, for which $\delta_{kl} = 1$ if $k = l$ and $\delta_{kl} = 0$ otherwise. The normalization of the eigenfunctions is actually arbitrary. We have assumed without loss of generality that $\|u_l\| = 1$, for all $l = 1, 2, \ldots, \infty$. It follows from (4.108) that

$$\|u_l\|_E^2 = a(u_l, u_l) = \lambda_l \tag{4.109}$$

and $a(u_k, u_l) = 0$ for $k \neq l$. Let $\mathcal{V}^h$ be either a standard finite element space (p-method) or a space of maximally smooth B-splines (k-method). The discrete counterpart of (4.108) is: Find $\lambda_l^h \in \mathbb{R}^+$ and $u_l^h \in \mathcal{V}^h$ such that for all $w^h \in \mathcal{V}^h$,

$$\lambda_l^h(w^h, u_l^h) = a(w^h, u_l^h) \tag{4.110}$$

The solution of (4.110) has similar properties to the solution of (4.108). Specifically, $0 < \lambda_1^h \leq \lambda_2^h \leq \ldots \leq \lambda_N^h$, where N is the dimension of $\mathcal{V}^h$, $(u_k^h, u_l^h) = \delta_{kl}$, $\|u_l^h\|_E^2 = a(u_l^h, u_l^h) = \lambda_l^h$, and $a(u_k^h, u_l^h) = 0$ if $k \neq l$. The comparison of $\{\lambda_l^h, u_l^h\}$ to $\{\lambda_l, u_l\}$ for all $l = 1, 2, \ldots, N$ is the key to gaining insight into the errors of the discrete approximations to the elliptic boundary-value problem and the parabolic and hyperbolic initial-value problems.

The fundamental global error analysis result for elliptic eigenvalue problems is the *Pythagorean eigenvalue error theorem*. It is simply derived and is done so on page 233 of Strang and Fix [111] The theorem is global in that it is applicable to each and every mode in the discrete approximation. Provided that $\|u_l^h\| = \|u_l\|$,

$$\frac{\lambda_l^h - \lambda_l}{\lambda_l} + \frac{\|u_l^h - u_l\|^2}{\|u_l\|^2} = \frac{\|u_l^h - u_l\|_E^2}{\|u_l\|_E^2}, \qquad \forall l = 1, 2, \ldots, N \qquad (4.111)$$

Note that the relative error in the lth eigenvalue and the square of the relative $(L^2(\Omega))^n$-norm error in the lth eigenfunction sum to equal the square of the relative energy-norm error in the lth eigenfunction. Due to the normalization introduced earlier, (4.111) can also be written as

$$\frac{\lambda_l^h - \lambda_l}{\lambda_l} + \|u_l^h - u_l\|^2 = \frac{\|u_l^h - u_l\|_E^2}{\lambda_l}, \qquad \forall l = 1, 2, \ldots, N \qquad (4.112)$$

See Fig. 4.26. We note that the first term in (4.112) is always non-negative as $\lambda_l^h \geq \lambda_l$, a consequence of the "minimax" characterization of eigenvalues (see [111], p. 223). It also immediately follows from (4.112) that

$$\lambda_l^h - \lambda_l \leq \|u_l^h - u_l\|_E^2 \qquad (4.113)$$

$$\|u_l^h - u_l\|^2 \leq \frac{\|u_l^h - u_l\|_E^2}{\lambda_l} \qquad (4.114)$$

We consider the elliptic eigenvalue problem for the second-order differential operator in one-dimension with homogeneous Dirichlet boundary conditions. The variational form of the problem is given by (4.108), in which

$$a(w, u_l) = \int_0^1 \frac{dw}{dx} \frac{du_l}{dx} dx \qquad (4.115)$$

$$(w, u_l) = \int_0^1 w u_l dx \qquad (4.116)$$

Fig. 4.26 Graphical representation of the Pythagorean eigenvalue error theorem

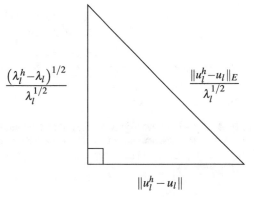

$$\frac{\left(\lambda_l^h - \lambda_l\right)^{1/2}}{\lambda_l^{1/2}}$$

$$\frac{\|u_l^h - u_l\|_E}{\lambda_l^{1/2}}$$

$$\|u_l^h - u_l\|$$

The eigenvalues are $\lambda_l = \pi^2 l^2$ and the eigenfunctions are $u_l = \sqrt{2}\sin(l\pi x)$, $l = 1, 2, \ldots, \infty$. Now, we will present the eigenvalue errors, rather than the eigenfrequency errors, and, in addition, $L^2(0, 1)$- and energy-norm eigenfunction errors. We will plot the various errors in a format that represents the Pythagorean eigenvalue error theorem budget. We will restrict our study to quadratic, cubic, and quartic finite elements and B-splines. In all cases, we assume linear geometric parametrizations and uniform meshes. Strictly speaking, for the k-method the results are only true for sufficiently large N, due to the use of open knot vectors, but in this case "sufficiently large" is not very large at all, say $N > 30$. For smaller spaces, the results change slightly. The results that we present here were computed using $N \approx 1000$ and, in [73], have been validated using a mesh convergence study and by comparing to analytical computations.

Let us begin with results for the quadratic k-method, i.e. C^1-continuous quadratic B-splines, presented in Fig. 4.27a. The results for the relative eigenvalue errors (red curve) follow the usual pattern that has been seen before. The squares of the eigenfunction errors in $L^2(0, 1)$ are also well-behaved (blue curve) with virtually no discernible error until about $l/N = 0.6$, and then monotonically increasing errors in the highest modes. The sums of the errors produce the squares of the relative

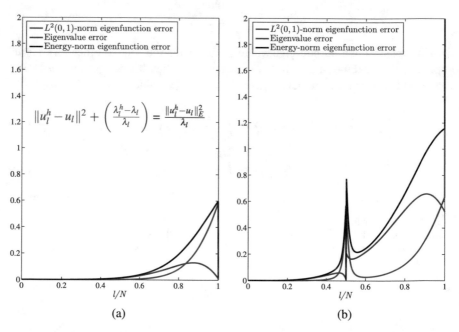

Fig. 4.27 Pythagorean eigenvalue error theorem budget for quadratic elements. (**a**) C^1-continuous B-splines; (**b**) C^0-continuous finite elements. The blue curves are $\|u_l^h - u_l\|^2$, the red curves are $(\lambda_l^h - \lambda_l)/\lambda_l$, and the black curves are $\|u_l^h - u_l\|_E^2/\lambda_l$. Note that $\|u_l\| = \|u_l^h\| = 1$, $\|u_l\|_E^2 = \lambda_l$, and $\|u_l^h\|_E^2 = \lambda_l^h$

energy-norm errors (black curve), as per the Pythagorean eigenvalue error theorem budget. There are no surprises here.

Next we compare with quadratic p-method, i.e., C^0-continuous quadratic finite elements in Fig. 4.27b. The pattern of eigenvalue errors (red curve), consisting of two branches, the acoustic branch for $l/N < 1/2$, and the optical branch for $l/N \geq 1/2$, is the one known from Sect. 4.5.1. However, the eigenfunction error in $L^2(0, 1)$ (blue curve) represents a surprise in that there is a large spike about $l/N = 1/2$, the transition point between the acoustic and optical branches. Again, the square of the energy-norm eigenfunction error term (black curve) is the sum, as per the budget. This is obviously not a happy result. It suggests that if modes in the neighborhood of $l/N = 1/2$ are participating in the solution of a boundary-value or initial-value problem, the results will be in significant error. The two unpleasant features of this result are (1) the large magnitude of the eigenfunction errors about $l/N = 1/2$ and (2) the fact that they occur at a relatively low mode number. That the highest modes are significantly in error is well-established for C^0-continuous finite elements, but that there are potential danger zones much earlier in the spectrum had not been recognized previously. The midpoint of the spectrum in one-dimension corresponds to the quarter point in two dimensions and the eighth point in three dimensions, and so one must be aware of the fact that the onset of inaccurate modes occurs much earlier in higher dimensions.

The spikes in the eigenfunction error spectrum for C^0-finite elements raise the question as to whether or not the eigenfunctions are representative of the best approximation to eigenfunctions in the vicinity of $l/N = 1/2$. To answer this question, we computed the $L^2(0, 1)$ best approximations of some of the exact eigenfunctions and plotted them in Fig. 4.28b. (They are indicated by ×.) The case for C^1-continuous quadratic B-splines is presented in Fig. 4.28a for comparison. For

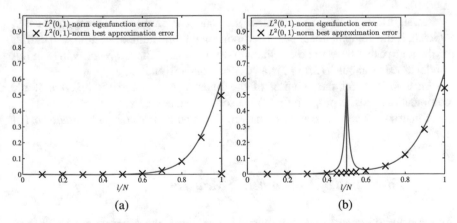

(a)   (b)

Fig. 4.28 Comparisons of eigenfunctions computed by the Galerkin method with $L^2(0, 1)$ best approximations of the exact eigenfunctions. (a) C^1-continuous quadratic B-splines; (b) C^0-continuous quadratic finite elements. The blue curves are $\|u_l^h - u_l\|^2$, where u_l^h is the Galerkin approximation of u_l, and the ×'s are $\|\widetilde{u}_l^h - u_l\|^2$, where $\widetilde{u}_l^h$ is the $L^2(0, 1)$ best approximation of u_l

this case there are almost no differences between the best approximation of the exact eigenfunctions and the computed eigenfunctions. However, for the C^0-continuous quadratic finite elements, the differences between the computed eigenfunctions and the $L^2(0, 1)$ best approximations of the exact eigenfunctions are significant, as can be seen in Fig. 4.28b. The spike is nowhere to be seen in the best approximation results. We conclude that the Galerkin formulation of the eigenvalue problem is simply not producing good approximations to the exact eigenfunctions about $l/N = 1/2$ in the finite element case.

For higher-order cases, in particular cubic and quartic, see [73] where it is shown that the essential observations made for the quadratic case persist. An investigation of the behavior of outlier frequencies and eigenfunctions is also presented in [73], along with discussion of the significance of eigenvalue and eigenfunction errors in the context of elliptic, parabolic and hyperbolic partial differential equations.

4.6 Computational Efficiency

High-degree high-regularity splines, and extensions, deliver higher accuracy per degree-of-freedom in comparison to C^0 finite elements but at a higher computational cost, when standard finite element implementation is adopted. In this section we present recent advances on the formation of the system matrix (Sects. 4.6.1 and 4.6.2), the solution of linear systems (Sect. 4.6.3) and the use a matrix-free approach (Sect. 4.6.4)

We consider, as a model case, the d-dimensional Poisson problem on a single-patch domain, and an isogeometric tensor-product space of degree p, continuity C^{p-1} and total dimension N, with $N \gg p$. This is the typical setting for the k-method.

An algorithm for the formation of the matrix is said to be (computationally) *efficient* if the computational cost is proportional to the number of non-zero entries of the matrix that have to be calculated (storage cost). The *stiffness matrix* in our model case has about $N(2p + 1)^d \approx CNp^d$ non-zero entries.

An algorithm for the solution of the linear system matrix is *efficient* if the computational cost is proportional to the solution size, i.e., N.

A matrix-free approach aims at an overall computational cost and storage cost of CN.

4.6.1 Formation of Isogeometric Matrices

When a finite element code architecture is adopted, the simplest approach is to use element-wise Gaussian quadrature and element-by-element assembling. Each elemental stiffness matrix has dimension $(p + 1)^{2d}$ and each entry is calculated by quadrature on $(p + 1)^d$ Gauss points. The total cost is $CN_{EL}p^{3d} \approx CNp^{3d}$, where N_{EL} is the number of elements and, for the k-method, $N_{EL} \approx N$.

A strategy to reduce the cost is to reduce the number of quadrature points. The paper [72] proposed to use *generalized* Gaussian rules for smooth spline integrands. These rules are not known analytically and need to be computed numerically (see also [9, 13, 14] and the recent paper [75] where the problem is effectively solved by a Newton method with continuation). Furthermore, *reduced* quadrature rules have been considered in [1, 101] and [65]. Another important step is to reduce the number of operations by arranging the computations in a way that exploits the tensor-product structure of multivariate splines: this is done by so-called sum factorization achieving a computational cost of CNp^{2d+1}, see [3].

Keeping the element-wise assembling loop is convenient, as it allows reusing available finite element routines. On the other hand, as the computation of each elemental stiffness matrix needs at least Cp^{2d} FLOPs (proportional to the elemental matrix size and assuming integration cost does not depend on p) the total cost is at least $CN_{EL}p^{2d} \approx CNp^{2d}$.

Further cost reduction is possible but only with a change of paradigm from element-wise assembling. This study has been recently initiated and two promising strategies have emerged.

One idea, in [84], is to use a low-rank expansion in order to approximate the stiffness matrix by a sum of R Kronecker type matrices that can be easily formed, thanks to their tensor-product structure. This approach has a computational cost of $CNRp^d$ FLOPs.

Another possibility, from [42], is based on two new concepts. The first is the use of a row loop instead of an element loop, and the second is the use of *weighted* quadrature. This will be discussed in the next section.

4.6.2 Weighted Quadrature

This idea has been proposed in [42]. Assume we want to compute integrals of the form:

$$\int_0^1 \widehat{B}_i(\zeta)\,\widehat{B}_j(\zeta)\,d\zeta, \qquad (4.117)$$

where $\{\widehat{B}_i\}_{i=1,\dots,n}$ are p-degree univariate B-spline basis functions. Consider for simplicity only the maximum regularity case, C^{p-1}, and for the moment a periodic uniform knot vector. Being in the context of Galerkin method, $\widehat{B}_i(\zeta)$ represents a test function and $\widehat{B}_j(\zeta)$ represents a trial function.

We are interested in a fixed point quadrature rule. In the lowest degree case, $p = 1$, exact integration is performed by a composite Cavalieri-Simpson rule:

$$\int_0^1 \widehat{B}_i(\zeta)\,\widehat{B}_j(\zeta)\,d\zeta = \mathbb{Q}^{CS}(\widehat{B}_i\,\widehat{B}_j) = \sum_q w_q^{CS}\widehat{B}_i(x_q^{CS})\,\widehat{B}_j(x_q^{CS}), \quad (4.118)$$

where x_q^{CS} are the quadrature points and w_q^{CS} the relative weights. In the above hypotheses the points x_q^{CS} are the knots and the midpoints of the knot-spans and $w_q^{CS} = \frac{h}{3}$ on knots and $w_q^{CS} = \frac{2h}{3}$ on midpoints.

Unbalancing the role of the test and the trial factors in (4.118), we can see it as a weighted quadrature:

$$\int_0^1 \widehat{B}_i(\zeta)\, \widehat{B}_j(\zeta)\, d\zeta = \mathbb{Q}_i^{WQ}(\widehat{B}_j) = \sum_q w_{q,i}^{WQ}\, \widehat{B}_j(x_{q,i}^{WQ}), \qquad (4.119)$$

where $x_q^{CS} = x_{q,i}^{WQ}$ and $w_{q,i}^{WQ} = \widehat{B}_i(x_{q,i}^{WQ})w_q^{CS}$. Because of the local support of the function $\widehat{B}_i$ only in three points the quadrature $\mathbb{Q}_i^{WQ}$ is non-zero and the weights are equal to $\frac{h}{3}$.

If we go to higher degree, we need more quadrature points in (4.118). For p-degree splines the integrand $\widehat{B}_i\widehat{B}_j$ is a piecewise polynomial of degree $2p$ and an element-wise integration requires $2p + 1$ equispaced points, or $p + 1$ Gauss points, or about $p/2$ points with generalized Gaussian integration (see [9, 31, 41, 72]). On the other hand, we can generalize (4.119) to higher degree still using as quadrature points only the knots and midpoints of the knot spans. Indeed this choice ensures that, for each basis function $\widehat{B}_i$, $i = 1, \ldots, n$, there are $2p + 1$ "active" quadrature points where $\widehat{B}_i$ is nonzero. Therefore we can compute the $2p + 1$ quadrature weights by imposing conditions for the $2p + 1$ B-splines $\widehat{B}_j$ that need to be exactly integrated. Clearly, the advantage of the weighted quadrature approach is that its computational complexity, i.e., the total number of quadrature points, is independent of p.

Given a weighted quadrature rule of the kind above, we are then interested in using it for the approximate calculation of integrals as:

$$\int_0^1 c(\zeta)\widehat{B}_i(\zeta)\, \widehat{B}_j(\zeta)\, d\zeta \approx \mathbb{Q}_i^{WQ}\left(c(\cdot)\widehat{B}_j(\cdot)\right) = \sum_q w_{q,i}^{WQ} c(x_{q,i}^{WQ})\widehat{B}_j(x_{q,i}^{WQ}).$$

$$(4.120)$$

For a non-constant function $c(\cdot)$, (4.120) is in general just an approximation.

We consider now the model reaction-diffusion problem

$$\begin{cases} -\nabla^2 u + u = f & \text{on} \quad \Omega, \\ \qquad\quad u = 0 & \text{on} \quad \partial\Omega, \end{cases} \qquad (4.121)$$

Its Galerkin approximation requires the stiffness matrix $\mathbb{S}$ and mass matrix $\mathbb{M}$. After change of variable we have $\mathbb{M} = \{m_{i,j}\} \in \mathbb{R}^{N \times N}$ with entries given by:

$$m_{i,j} = \int_{\widehat{\Omega}} \widehat{B}_i\, \widehat{B}_j\, \det\widehat{\mathbf{D}}F\, d\zeta.$$

For notational convenience we write:

$$m_{i,j} = \int_{\widehat{\Omega}} \widehat{B}_i(\zeta)\, \widehat{B}_j(\zeta)\, c(\zeta)\, d\zeta \ . \tag{4.122}$$

In more general cases, the factor c incorporates the coefficient of the equation and, for NURBS functions, the polynomial denominator. Similarly for the stiffness matrix $\mathbb{S} = \{s_{i,j}\} \in \mathbb{R}^{N \times N}$ we have:

$$s_{i,j} = \int_{\widehat{\Omega}} \left(\widehat{\mathbf{D}}F^{-T} \widehat{\nabla} \widehat{B}_i \right)^T \left(\widehat{\mathbf{D}}F^{-T} \widehat{\nabla} \widehat{B}_j \right) \det \widehat{\mathbf{D}}F \, d\zeta$$

$$= \int_{\widehat{\Omega}} \widehat{\nabla} \widehat{B}_i^T \left([\widehat{\mathbf{D}}F^{-1} \widehat{\mathbf{D}}F^{-T}] \det \widehat{\mathbf{D}}F \right) \widehat{\nabla} \widehat{B}_j \, d\zeta$$

which we write in compact form:

$$s_{i,j} = \sum_{l,m=1}^{d} \int_{\widehat{\Omega}} \left(\widehat{\nabla} \widehat{B}_i(\zeta) \right)_l c_{l,m}(\zeta) \left(\widehat{\nabla} \widehat{B}_j(\zeta) \right)_m \, d\zeta. \tag{4.123}$$

Here we have denoted by $\{c_{l,m}(\zeta)\}_{l,m=1,\dots,d}$ the following matrix:

$$c_{l,m}(\zeta) = \left\{ \left[\widehat{\mathbf{D}}F^{-1}(\zeta) \widehat{\mathbf{D}}F^{-T}(\zeta) \right] \det \widehat{\mathbf{D}}F(\zeta) \right\}_{l,m}. \tag{4.124}$$

The number of non-zero elements N_{NZ} of $\mathbb{M}$ and $\mathbb{S}$ depends on the polynomial degree p and the required regularity r. We introduce the following sets:

$$\mathscr{I}_{l,i_l} = \left\{ j_l \in \{1, \dots, n_l\} \ s.t. \ \widehat{B}_{i_l} \cdot \widehat{B}_{j_l} \neq 0 \right\}, \qquad \mathscr{I}_i = \prod_{l=1}^{d} \mathscr{I}_{l,i_l} \tag{4.125}$$

We have $\#\mathscr{I}_{l,i} \leq (2p+1)$ and $N_{NZ} = O(N\, p^d)$. In particular, with maximal regularity in the case $d=1$ one has $N_{NZ} = (2p+1)N - p(p+1)$.

Consider the calculation of the mass matrix. The first step is to write the integral in a nested way, as done in [3]:

$$m_{i,j} = \int_{\widehat{\Omega}} \widehat{B}_i(\zeta) \widehat{B}_j(\zeta) c(\zeta)\, d\zeta$$

$$= \int_0^1 \widehat{B}_{i_1}(\zeta_1) \widehat{B}_{j_1}(\zeta_1) \left[\int_0^1 \widehat{B}_{i_2}(\zeta_2) \widehat{B}_{j_2}(\zeta_2) \cdots \right.$$

$$\left. \left[\int_0^1 \widehat{B}_{i_d}(\zeta_d) \widehat{B}_{j_d}(\zeta_d) c(\zeta)\, d\zeta_d \right] \cdots d\zeta_2 \right] d\zeta_1$$

The idea in is to isolate the *test function* $\widehat{B}_{i_l}$ univariate factors in each univariate integral and to consider it as a weight for the construction of the weighted quadrature (WQ) rule. This leads to a quadrature rule for each i_l that is:

$$
\begin{aligned}
m_{i,j} \approx \tilde{m}_{i,j} &= \mathbb{Q}_i^{WQ}\left(\widehat{B}_j(\zeta)c(\zeta)\right) = \mathbb{Q}_i\left(\widehat{B}_j(\zeta)c(\zeta)\right) \\
&= \mathbb{Q}_{i_1}\left(\widehat{B}_{j_1}(\zeta_1)\mathbb{Q}_{i_2}\left(\cdots\mathbb{Q}_{i_d}\left(\widehat{B}_{j_d}(\zeta_d)c(\zeta)\right)\right)\right) .
\end{aligned}
\tag{4.126}
$$

Notice that we drop from now on the label WQ used in the introduction in order to simplify notation. The key ingredients for the construction of the quadrature rules that preserve the optimal approximation properties are the exactness requirements. Roughly speaking, exactness means that in (4.126) we have $m_{i,j} = \tilde{m}_{i,j}$ whenever c is a constant coefficient. When the stiffness term is considered, also terms with derivatives have to be considered.

We introduce the notation:

$$
\begin{aligned}
\mathbb{I}_{l,i_l,j_l}^{(0,0)} &:= \int_0^1 \widehat{B}_{i_l}(\zeta_l)\widehat{B}_{j_l}(\zeta_l)\,d\zeta_l \\
\mathbb{I}_{l,i_l,j_l}^{(1,0)} &:= \int_0^1 \widehat{B}'_{i_l}(\zeta_l)\widehat{B}_{j_l}(\zeta_l)\,d\zeta_l \\
\mathbb{I}_{l,i_l,j_l}^{(0,1)} &:= \int_0^1 \widehat{B}_{i_l}(\zeta_l)\widehat{B}'_{j_l}(\zeta_l)\,d\zeta_l \\
\mathbb{I}_{l,i_l,j_l}^{(1,1)} &:= \int_0^1 \widehat{B}'_{i_l}(\zeta_l)\widehat{B}'_{j_l}(\zeta_l)\,d\zeta_l
\end{aligned}
\tag{4.127}
$$

For each integral in (4.127) we define a quadrature rule: we look for

- points $\tilde{x}_q = (\tilde{x}_{1,q_1}, \tilde{x}_{2,q_2}, \ldots, \tilde{x}_{d,q_d})$ with $q_l = 1, \ldots n_{\mathrm{QP},l}$, with N_{QP} is $\#\{\tilde{x}\} = \prod_{l=1}^d n_{\mathrm{QP},l}$;
- for each index $i_l = 1, \ldots, n_{\mathrm{DOF},l}$; $l = 1, \ldots, d$, four quadrature rules such that:

$$
\begin{aligned}
\mathbb{Q}_{i_l}^{(0,0)}(f) &:= \sum_{q_l=1}^{n_{\mathrm{QP},l}} w_{l,i_l,q_l}^{(0,0)} f(\tilde{x}_{l,q_l}) \approx \int_0^1 f(\zeta_l)\widehat{B}_{i_l}(\zeta_l)d\zeta_l ; \\
\mathbb{Q}_{i_l}^{(1,0)}(f) &:= \sum_{q_l=1}^{n_{\mathrm{QP},l}} w_{l,i_l,q_l}^{(1,0)} f(\tilde{x}_{l,q_l}) \approx \int_0^1 f(\zeta_l)\widehat{B}_{i_l}(\zeta_l)d\zeta_l ; \\
\mathbb{Q}_{i_l}^{(0,1)}(f) &:= \sum_{q_l=1}^{n_{\mathrm{QP},l}} w_{l,i_l,q_l}^{(0,1)} f(\tilde{x}_{l,q_l}) \approx \int_0^1 f(\zeta_l)\widehat{B}'_{i_l}(\zeta_l)d\zeta_l ; \\
\mathbb{Q}_{i_l}^{(1,1)}(f) &:= \sum_{q_l=1}^{n_{\mathrm{QP},l}} w_{l,i_l,q_l}^{(1,1)} f(\tilde{x}_{l,q_l}) \approx \int_0^1 f(\zeta_l)\widehat{B}'_{i_l}(\zeta_l)d\zeta_l .
\end{aligned}
\tag{4.128}
$$

fulfilling the exactness requirement:

$$
\begin{aligned}
\mathbb{Q}_{i_l}^{(0,0)}(\widehat{B}_{j_l}) &= \mathbb{I}_{l,i_l,j_l}^{(0,0)} \\
\mathbb{Q}_{i_l}^{(1,0)}(\widehat{B}'_{j_l}) &= \mathbb{I}_{l,i_l,j_l}^{(1,0)} \\
\mathbb{Q}_{i_l}^{(0,1)}(\widehat{B}_{j_l}) &= \mathbb{I}_{l,i_l,j_l}^{(0,1)} \quad, \quad \forall j_l \in \mathscr{I}_{l,i_l}\,. \\
\mathbb{Q}_{i_l}^{(1,1)}(\widehat{B}'_{j_l}) &= \mathbb{I}_{l,i_l,j_l}^{(1,1)}
\end{aligned}
\tag{4.129}
$$

We also require that the quadrature rules $\mathbb{Q}_{i_l}^{(\cdot,\cdot)}$ have support included in the support of $\widehat{B}_{i_l}$, that is

$$
q_l \notin \mathscr{Q}_{l,i_l} \Rightarrow w_{l,i_l,q_l}^{(\cdot,\cdot)} = 0\,.
\tag{4.130}
$$

where $\mathscr{Q}_{l,i_l} := \{q_l \in 1, \ldots, n_{\mathsf{QP},l}$ s.t. $\widetilde{x}_{l,q_l} \in \operatorname{supp}\left(\widehat{B}_{i_l}\right)\}$; recall that here the support of a function is considered an open set. Correspondingly, we introduce the set of multi-indexes $\mathscr{Q}_i := \prod_{l=1}^{d} \mathscr{Q}_{l,i_l}$.

Once the points $\widetilde{x}_q$ are fixed, the quadrature rules have to be determined by the exactness requirements, that are a system of linear equations of the unknown weights (each of the (4.129)). For that we require

$$
\#\mathscr{Q}_{l,i_l} \geq \#\mathscr{I}_{l,i_l}\,.
\tag{4.131}
$$

See[42] for a discussion on the well-posedness of the linear systems for the weights.

The construction of a global grid of quadrature points is done in order to save computations. For the case of maximum C^{p-1} regularity considered here, the choice for quadrature points of [42] is endpoints (knots) and midpoints of all internal knot-spans, while for the boundary knot-spans (i.e. those that are adjacent to the boundary of the parameter domain $\widehat{\Omega}$) we take $p+1$ equally spaced points. Globally $N_{\mathsf{QP}} \approx 2^d N_{\mathsf{EL}} = O(N)$ considering only the dominant term.

When all the quadrature rules are available we can write the computation of the approximate mass matrix following (4.126), where the quadrature rules $\mathbb{Q}_{i_1}^{(0,0)}, \ldots, \mathbb{Q}_{i_d}^{(0,0)}$ are used. Similar formulae and algorithms can be written for the stiffness matrix. In that case, all the integrals are approximated separately, and all the quadrature rules $\mathbb{Q}_{i_l}^{(\cdot,\cdot)}$ are necessary.

The mass matrix formation algorithm is mainly a loop over all rows i, for each i we consider the calculation of

$$
\widetilde{m}_{i,j} = \sum_{q \in \mathscr{Q}_i} w_{i,q}^{(0,0)} c(\widetilde{x}_q) \widehat{B}_j\left(\widetilde{x}_q\right).
\tag{4.132}
$$

where $w_{i,q}^{(0,0)} = w_{1,i_1,q_1}^{(0,0)} \cdots w_{d,i_d,q_d}^{(0,0)}$.

The computational cost of (4.132) is minimised by a sum factorization approach. Nota that (4.132) can be rearranged as in (4.126) to obtain the following sequence of nested summations:

$$\tilde{m}_{i,j} = \sum_{q_1 \in \mathcal{Q}_{1,i_1}} w_{1,i_1,q_1}^{(0,0)} \widehat{B}_{j_1}(x_{1,q_1}) \left(\sum_{q_2 \in \mathcal{Q}_{2,i_2}} \cdots \right. \tag{4.133}$$

$$\left. \sum_{q_d \in \mathcal{Q}_{d,i_d}} w_{d,i_d,q_d}^{(0,0)} \widehat{B}_{j_d}(x_{d,q_d}) c\left(x_{1,q_1}, \ldots, x_{d,q_d}\right) \right).$$

To write (4.133) in a more compact form, we introduce the notion of matrix-tensor product. Let $\mathcal{X} = \left\{ x_{k_1,\ldots,k_d} \right\} \in \mathbb{R}^{n_1 \times \ldots \times n_d}$ be a d−dimensional tensor, and let $m \in \{1, \ldots, d\}$. The m−mode product of $\mathcal{X}$ with a matrix $A = \left\{ a_{i,j} \right\} \in \mathbb{R}^{t \times n_m}$, denoted with $\mathcal{X} \times_m A$, is a tensor of dimension $n_1 \times \ldots \times n_{m-1} \times t \times n_{m+1} \times \ldots \times n_d$, with components

$$(\mathcal{X} \times_m A)_{k_1,\ldots,k_d} = \sum_{j=1}^{n_m} a_{k_m,j} \, x_{k_1,\ldots k_{m-1},j,k_{m+1},\ldots k_d}.$$

For $l = 1, \ldots, d$ and $i_l = 1, \ldots, n_l$ we define the matrices

$$\mathbf{B}^{(l,i_l)} = \left(\widehat{B}_{j_l}(x_{l,q_l}) \right)_{j_l \in \mathcal{I}_{l,i_l}, q_l \in \mathcal{Q}_{l,i_l}}, \qquad \mathbf{W}^{(l,i_l)} = \mathrm{diag}\left(\left(w_{l,i_l,q_l}^{(0,0)} \right)_{q_l \in \mathcal{Q}_{l,i_l}} \right),$$

where $\mathrm{diag}(v)$ denotes the diagonal matrix obtained by the vector v. We also define, for each index i, the d−dimensional tensor

$$\mathcal{C}_i = c(\tilde{\mathbf{x}}_{\mathcal{Q}_i}) = \left(c(\tilde{x}_{1,q_1}, \ldots, \tilde{x}_{d,q_d}) \right)_{q_1 \in \mathcal{Q}_{1,i_1}, \ldots, q_d \in \mathcal{Q}_{d,i_d}}.$$

Using the above notations, we have

$$\tilde{m}_{i,\mathcal{I}_i} = \mathcal{C}_i \times_d \left(\mathbf{B}^{(d,i_d)} \mathbf{W}^{(d,i_d)} \right) \times_{d-1} \cdots \times_1 \left(\mathbf{B}^{(1,i_1)} \mathbf{W}^{(1,i_1)} \right). \tag{4.134}$$

Since with this choice of the quadrature points $\#\mathcal{Q}_{l,i_l}$ and $\#\mathcal{I}_{l,i_l}$ are both $O(p)$, the computational cost associated with (4.134) is $O(p^{d+1})$ FLOPs. Note that $\tilde{m}_{i,\mathcal{I}_i}$ includes all the nonzeros entries of the i-th row of $\tilde{\mathbb{M}}$. Hence if we compute it for each $i = 1, \ldots, N$ the total cost amounts to $O(N\, p^{d+1})$ FLOPs. This approach is summarized in Algorithm 1.

From [42], we report CPU time results for the formation on a single patch domain of mass matrices. Comparison is made with GeoPDEs 3.0, the optimized but SGQ-based MATLAB isogeometric library developed by Rafael Vázquez, see[124]. In Fig. 4.29 we plot the time needed for the mass matrix formation up to degree $p = 10$ with $N = 20^3$. The tests confirm the superior performance of the proposed row-loop WQ-based algorithm vs SGQ. In the case $p = 10$ GeoPDEs takes more than 62 h to

Input: Quadrature rules, evaluations of coefficients
1 **for** $i = 1, \ldots, N$ **do**
2   Set $\mathscr{C}_i^{(0)} := c(\widetilde{x}_{\mathscr{Q}_i})$;
3   **for** $l = d, d-1, \ldots, 1$ **do**
4   Load the quadrature rule $\mathbb{Q}_{i_l}^{(0,0)}$ and form the matrices $\mathrm{B}^{(l,i_l)}$ and $\mathrm{W}^{(l,i_l)}$;
5   Compute $\mathscr{C}_i^{(d+1-l)} = \mathscr{C}_i^{(d-l)} \times_l \left(\mathrm{B}^{(l,i_l)}\mathrm{W}^{(l,i_l)}\right)$;
6   **end**
7   Store $\widetilde{m}_{i,\mathscr{I}_i} = \mathscr{C}_i^{(d)}$;
8 **end**

Algorithm 1: Construction of mass matrix by sum-factorization

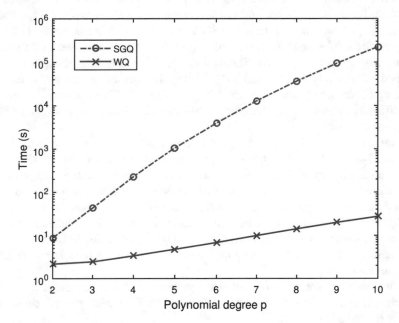

Fig. 4.29 Time for mass matrix assembly in the framework of isogeometric-Galerkin method with maximal regularity on a single patch domain of 20^3 elements. The comparison is between the WQ approach and the SGQ as implemented in GeoPDEs 3.0 [125]

form the mass matrix while the proposed algorithm needs only 27 s, so the use high degrees is possible with WQ.

4.6.3 Linear Solvers and Preconditioners

The study of the computational efficiency of linear solvers for isogeometric discretizations has been initiated in the papers [46, 47], where it has been shown that the algorithms used with the finite element method suffer of performance degradation when used to solve isogeometric linear systems. Consider, for example, a Lagrangian finite element method with polynomial degree p and N degrees-of-freedom, in 3D, for a Poisson model problem:. As shown in [46], a multifrontal direct solver requires $O(N^2)$ FLOPs (under the assumption $N > p^9$) to solve the resulting linear system. If, instead, we consider the isogeometric k-method with C^{p-1} p-degree splines and N degrees-of-freedom, the same direct solver requires $O(N^2 p^3)$ FLOPs, i.e., p^3 times more than in the finite element case. The memory required is also higher for the k-method.

Iterative solvers have attracted more attention in the isogeometric community since they allow, though it is not trivial, optimal computational cost. The effort has been primarily on the development of preconditioners for the Poisson model problem, for arbitrary degree and continuity splines. As reported in [47], standard algebraic preconditioners (Jacobi, SSOR, incomplete factorization) commonly adopted for finite elements exhibit reduced performance when used in the context of the isogeometric k-method. Standard multilevel and multigrid approaches are studied respectively in [39] and [60], while advances in the theory of domain-decomposition based solvers are given in, e.g., [24, 28]. These papers also confirm the difficulty in achieving both robustness and computational efficiency for the high-degree k-method.

More sophisticated multigrid preconditioners have been proposed in the recent papers [54] and [67]. The latter, in particular, contains a proof of robustness, based on the theory of [115]. The two works are based on the following common ingredients: specific spectral properties of the discrete operator of the isogeometric k-method and the tensor-product structure of isogeometric spaces.

The tensor-product structure of multivariate spline space is exploited in [61, 96], based on approaches that have been developed for the so-called Sylvester equation. The tensor product structure of splines spaces yields to a Kronecker structure of isogeometric matrices.

We first recall the notation and basic properties of the Kronecker product of matrices. Let $A \in \mathbb{R}^{n_a \times n_a}$, and $B \in \mathbb{R}^{n_b \times n_b}$. The Kronecker product between A and B is defined as

$$
A \otimes B = \begin{bmatrix} a_{11}B & \dots & a_{1n_a}B \\ \vdots & \ddots & \vdots \\ a_{n_a 1}B & \dots & a_{n_a n_a}B \end{bmatrix} \in \mathbb{R}^{n_a n_b \times n_a n_b},
$$

where a_{ij}, $i, j = 1, \dots n_a$, denote the entries of A. The Kronecker product is an associative operation, and it is bilinear with respect to matrix sum and scalar multiplication. Some properties of the Kronecker product that will be useful in the following.

• It holds

$$
(A \otimes B)^T = A^T \otimes B^T. \tag{4.135}
$$

- If C and D are matrices of conforming order, then

$$(A \otimes B)(C \otimes D) = (AC \otimes BD). \tag{4.136}$$

- For any matrix $X \in \mathbb{R}^{n_a \times n_b}$ we denote with $\varepsilon c(X)$ the vector of $\mathbb{R}^{n_a n_b}$ obtained by "stacking" the columns of X. Then if A, B and X are matrices of conforming order, and $x = \varepsilon c(X)$, it holds

$$(A \otimes B)x = \varepsilon c(BXA^T). \tag{4.137}$$

The last property can be used to cheaply compute matrix-vector products with a matrix having Kronecker structure. Indeed, it shows that computing a matrix-vector product with $A \otimes B$ is equivalent to computing n_b matrix-vector products with A and n_a matrix-vector products with B. Note in particular that $A \otimes B$ does not have to be formed.

Consider the Laplace operator with constant coefficients, on the square $[0, 1]^2$, then the tensor-product spline Galerkin discretization leads to the system

$$(K_1 \otimes M_2 + M_1 \otimes K_2)u = b \tag{4.138}$$

where K_ℓ and M_ℓ denote the univariate stiffness and mass matrices in the ℓ direction, $\ell = 1, 2$, and $\otimes$ is the Kronecker product. For simplicity, we assume that all the univariate matrices have the same order, which we denote with n. Note in particular that $N = n^2$.

Observe that in general, for variable coefficients, general elliptic problems, non-trivial and possibly multipatch geometry parametrization, the isogeometric system is not as in (4.138). In this case, a fast solver for (4.138) plays the role of a preconditioner. At each iterative step, the preconditioner takes the form

$$(K_1 \otimes \mathcal{M}_2 + \mathcal{M}_1 \otimes K_2)s = r. \tag{4.139}$$

Using relation (4.137), we can rewrite this equation in matrix form

$$\mathcal{M}_2 S K_1 + K_2 S \mathcal{M}_1 = R, \tag{4.140}$$

where $\varepsilon c(S) = s$ and $\varepsilon c(R) = r$. Equation (4.140) takes the name of (generalized) Sylvester equation. Due to its many applications, the literature dealing with Sylvester equation (and its variants) is vast, and a number of methods have been proposed for its numerical solution. We refer to [110] for a recent survey on this subject.

Following [96], we consider the fast diagonalization (FD) method which is a direct solver, that is, $s = \mathcal{P}^{-1}r$ is computed exactly. It was first presented in 1964 by Lynch, Rice and Thomas [83] as a method for solving elliptic partial differential equations discretized with finite differences. This approach was extended to a

general Sylvester equation involving nonsymmetric matrices by Bartels and Stewart
in 1972 [11], although this is not considered here.

We consider the generalized eigendecomposition of the matrix pencils (K_1, M_1)
and (K_2, M_2), namely

$$K_1 U_1 = M_1 U_1 D_1 \qquad K_2 U_2 = M_2 U_2 D_2, \tag{4.141}$$

where D_1 and D_2 are diagonal matrices whose entries are the eigenvalues of $M_1^{-1} K_1$
and $M_2^{-1} K_2$, respectively, while U_1 and U_2 satisfy

$$U_1^T M_1 U_1 = I, \qquad U_2^T M_2 U_2 = I,$$

which implies in particular $U_1^{-T} U_1^{-1} = M_1$ and $U_2^{-T} U_2^{-1} = M_2$, and also,
from (4.141), $U_1^{-T} D_1 U_1^{-1} = K_1$ and $U_2^{-T} D_2 U_2^{-1} = K_2$. Therefore we factorize
$\mathscr{P}$ in (4.139) as follows:

$$(U_1 \otimes U_2)^{-T} (D_1 \otimes I + I \otimes D_2) (U_1 \otimes U_2)^{-1} s = r,$$

and adopt the following strategy:

- Compute the generalized eigendecompositions (4.141)
- Compute $\widetilde{r} = (U_1 \otimes U_2)^T r$
- Compute $\widetilde{s} = (D_1 \otimes I + I \otimes D_2)^{-1} \widetilde{r}$
- Compute $s = (U_1 \otimes U_2)\widetilde{s}$

Algorithm 2: FD direct method (2D)

The exact cost of the eigendecompositions in line 1 depends on the algorithm
employed. A simple approach is to first compute the Cholesky factorization $M_1 =
LL^T$ and the symmetric matrix $\widetilde{K}_1 = L^{-1} K_1 L^{-T}$. Since M_1 and K_1 are banded,
the cost of these computations is $O(pn^2)$ FLOPs. The eigenvalues of $\widetilde{K}_1$ are the
same of (4.141), and once the matrix $\widetilde{U}_1$ of orthonormal eigenvectors is computed
then one can compute $U_1 = L^{-T} \widetilde{U}_1$, again at the cost of $O(pn^2)$ FLOPs. Being $\widetilde{U}_1$
orthogonal, then $U_1^T M_1 U_1 = I_n$. If the eigendecomposition of $\widetilde{K}_1$ is computed
using a divide-and-conquer method, the cost of this operation is roughly $4n^3$
FLOPs. We remark that the divide-and-conquer approach is also very suited for
parallelization. In conclusion, by this approach, line 1 requires roughly $8n^3$ FLOPs.

Lines 2 and 4 each involve a matrix-vector product with a matrix having
Kronecker structure, and each step is equivalent (see (4.137)) to 2 matrix products
involving dense $n \times n$ matrices. The total computational cost of both steps is $8n^3$
FLOPs. Line 3 is just a diagonal scaling, and its $O(n^2)$ cost is negligible. We
emphasize that the overall computational cost of Algorithm 2 is independent of p.

If we apply Algorithm 2 as a preconditioner, then Step 1 may be performed
only once, since the matrices involved do not change throughout the CG iteration.

In this case the main cost can be quantified in approximately $8n^3$ FLOPs per CG iteration. The other main computational effort of each CG iteration is the residual computation, that is the product of the system matrix $\mathscr{A}$ by a vector, whose cost in FLOPs is twice the number of nonzero entries of $\mathscr{A}$, that is approximately $2(2p+1)^2 n^2$. In conclusion, the cost ratio between the preconditioner application and the residual computation is $O(n/p^2)$.

When $d = 3$, Eq. (4.139) takes the form

$$(K_1 \otimes M_2 \otimes M_3 + M_1 \otimes K_2 \otimes M_3 + M_1 \otimes M_2 \otimes K_3)\, s = r, \qquad (4.142)$$

where, as in the 2D case, we assume that all the univariate matrices have order n (and hence $N = n^3$).

The FD method above admits a straightforward generalization to the 3D case. We consider the generalized eigendecompositions

$$K_1 U_1 = M_1 U_1 D_1, \qquad K_2 U_2 = M_2 U_2 D_2, \qquad K_3 U_3 = M_3 U_3 D_3, \qquad (4.143)$$

with D_1, D_2, D_3 diagonal matrices and

$$U_1^T M_1 U_1 = I, \qquad U_2^T M_2 U_2 = I, \qquad U_3^T M_3 U_3 = I.$$

Then, (4.142) can be factorized as

$$(U_1 \otimes U_2 \otimes U_3)^{-1} (D_1 \otimes I \otimes I + I \otimes D_2 \otimes I + I \otimes I \otimes D_3) (U_1 \otimes U_2 \otimes U_3)^{-T}$$

$$s = r,$$

which suggests the following algorithm.

- Compute the generalized eigendecompositions (4.143)
- Compute $\widetilde{r} = (U_1 \otimes U_2 \otimes U_3) r$
- Compute $\widetilde{s} = (D_1 \otimes I \otimes I + I \otimes D_2 \otimes I + I \otimes I \otimes D_3)^{-1} \widetilde{r}$
- Compute $s = (U_1 \otimes U_2 \otimes U_3)^T \widetilde{s}$

Algorithm 3: FD direct method (3D)

Lines 1 and 3 require $O(n^3)$ FLOPs. Lines 2 and 4, as can be seen by nested applications of formula (4.137), are equivalent to performing a total of 6 products between dense matrices of size $n \times n$ and $n \times n^2$. Thus, neglecting lower order terms the overall computational cost of Algorithm 3 is $12n^4$ FLOPs.

The FD method is even more appealing in the 3D case than it was in the 2D case, for at least two reasons. First, the computational cost associated with the preconditioner setup, that is the eigendecomposition, is negligible. This means that the main computational effort of the method consists in a few (dense) matrix-matrix products, which are level 3 BLAS operations and typically yield high efficiency thanks to a dedicated implementation on modern computers by optimized usage of the memory cache hierarchy Second, in a preconditioned CG iteration the cost for applying the preconditioner has to be compared with the cost of the residual computation (a matrix-vector product with $\mathscr{A}$) which can be quantified in approximately $2(2p + 1)^3 n^3$ for 3D problems, resulting in a FLOPs ratio of the preconditioner application to residual computation of $O(n/p^3)$. However in numerical tests we will see that, for all cases of practical interest in 3D, the computational time used by the preconditioner application is far lower that the residual computation itself. This is because the computational time depends not only on the FLOPs count but also on the memory usage and, as mentioned above, dense matrix-matrix multiplications greatly benefit of modern computer architecture.

We report some 3D single-patch numerical tests from [96]. We consider a two domains: the first one is a thick quarter of ring; note that this solid has a trivial geometry on the third direction. The second one is the solid of revolution obtained by the 2D quarter of ring. Specifically, we performed a $\pi/2$ revolution around the axis having direction $(0, 1, 0)$ and passing through $(-1, -1, -1)$. We emphasize that here the geometry is nontrivial along all directions.

We consider a standard Incomplete Cholesky (IC) preconditioner (no reordering is used in this case, as the resulting performance is better than when using the standard reorderings available in MATLAB).

In Table 4.1 we report the results for the thick quarter ring while in Table 4.2 we report the results for the revolved ring. The symbol "*" denotes the cases in which even assembling the system matrix $\mathscr{A}$ was unfeasible due to memory limitations. From these results, we infer that most of the conclusions drawn for the 2D case

Table 4.1 Thick quarter of ring domain

h^{-1}	CG + $\mathscr{P}$ iterations/time (s)				
	$p = 2$	$p = 3$	$p = 4$	$p = 5$	$p = 6$
32	26/0.19	26/0.38	26/0.75	26/1.51	26/2.64
64	27/1.43	27/3.35	27/6.59	27/12.75	27/21.83
128	28/14.14	28/32.01	28/61.22	*	*
h^{-1}	CG + $\mathscr{P}_J$ iterations/time (s)				
	$p = 2$	$p = 3$	$p = 4$	$p = 5$	$p = 6$
32	26 (7)/0.88	26 (7)/1.20	26 (7)/1.71	26 (7)/2.62	27 (8)/4.08
64	27 (7)/7.20	27 (8)/10.98	27 (8)/14.89	27 (8)/21.81	27 (8)/30.56
128	28 (8)/99.01	28 (8)/98.39	28 (8)/143.45	*	*
h^{-1}	CG + IC iterations/time (s)				
	$p = 2$	$p = 3$	$p = 4$	$p = 5$	$p = 6$
32	21/0.37	15/1.17	12/3.41	10/9.43	9/24.05
64	37/4.26	28/13.23	22/33.96	18/88.94	16/215.31
128	73/65.03	51/163.48	41/385.54	*	*

Performance of CG preconditioned by the direct method (upper table), by ADI (middle table) and by Incomplete Cholesky (lower table)

Table 4.2 Revolved quarter of ring domain

h^{-1}	CG + $\mathscr{P}$ iterations/time (s)				
	$p = 2$	$p = 3$	$p = 4$	$p = 5$	$p = 6$
32	40/0.27	41/0.63	41/1.24	42/2.38	42/4.13
64	44/2.30	44/5.09	45/10.75	45/20.69	45/35.11
128	47/23.26	47/55.34	47/101.94	*	*
h^{-1}	CG + $\mathscr{P}_J$ iterations/time (s)				
	$p = 2$	$p = 3$	$p = 4$	$p = 5$	$p = 6$
32	40 (7)/1.39	41 (7)/1.93	41 (7)/2.67	42 (7)/4.17	42 (8)/6.25
64	44 (7)/11.82	44 (8)/16.96	45 (8)/24.31	45 (8)/35.76	45 (8)/49.89
128	47 (8)/170.69	47 (8)/168.45	47 (9)/239.07	*	*
h^{-1}	CG + IC iterations/time (s)				
	$p = 2$	$p = 3$	$p = 4$	$p = 5$	$p = 6$
32	24/0.44	18/1.28	15/3.61	12/9.63	11/24.57
64	47/5.19	35/14.95	28/37.33	24/94.08	20/222.09
128	94/81.65	71/211.53	57/464.84	*	*

Performance of CG preconditioned by the direct method (upper table), by ADI (middle table) and by Incomplete Cholesky (lower table)

Table 4.3 Percentage of time spent in the application of the 3D FD preconditioner with respect to the overall CG time

h^{-1}	$p = 2$	$p = 3$	$p = 4$	$p = 5$	$p = 6$
32	25.60	13.34	7.40	4.16	2.44
64	22.69	11.26	5.84	3.32	1.88
128	25.64	13.09	6.92	*	*

Revolved ring domain

still hold in 3D. In particular, both Sylvester-based preconditioners yield a better performance than the IC preconditioner, especially for small h.

Somewhat surprisingly, however, the CPU times show a stronger dependence on p than in the 2D case, and the performance gap between the ADI and the FD approach is not as large as for the cube domain. This is due to the cost of the residual computation in the CG iteration (a sparse matrix-vector product, costing $O(p^3 n^3)$ FLOPs). This step represents now a significant computational effort in the overall CG performance. In fact, our numerical experience shows that the 3D FD method is so efficient that the time spent in the preconditioning step is often negligible w.r.t. the time required for the residual computation. This effect is clearly shown in Table 4.3, where we report the percentage of time spent in the application of the preconditioner when compared with the overall time of CG, in the case of the revolved ring domain. Interestingly, this percentage is almost constant w.r.t. h up to the finest discretization level, corresponding to about two million degrees-of-freedom.

For conforming multi-patch parametrization, we can easily combine the approaches discussed above with an overlapping Schwarz preconditioner. For details, see [96]. Extension of this approach to nonconforming discretizations would require the use of nonconforming DD preconditioners (e.g., [81]) instead of an overlapping Schwarz preconditioner.

4.6.4   Matrix-Free Computationally-Efficient k-Refinement

The techniques of Sects. 4.6.2 and 4.6.3 are still not enough to achieve the full potential of the k-method and motivate the k-refinement from the point of view of computational efficiency. Matrix operations are too slow and the matrix storage itself poses restrictions to degree elevation. Therefore in [97] the idea of forming and storing the needed matrices is abandoned and, still relying on weighted quadrature, a matrix-free approach is developed. In such a case, the system matrix is available only as a function that computes matrix-vector products. This is exactly what is needed by an iterative solver. Matrix-free approaches have been use in high-order methods based on a tensor construction like spectral elements (see [121]) and have been recently extended to hp-finite elements [2, 80]. They are commonly used in non-linear solvers, parallel implementations, typically for application that are computationally demanding, for example in computational-fluid-dynamics [74, 92].

The cost to initialize the matrix-free approach is only $O(N)$ FLOPs, while the computation of matrix-vector products costs only $O(Np)$ FLOPs. Moreover, the memory required by this approach is just $O(N)$, i.e., it is proportional to the number of degrees of freedom. On the other hand, in 3D the memory required to store the matrix would be $O(Np^3)$, and the cost to compute standard matrix-vector products would be $O(Np^3)$ FLOPs. It is important to remark that, while in some cases the reduction in storage is the major motivation of the matrix-free approach, in this case framework both FLOPs and memory savings are fundamental in order to make the use of the high-degree k-method possible and advantageous. We emphasize that other matrix-free approaches which rely on more standard quadrature rules (e.g. Gaussian quadrature) require $O(Np^4)$ FLOPs to compute matrix-vector products.

The innovative implementation described below is, in the case of the k-method (the isogeometric method based on splines or NURBS, etc., with maximum regularity), orders of magnitude faster than the standard implementation inherited by finite elements. The speedup on a mesh of 256^3 elements is 13 times for degree $p = 1$, 44 times for degree $p = 2$, while higher degrees can not be handled in the standard framework. Indeed, in the standard implementation, higher degrees are beyond the memory constraints of nowadays workstations, while they are easily allowed in the new framework. This has the upshot: it gives, for the first time, clear evidence of the superiority of the high-degree k-method with respect to low-degree isogeometric discretizations in terms of computational efficiency.

This approach has been also studied, implemented and tested in an innovative environment and hardware for dataflow computing, in the thesis [122].

For brevity we only present here the weighted quadrature matrix-free algorithm for the mass matrix multiplication. Let $\widetilde{\mathbb{M}}$ be the approximation of $\mathbb{M}$ obtained with weighted quadrature, as described in Sect. 4.6.2. We use however indices instead of multi-indices, for the sake of simplicity. We want to compute the vector $\widetilde{\mathbb{M}}v$, where $v \in \mathbb{R}^N$ is a given vector.

For $i = 1, \ldots, N$, we observe that

$$\left(\widetilde{\mathbb{M}}v\right)_i = \sum_{j=1}^{N} \widetilde{m}_{ij} v_j = \sum_{j=1}^{N} \sum_{q=1}^{N_{QP}} w_{i,q} c(\mathbf{x}_q) \widehat{B}_j(\mathbf{x}_q) v_j$$

$$= \sum_{q=1}^{N_{QP}} w_{i,q} c(\mathbf{x}_q) \left(\sum_{j=1}^{N} \widehat{B}_j(\mathbf{x}_q) v_j \right),$$

where we have used the definition of $\widetilde{m}_{ij}$ from (4.133). If we define $v_h = \sum_{j=1}^{N} v_j \widehat{B}_j$, we have then the obvious relation

$$\left(\widetilde{\mathbb{M}}v\right)_i = \sum_{q=1}^{N_{QP}} w_{i,q} \, c(\mathbf{x}_q) v_h(\mathbf{x}_q) = \mathbb{Q}_i^{WQ} \left(c(\cdot)v_h(\cdot)\right). \tag{4.144}$$

Above, we see that weighted-quadrature is well suited for a direct calculation of the i-th entry of $\widetilde{\mathbb{M}}v$: this is just equivalent to approximating the integral of the function $c \, v_h$ using the i-th quadrature rule.

Then $\widetilde{\mathbb{M}}v$ can be computed with the following steps:

1. Compute $\widetilde{v} \in \mathbb{R}^{N_{QP}}$, with $\widetilde{v}_q := v_h(\mathbf{x}_q)$, $q = 1, \ldots, N_{QP}$.
2. Compute $\widetilde{\widetilde{v}} \in \mathbb{R}^{N_{QP}}$, with $\widetilde{\widetilde{v}}_q := c(\mathbf{x}_q) \cdot v_h(\mathbf{x}_q)$, $q = 1, \ldots, N_{QP}$.
3. Compute $\left(\widetilde{\mathbb{M}}v\right)_i = \sum_{q=1}^{N_{QP}} w_{i,q} \widetilde{\widetilde{v}}_q$, $i = 1, \ldots, N$.

This algorithm, and in particular steps 1 and 3, can be performed efficiently by exploiting the tensor structure of the basis functions and of the weights. In order to make this fact apparent, we now derive a matrix expression for the above algorithm. Consider the matrix of B-spline values $\mathscr{B} \in \mathbb{R}^{N_{QP} \times N}$, with $\mathscr{B}_{qj} := \widehat{B}_j(\mathbf{x}_q)$, $q = 1, \ldots, N_{QP}$, $j = 1, \ldots, N$, which can be written as

$$\mathscr{B} = \mathbf{B}_d \otimes \ldots \otimes \mathbf{B}_1, \tag{4.145}$$

where

$$(\mathbf{B}_l)_{q_l j_l} = \widehat{B}_{l,j_l}(x_{l,q_l}) \qquad q_l = 1, \ldots, n_q, \; j_l = 1, \ldots, n. \tag{4.146}$$

We also consider the matrix of weights $\mathscr{W} \in \mathbb{R}^{N \times N_{QP}}$, with $\mathscr{W}_{iq} := w_{i,q}$, $i = 1, \ldots, N$, $q = 1, \ldots, N_{QP}$. Thanks to the tensor structure of the weights, it holds

$$\mathscr{W} = \mathbf{W}_d \otimes \ldots \otimes \mathbf{W}_1 \tag{4.147}$$

where

$$(\mathbf{W}_l)_{i_l q_l} = w_{l,i_l,q_l}, \qquad i_l = 1, \ldots, n, \; q_l = 1, \ldots, n_q.$$

Finally we introduce the diagonal matrix of coefficient values

$$\mathscr{D} := \mathrm{diag}\left(\left\{c(\mathbf{x}_q)\right\}_{q=1,\ldots,N_{\mathrm{QP}}}\right). \tag{4.148}$$

Then for every $i, j = 1, \ldots, N$ we infer that

$$\widetilde{\mathsf{M}}_{ij} = \sum_{q=1}^{N_{\mathrm{QP}}} w_{i,q} c(\mathbf{x}_q) \widehat{B}_j(\mathbf{x}_q) = \sum_{q=1}^{N_{\mathrm{QP}}} \mathscr{W}_{iq} \mathscr{D}_{qq} \mathscr{B}_{qj} = (\mathscr{W}\mathscr{D}\mathscr{B})_{ij}.$$

Thus it holds

$$\widetilde{\mathsf{M}} = \mathscr{W}\mathscr{D}\mathscr{B} \tag{4.149}$$

The factorization above of $\widetilde{\mathsf{M}}$ justifies Algorithm 4, which computes efficiently the matrix-vector product.

We now analyze Algorithm 4 in terms of memory usage and of computational cost, where we distinguish between setup cost and application cost. The initialization of Algorithm 4 requires the computation and storage of the coefficient values $c(\mathbf{x}_q)$, $q = 1, \ldots, N_{\mathrm{QP}}$, and of the (sparse) matrices $\mathrm{W}_l \in \mathbb{R}^{n \times n_q}$ and $\mathrm{B}_l \in \mathbb{R}^{n_q \times n}$, for $l = 1, \ldots, d$. The latter part, which involves only the computation and storage of univariate function values and weights, has negligible requirements both in terms of memory and arithmetic operations. The computational cost of the evaluation of the coefficients $c(\mathbf{x}_q)$ is problem dependent. For example, when $c(\boldsymbol{\xi}) = \det(J_{FR}(\boldsymbol{\xi}))$ and FR is a spline/NURBS parametrization of degree lower than the one of the isogeometric space, as it happens in the numerical benchmarks of the isogeometric k-method, one can assume this cost is $O(N)$ FLOPs, i.e., independent of p. In general, the storage of such coefficients clearly requires $N_{\mathrm{QP}} \approx 2^d N = O(N)$ memory.[3] We emphasize that this memory requirement is completely independent of p; this is a great improvement if we consider that storing the whole mass matrix

---

Initialization: Compute and store the matrices $\mathscr{D}$, B_l and W_l, for $l = 1, \ldots, d$.
Input   : Vector $v \in \mathbb{R}^N$.
1 Compute $\widetilde{v} = (\mathrm{B}_d \otimes \ldots \otimes \mathrm{B}_1)\, v$;
2 Compute $\widetilde{\widetilde{v}} = \mathscr{D}\widetilde{v}$;
3 Compute $w = (\mathrm{W}_d \otimes \ldots \otimes \mathrm{W}_1)\widetilde{\widetilde{v}}$;
Output   : Vector $w = \widetilde{\mathsf{M}}v \in \mathbb{R}^N$.

Algorithm 4: Matrix-free product (mass)

---

[3]It is possible to further reduce the memory requirements at the cost of increasing the number of computations. Indeed, note that it is not necessary to store the whole $\mathscr{D}$, $\widetilde{v}$ and $\widetilde{\widetilde{v}}$ since w in Algorithm 4 can be computed component by component with on-the-fly calculation of the portion of $\mathscr{D}$, $\widetilde{v}$ and $\widetilde{\widetilde{v}}$ that is needed).

would require roughly $(2p+1)^d N = O(Np^d)$ memory. As for the application cost, Step 2 only requires N FLOPs. Using the properties of Kronecker product and the fact that nnz $\left(\mathbf{B}^{(l)}\right) \approx 2pn, l = 1, \ldots, d$, we find that the number of FLOPs required by Step 1 is

$$4pn\left(n^{d-1} + 2n^{d-1} + \ldots + 2^{d-1}n^{d-1}\right) \leq 2^{d+2}Np = O(Np).$$

Approximately the same number of operations is required for Step 3. Hence we conclude that the total application cost of Algorithm is $O(Np)$ FLOPs. This should be compared with the $O(Np^d)$ cost of the standard matrix-vector product.

Similar conclusions hold for the stiffness matrix, though it requires a different treatment of the different derivatives, in the spirit of the weighted quadrature.

Now we report some numerical tests of this approach, from [97], considering a Poisson problem on a mesh of 256^3 elements, on a thick quarter of annulus as in Fig. 4.30 (left). For the sake of simplicity, a uniform mesh is considered but all the algorithms do not take any advantage of it and work on non-uniform meshes.

The problem solution is an oscillating manufactured solution, namely

$$u(x, y, z) = \sin(5\pi x) \sin(5\pi y) \sin(5\pi z) \left(x^2 + y^2 - 1\right)\left(x^2 + y^2 - 4\right).$$

$$(4.150)$$

In the tests we see that the k-refinement, whose use has always been discouraged by its prohibitive computational cost, becomes very appealing in the present setting.

For different values of h and p we report the total computation time (setup and solution of the system) and the error $\|u - \widetilde{u}_h\|_{H^1}$, where $\widetilde{u}_h \in V_h$ is the function associated with the approximate solution of the linear system (using BiCGStab and the preconditioner of Sect. 4.6.3). Results are shown in Table 4.4 and in Fig. 4.31.

There is a minimal mesh resolution which is required to allow k-refinement convergence. This depends on the solution, which is in the example (4.150) a simple

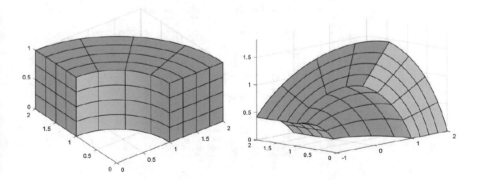

Fig. 4.30 Thick ring and revolved ring domains

Table 4.4 H^1 error and total computation time for the matrix-free WQ strategy

p	h				
	H^1 relative error/total computation time (s)				
	2^{-4}	2^{-5}	2^{-6}	2^{-7}	2^{-8}
1	$5.8 \times 10^{-1}/2.4 \times 10^{-1}$	$2.8 \times 10^{-1}/9.9 \times 10^{-1}$	$1.4 \times 10^{-1}/7.4 \times 10^{0}$	$6.8 \times 10^{-2}/6.2 \times 10^{1}$	$3.4 \times 10^{-2}/5.1 \times 10^{2}$
2	$5.3 \times 10^{-1}/2.5 \times 10^{-1}$	$7.1 \times 10^{-2}/1.1 \times 10^{0}$	$1.2 \times 10^{-2}/8.8 \times 10^{0}$	$2.6 \times 10^{-3}/7.6 \times 10^{1}$	$6.2 \times 10^{-4}/6.9 \times 10^{2}$
3	$4.5 \times 10^{-1}/2.7 \times 10^{-1}$	$3.3 \times 10^{-2}/1.3 \times 10^{0}$	$2.5 \times 10^{-3}/1.1 \times 10^{1}$	$2.7 \times 10^{-4}/9.3 \times 10^{1}$	$3.2 \times 10^{-5}/8.2 \times 10^{2}$
4	$5.1 \times 10^{-1}/3.0 \times 10^{-1}$	$1.4 \times 10^{-2}/1.6 \times 10^{0}$	$3.8 \times 10^{-4}/1.2 \times 10^{1}$	$1.8 \times 10^{-5}/1.1 \times 10^{2}$	$1.0 \times 10^{-6}/9.9 \times 10^{2}$
5	$4.4 \times 10^{-1}/3.4 \times 10^{-1}$	$6.8 \times 10^{-3}/1.8 \times 10^{0}$	$7.1 \times 10^{-5}/1.5 \times 10^{1}$	$1.5 \times 10^{-6}/1.3 \times 10^{2}$	$4.3 \times 10^{-8}/1.2 \times 10^{3}$
6	$4.9 \times 10^{-1}/3.8 \times 10^{-1}$	$3.3 \times 10^{-2}/2.1 \times 10^{0}$	$1.3 \times 10^{-5}/1.7 \times 10^{1}$	$1.2 \times 10^{-7}/1.6 \times 10^{2}$	$1.6 \times 10^{-9}/1.4 \times 10^{3}$
7	$4.1 \times 10^{-1}/4.2 \times 10^{-1}$	$1.7 \times 10^{-3}/2.4 \times 10^{0}$	$2.5 \times 10^{-6}/2.0 \times 10^{1}$	$1.1 \times 10^{-8}/1.9 \times 10^{2}$	$6.7 \times 10^{-11}/1.6 \times 10^{3}$
8	$4.7 \times 10^{-1}/4.7 \times 10^{-1}$	$9.2 \times 10^{-4}/2.7 \times 10^{0}$	$5.1 \times 10^{-7}/2.3 \times 10^{1}$	$9.3 \times 10^{-10}/2.0 \times 10^{2}$	$2.8 \times 10^{-12}/1.8 \times 10^{3}$

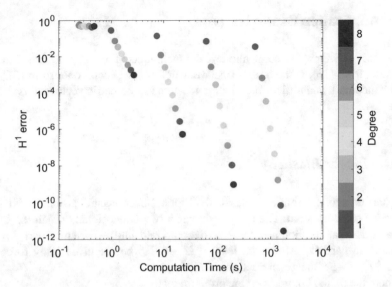

Fig. 4.31 Representation in the time-error plane of the results shown in Table 4.4

oscillating function with wavelength $1/5$ on a domain with diameter 3. Indeed, there is no approximation (i.e., the relative approximating error remains close to 1) for meshes of 16^3 elements or coarser, for any p. Convergence begin at a resolution of 32^3 elements.

The computation time of the proposed matrix-free method grows almost linearly with respect to $N = \left(\frac{1}{h}\right)^3$ (note that the growth is slower between the two coarser discretization level, where apparently we are still in the pre-asymptotic regime). Time dependence on p is also very mild: the computation time for $p = 8$ is $1/3$ the one for $p = 2$, keeping the same mesh resolution. The time growth with respect to N and p is due not only to the increased cost for system setup, matrix-vector product and application of the preconditioner, but also to the increased number of iterations. In turn, the number of iterations grows not because of a worsening of the preconditioner's quality (according to the results in [96, 117]) but because of a smaller discretization error, which corresponds to a more stringent stopping criterion.

The higher the degree, the higher the computational efficiency of the k-method. This is clearly seen in Fig. 4.31 where the red dots (associated to $p = 8$, the highest degree in our experiments), are at the bottom of the error vs. computation time plot.

The k-refinement is superior to low-degree h-refinement given a target accuracy: for example, for a relative accuracy of order 10^{-3}, we can select degree $p = 8$ on a mesh of 32^3 elements or $p = 2$ on a mesh of 256^3 elements: the former approximation is obtained in 2.7 s while the latter takes about 690 s on our workstation, with speedup factor higher than 250.

4.7 Application Examples

In this section we present some numerical benchmarks of model problems. The first example, from [15], is the one of linear elasticity. The second one, from [57], is a fluid benchmark and utilizes the divergence-free isogeometric vector fields defined in Sect. 4.3.2.

4.7.1 Linear Elasticity

We start by considering the classical elliptic linear elastic problem. First we introduce some notation. The body occupies a two-dimensional domain $\Omega \subset \mathbb{R}^2$. We assume that the boundary $\partial\Omega$ is decomposed into a Dirichlet part Γ_D and a Neumann part Γ_N. Moreover, let $\mathbf{f} : \Omega \to \mathbb{R}^d$ be the given body force and $\mathbf{g} : \Omega : \Gamma_N \to \mathbb{R}^d$ the given traction on Γ_N.

Then, the mixed boundary-value problem reads

$$\begin{cases} \operatorname{div} \mathbb{C}\boldsymbol{\varepsilon}(\mathbf{u}) + \mathbf{f} = \mathbf{0} & \text{in } \Omega \\ \mathbf{u} = \mathbf{0} & \text{on } \Gamma_D \\ \mathbb{C}\boldsymbol{\varepsilon}(\mathbf{u}) \cdot \mathbf{n} = \mathbf{g} & \text{on } \Gamma_N, \end{cases} \tag{4.151}$$

where $\mathbf{u}$ is the body displacement and $\boldsymbol{\varepsilon}(\mathbf{u})$ its symmetric gradient, $\mathbf{n}$ is the unit outward normal at each point of the boundary and the fourth-order tensor $\mathbb{C}$ satisfies

$$\mathbb{C}\mathbf{w} = 2\mu \left[\mathbf{w} + \frac{\nu}{1 - 2\nu} \operatorname{tr}(\mathbf{w})\mathbf{I} \right] \tag{4.152}$$

for all second-order tensors $\mathbf{w}$, where tr represents the trace operator and $\mu > 0$, $0 \le \nu < 1/2$ are, respectively, the shear modulus and Poisson's ratio. The stress, $\boldsymbol{\sigma}$, is given by Hooke's law, $\boldsymbol{\sigma} = \mathbb{C}\boldsymbol{\varepsilon}$.

Assuming for simplicity a regular loading $\mathbf{f} \in [L^2(\Omega)]^2$ and $\mathbf{g} \in [L^2(\Gamma_N)]^2$, we introduce also

$$< \boldsymbol{\psi}, \mathbf{v} > = (\mathbf{f}, \mathbf{v})_\Omega + (\mathbf{g}, \mathbf{v})_{\Gamma_N} \qquad \forall \mathbf{v} \in [H^1(\Omega)]^d, \tag{4.153}$$

where $(\,,\,)_\Omega$, $(\,,\,)_{\Gamma_N}$ indicate, as usual, the L^2 scalar products on Ω and Γ_N, respectively. The variational form of problem (4.151) then reads: find $\mathbf{u} \in [H^1_{\Gamma_D}(\Omega)]^d$ such that

$$(\mathbb{C}\boldsymbol{\varepsilon}(\mathbf{u}), \boldsymbol{\varepsilon}(\mathbf{v}))_\Omega = < \boldsymbol{\psi}, \mathbf{v} > \qquad \forall \mathbf{v} \in [H^1_{\Gamma_D}(\Omega)]^d \tag{4.154}$$

To solve (4.151), we introduce an isogeometric vector space V_h as defined in Sect. 4.3.1 and look for the Galerkin isogeometric approximation $\mathbf{u}_h \in V_h$ such that

$$(\mathbb{C}\boldsymbol{\varepsilon}(\mathbf{u}), \boldsymbol{\varepsilon}(\mathbf{v}))_\Omega = < \boldsymbol{\psi}, \mathbf{v} > \qquad \forall \mathbf{v} \in \boldsymbol{V}_h, \tag{4.155}$$

where

$$\boldsymbol{V}_h = [V_h]^d \cap [H^1_{\Gamma_D}(\Omega)]^d. \tag{4.156}$$

This is an elliptic problem, then a Galerkin method returns the best approximation in the energy norm. The order of convergence of the numerical error $\mathbf{u} - \mathbf{u}_h$ follows from the approximation properties of isogeometric spaces, see Sect. 4.4.

We will see this for the model of an infinite plate with a hole, modeled by a finite quarter plate. The exact solution [64, pp. 120–123], evaluated at the boundary of the finite quarter plate, is applied as a Neumann boundary condition. The setup is illustrated in Fig. 4.32. T_x is the magnitude of the applied stress at infinity, R is the radius of the traction-free hole, L is the length of the finite quarter plate, E is Young's modulus, and ν is Poisson's ratio. The rational quadratic basis is the minimum order capable of exactly representing a circle.

The first six meshes used in the analysis are shown in Fig. 4.33. The cubic and quartic NURBS are obtained by order elevation of the quadratic NURBS on the coarsest mesh (for details of the geometry and mesh construction, see [70]). Continuity of the basis is C^{p-1} everywhere, except along the line which joins the center of the circular edge with the upper left-hand corner of the domain. There it is C^1 as is dictated by the coarsest mesh employing rational quadratic parametrization. In this example, the geometry parametrization is singular at the upper left-hand corner of the domain. Convergence results in the L^2-norm of stresses (which is equivalent to the H^1-seminorm of the displacements) are shown in Fig. 4.34. As can be seen, the L^2-convergence rates of stress for quadratic, cubic, and quartic

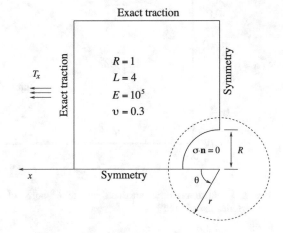

Fig. 4.32 Elastic plate with a circular hole: problem definition

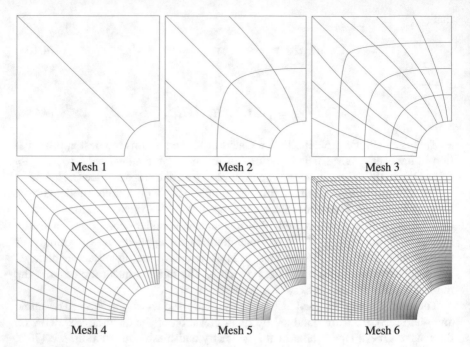

Fig. 4.33 Elastic plate with circular hole. Meshes produced by h-refinement (knot insertion)

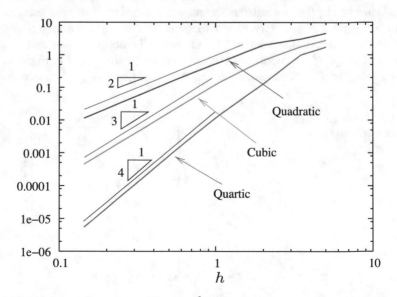

Fig. 4.34 Elasticity: error measured in the L^2-norm of stress vs. mesh parameter (optimal convergence rates in red)

NURBS are 2, 3, and 4, respectively, hence optimal in all cases, consistent with the approximation estimates described in Sect. 4.4.

4.7.2 Steady Navier-Stokes Problem

We consider now the steady Navier-Stokes Problem. The fluid occupies the domain $\Omega \subset \mathbb{R}^3$. We assume that the boundary $\partial\Omega = \Gamma_D$ for simplicity and take $\mathbf{f} : \Omega \to \mathbb{R}^3$ as the external driving force. Then, the problem reads

$$\begin{cases} \operatorname{div}(\mathbf{u} \otimes \mathbf{u}) - \operatorname{div}(2\nu\varepsilon(\mathbf{u})) + \nabla p = \mathbf{f} & \text{in } \Omega \\ \operatorname{div}\mathbf{u} = 0 & \text{in } \Omega \\ \mathbf{u} \cdot \mathbf{n} & \text{on } \partial\Omega, \end{cases} \tag{4.157}$$

where $\mathbf{u}$ is the fluid velocity, p is the pressure, ν is the kinematic viscosity and $\varepsilon(\mathbf{u})$ is the symmetric gradient operator.

The variational form of (4.157) reads as follows: find $\mathbf{u} \in [H_0^1(\Omega)]^d$ and $p \in L_0^2(\Omega)$ such that

$$(2\nu\varepsilon(\mathbf{u}), \varepsilon(\mathbf{v}))_\Omega - (\mathbf{u} \otimes \mathbf{u}, \nabla\mathbf{v})_\Omega - (p, \operatorname{div}\mathbf{u})_\Omega + (q, \operatorname{div}\mathbf{v})_\Omega$$
$$= (\mathbf{f}, \mathbf{v})_\Omega, \quad \forall \mathbf{u} \in [H_0^1(\Omega)]^d, q \in L_0^2(\Omega), \tag{4.158}$$

where $L_0^2(\Omega)$ is the subspace of $L^2(\Omega)$ functions having zero average on Ω. At the discrete level, we are going to adopt a divergence-free (X_h^2, X_h^3) isogeometric discretization for the velocity-pressure pair, as defined in Sect. 4.3.2. In this case, only the Dirichlet boundary condition on the normal velocity component (i.e., no-penetration condition) can be imposed strongly (see [38]) while the other boundary conditions, including the Dirichlet boundary condition on the tangential velocity component, have to be imposed weakly, for example by Nitsche's method, as studied in [57]. For that, we introduce the space $H_\mathbf{n}^1(\Omega) = \{\mathbf{w} \in [H^1(\Omega)]^d \text{ such that } \mathbf{w}\cdot\mathbf{n} = 0 \text{ on } \partial\Omega\}$, and the discrete variational formulation is: find $\mathbf{u}_h \in X_h^2 \cap H_\mathbf{n}^1(\Omega)$ and $p_h \in X_h^3 \cap L_0^2(\Omega)$ such that

$$(2\nu\varepsilon(\mathbf{u}_h), \varepsilon(\mathbf{v}_h))_\Omega - (\mathbf{u}_h \otimes \mathbf{u}_h, \nabla\mathbf{v}_h)_\Omega - (p_h, \operatorname{div}\mathbf{u}_h)_\Omega + (q_h, \operatorname{div}\mathbf{v}_h)_\Omega$$

$$- \sum_{F \subset \partial\Omega} \int_F \left((\varepsilon(\mathbf{u}_h)\,\mathbf{n}) \cdot \mathbf{v}_h + (\varepsilon(\mathbf{v}_h)\,\mathbf{n}) \cdot \mathbf{u}_h - \frac{C_{pen}}{h_F} \mathbf{u}_h \cdot \mathbf{v}_h \right) ds$$

$$= (\mathbf{f}, \mathbf{v}_h)_\Omega, \qquad\qquad \forall \mathbf{v}_h \in X_h^2 \cap H_\mathbf{n}^1(\Omega), q_h \in X_h^3 \cap L_0^2(\Omega). \tag{4.159}$$

where $F \subset \partial \Omega$ denote the faces (in three dimensions) of the Bézier elements that are on the boundary of Ω and $C_{pen} > 0$ is a suitable penalty constant.

We consider a simple configuration (see [57, Section 8.2]), with $\Omega = [0, 1]^3$ and select in (4.32) the polynomial degrees $p_1 = p_2 = p_3 = 2$ and $p_1 = p_2 = p_3 = 3$. Selecting knots with single multiplicity, the former choice X_h^2 is formed by linear-quadratic splines and X_h^3 is formed by trilinear splines, which is the minimum degree required to have $X_h^2 \in [H^1(\Omega)]^d$. The right-hand side is set up in order to give the exact solution:

$$\mathbf{u}_h = \mathbf{curl} \begin{bmatrix} x(x-1)y^2(y-1)^2 z^2(z-1)^2 \\ 0 \\ x^2(x-1)^2 y^2(y-1)^2 z(z-1) \end{bmatrix} \quad ; \quad p = \sin(\pi x)\sin(\pi y) - \frac{4}{\pi^2}.$$

Streamlines associated with the exact solution are plotted in Fig. 4.35. The convergence rates are shown in Figs. 4.36 and 4.37 for Reynolds number $Re = 1$. Optimal convergence is obtained for both velocity and pressure. We remark that the discrete velocity is point-wise divergence-free, because of (4.38).

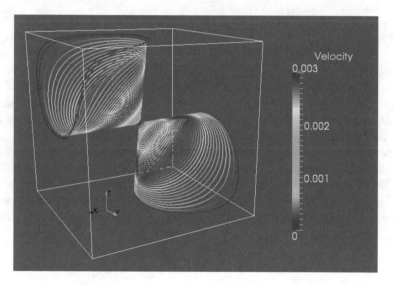

Fig. 4.35 Vortex manufactured solution: Flow velocity streamlines colored by velocity magnitude (from [57])

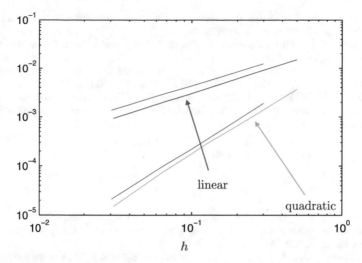

Fig. 4.36 Navier-Stokes: error measured in the H^1-norm of velocity vs. mesh-size h. The optimal convergence rates in red (from [57])

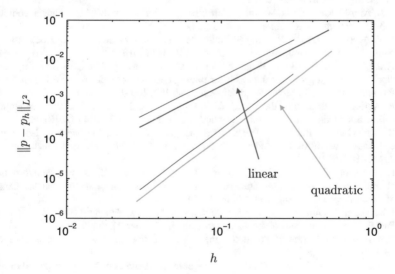

Fig. 4.37 Navier-Stokes: error measured in the L^2-norm of pressure vs. mesh-size h. Optimal convergence rates in red (from [57])

Acknowledgements This chapter is a review of the mathematical results available on IGA. The presentation follows in particular [25, 48, 50, 57, 71, 73, 96]. We thank our colleagues and coauthors: Yuri Bazilevs, Lourenco Beirão da Veiga, Annalisa Buffa, Francesco Calabrò, Annabelle Collin, Austin Cottrell, John Evans, Alessandro Reali, Thomas Takacs, and Rafael Vázquez. Thomas J.R. Hughes was partially supported by the Office of Naval Research (Grant Nos. N00014-17-1-2119, N00014-17-1-2039, and N00014-13-1-0500), and by the Army Research Office (Grant No. W911NF-13-1-0220). Giancarlo Sangalli and Mattia Tani were partially supported by the European Research Council (ERC Consolidator Grant n.616563 *HIGEOM*). This support is gratefully acknowledged.

References

1. C. Adam, T.J.R. Hughes, S. Bouabdallah, M. Zarroug, H. Maitournam, Selective and reduced numerical integrations for NURBS-based isogeometric analysis. Comput. Methods Appl. Mech. Eng. **284**, 732–761 (2015)
2. M. Ainsworth, O. Davydov, L.L. Schumaker, Bernstein-Bézier finite elements on tetrahedral-hexahedral-pyramidal partitions. Comput. Methods Appl. Mech. Eng. **304**, 140–170 (2016)
3. P. Antolin, A. Buffa, F. Calabrò, M. Martinelli, G. Sangalli, Efficient matrix computation for tensor-product isogeometric analysis: the use of sum factorization. Comput. Methods Appl. Mech. Eng. **285**, 817–828 (2015)
4. D.N. Arnold, R.S. Falk, R. Winther, Finite element exterior calculus, homological techniques, and applications. Acta Numer. **15**, 1–155 (2006)
5. D.N. Arnold, R.S. Falk, R. Winther, Finite element exterior calculus: from Hodge theory to numerical stability. Bull. Am. Math. Soc. **47**, 281–354 (2010)
6. F. Auricchio, L. Beirão da Veiga, A. Buffa, C. Lovadina, A. Reali, G. Sangalli, A fully "locking-free" isogeometric approach for plane linear elasticity problems: a stream function formulation. Comput. Methods Appl. Mech. Eng. **197**, 160–172 (2007)
7. F. Auricchio, L. Beirão da Veiga, T.J.R. Hughes, A. Reali, G. Sangalli, Isogeometric collocation methods. Math. Models Methods Appl. Sci. **20**, 2075–2107 (2010)
8. F. Auricchio, L. Beirão da Veiga, T.J.R. Hughes, A. Reali, G. Sangalli, Isogeometric collocation for elastostatics and explicit dynamics. Comput. Methods Appl. Mech. Eng. **249–252**, 2–14 (2012)
9. F. Auricchio, F. Calabrò, T.J.R. Hughes, A. Reali, G. Sangalli, A simple algorithm for obtaining nearly optimal quadrature rules for NURBS-based isogeometric analysis. Comput. Methods Appl. Mech. Eng. **249**, 15–27 (2012)
10. I. Babuska, W. Gui, The h, p and h-p versions of the finite element method in 1 dimension. Part II. The error analysis of the h and h-p versions. Numer. Math. **49**, 613–658 (1986)
11. R.H. Bartels, G.W. Stewart, Solution of the matrix equation $AX + XB = C$. Commun. ACM **15**, 820–826 (1972)
12. A. Bartezzaghi, L. Dedè, A. Quarteroni, Isogeometric analysis of high order partial differential equations on surfaces. Comput. Methods Appl. Mech. Eng. **295**, 446–469 (2015)
13. M. Bartoň, V.M. Calo, Gaussian quadrature for splines via homotopy continuation: rules for C^2 cubic splines. J. Comput. Appl. Math. **296**, 709–723 (2016)
14. M. Bartoň, V.M. Calo, Optimal quadrature rules for odd-degree spline spaces and their application to tensor-product-based isogeometric analysis. Comput. Methods Appl. Mech. Eng. **305**, 217–240 (2016)
15. Y. Bazilevs, L. Beirão da Veiga, J.A. Cottrell, T.J.R. Hughes, G. Sangalli, Isogeometric analysis: approximation, stability and error estimates for h-refined meshes. Math. Models Methods Appl. Sci. **16**, 1031–1090 (2006)

16. Y. Bazilevs, V. Calo, J.A. Cottrell, J.A. Evans, T.J.R. Hughes, S. Lipton, M.A. Scott, T.W. Sederberg, Isogeometric analysis using T-splines. Comput. Methods Appl. Mech. Eng. **199**, 229–263 (2010)
17. Y. Bazilevs, M.C. Hsu, M.A. Scott, Isogeometric fluid–structure interaction analysis with emphasis on non-matching discretizations, and with application to wind turbines. Comput. Methods Appl. Mech. Eng. **249**, 28–41 (2012)
18. E. Beeker, Smoothing of shapes designed with free-form surfaces. Comput. Aided Des. **18**, 224–232 (1986)
19. L. Beirão da Veiga, A. Buffa, J. Rivas, G. Sangalli, Some estimates for h-p-k-refinement in isogeometric analysis. Numer. Math. **118**, 271–305 (2011)
20. L. Beirão da Veiga, A. Buffa, D. Cho, G. Sangalli, Analysis-suitable t-splines are dual-compatible. Comput. Methods Appl. Mech. Eng. **249–252**, 42–51 (2012)
21. L. Beirão da Veiga, A. Buffa, C. Lovadina, M. Martinelli, G. Sangalli, An isogeometric method for the Reissner-Mindlin plate bending problem. Comput. Methods Appl. Mech. Eng. **209–212**, 45–53 (2012)
22. L. Beirão da Veiga, D. Cho, G. Sangalli, Anisotropic NURBS approximation in isogeometric analysis. Comput. Methods Appl. Mech. Eng. **209–212**, 1–11 (2012)
23. L. Beirão da Veiga, A. Buffa, G. Sangalli, R. Vázquez, Analysis-suitable T-splines of arbitrary degree: definition, linear independence and approximation properties. Math. Models Methods Appl. Sci. **23**, 1979–2003 (2013)
24. L. Beirão da Veiga, D. Cho, L.F. Pavarino, S. Scacchi, BDDC preconditioners for isogeometric analysis. Math. Models Methods Appl. Sci. **23**, 1099–1142 (2013)
25. L. Beirão da Veiga, A. Buffa, G. Sangalli, R. Vázquez, Mathematical analysis of variational isogeometric methods. Acta Numer. **23**, 157–287 (2014)
26. Benson, D.J., Y. Bazilevs, M.C. Hsu, T.J.R. Hughes, A large deformation, rotation-free, isogeometric shell. Comput. Methods Appl. Mech. Eng. **200**, 1367–1378 (2011)
27. M. Bercovier, T. Matskewich, Smooth Bezier surfaces over arbitrary quadrilateral meshes (2014, preprint). arXiv:1412.1125
28. M. Bercovier, I. Soloveichik, Overlapping non matching meshes domain decomposition method in isogeometric analysis (2015, preprint). arXiv:1502.03756
29. M.J. Borden, C.V. Verhoosel, M.A. Scott, T.J.R. Hughes, C.M. Landis, A phase-field description of dynamic brittle fracture. Comput. Methods Appl. Mech. Eng. **217–220**, 77–95 (2012)
30. M.J. Borden, T.J.R. Hughes, C.M. Landis, C.V. Verhoosel, A higher-order phase-field model for brittle fracture: formulation and analysis within the isogeometric analysis framework. Comput. Methods Appl. Mech. Eng. **273**, 100–118 (2014)
31. J. Bremer, Z. Gimbutas, V. Rokhlin, A nonlinear optimization procedure for generalized Gaussian quadratures. SIAM J. Sci. Comput. **32**, 1761–1788 (2010)
32. A. Bressan, Isogeometric regular discretization for the Stokes problem. IMA J. Numer. Anal. **31**, 1334–1356 (2011)
33. A. Bressan, G. Sangalli, Isogeometric discretizations of the Stokes problem: stability analysis by the macroelement technique. IMA J. Numer. Anal. **33**, 629–651 (2013)
34. E. Brivadis, A. Buffa, B. Wohlmuth, L. Wunderlich, Isogeometric mortar methods. Comput. Methods Appl. Mech. Eng. **284**, 292–319 (2015)
35. F. Buchegger, B. Jüttler, A. Mantzaflaris, Adaptively refined multi-patch B-splines with enhanced smoothness. Appl. Math. Comput. **272**, 159–172 (2016)
36. A. Buffa, C. Giannelli, Adaptive isogeometric methods with hierarchical splines: optimality and convergence rates. Math. Models Methods Appl. Sci. **27**, 2781 (2017)
37. A. Buffa, C. de Falco, G. Sangalli, Isogeometric analysis: stable elements for the 2D Stokes equation. Int. J. Numer. Methods Fluids **65**, 1407–1422 (2011)
38. A. Buffa, J. Rivas, G. Sangalli, R. Vázquez, Isogeometric discrete differential forms in three dimensions. SIAM J. Numer. Anal. **49**, 818–844 (2011)
39. A. Buffa, H. Harbrecht, A. Kunoth, G. Sangalli, BPX-preconditioning for isogeometric analysis. Comput. Methods Appl. Mech. Eng. **265**, 63–70 (2013)

40. A. Buffa, G. Sangalli, C. Schwab, Exponential convergence of the hp version of isogeometric analysis in 1D, in *Spectral and High Order Methods for Partial Differential Equations–ICOSAHOM 2012*, ed. by M. Azaiez, H. El Fekih, J.S. Hesthaven. Lecture Notes in Computational Science and Engineering, vol. 95 (Springer, Berlin, 2014), pp. 191–203

41. F. Calabrò, C. Manni, F. Pitolli, Computation of quadrature rules for integration with respect to refinable functions on assigned nodes. Appl. Numer. Math. **90**, 168–189 (2015)

42. F. Calabrò, G. Sangalli, M. Tani, Fast formation of isogeometric Galerkin matrices by weighted quadrature. Comput. Methods Appl. Mech. Eng. **316**, 606–622 (2017)

43. F. Cirak, M.J. Scott, E.K. Antonsson, M. Ortiz, P. Schröder, Integrated modeling, finite-element analysis, and engineering design for thin-shell structures using subdivision. Comput. Aided Des. **34**, 137–148 (2002)

44. E. Cohen, R. Riesenfeld, G. Elber, *Geometric Modeling with Splines: An Introduction*, vol. 1 (AK Peters, Wellesley, 2001)

45. E. Cohen, T. Martin, R.M. Kirby, T. Lyche, R.F. Riesenfeld, Analysis-aware modeling: understanding quality considerations in modeling for isogeometric analysis. Comput. Methods Appl. Mech. Eng. **199**, 334–356 (2010)

46. N. Collier, D. Pardo, L. Dalcin, M. Paszynski, V.M. Calo, The cost of continuity: a study of the performance of isogeometric finite elements using direct solvers. Comput. Methods Appl. Mech. Eng. **213**, 353–361 (2012)

47. N. Collier, L. Dalcin, D. Pardo, V.M. Calo, The cost of continuity: performance of iterative solvers on isogeometric finite elements. SIAM J. Sci. Comput. **35**, A767–A784 (2013)

48. A. Collin, G. Sangalli, T. Takacs, Analysis-suitable G^1 multi-patch parametrizations for C^1 isogeometric spaces. Comput. Aided Geom. Des. **47**, 93–113 (2016)

49. J.A. Cottrell, A. Reali, Y. Bazilevs, T.J.R. Hughes, Isogeometric analysis of structural vibrations. Comput. Methods Appl. Mech. Eng. **195**, 5257–5296 (2006)

50. J.A. Cottrell, T.J.R. Hughes, Y. Bazilevs, *Isogeometric Analysis: Toward Integration of CAD and FEA* (Wiley, Hoboken, 2009)

51. C. de Boor, *A Practical Guide to Splines*, revised edn. (Springer, New York, 2001)

52. R. Dimitri, L. De Lorenzis, M. Scott, P. Wriggers, R. Taylor, G. Zavarise, Isogeometric large deformation frictionless contact using T-splines. Comput. Methods Appl. Mech. Eng. **269**, 394–414 (2014)

53. T. Dokken, T. Lyche, K.F. Pettersen, Polynomial splines over locally refined box-partitions. Comput. Aided Geom. Des. **30**, 331–356 (2013)

54. M. Donatelli, C. Garoni, C. Manni, S. Serra-Capizzano, H. Speleers, Robust and optimal multi-iterative techniques for IgA Galerkin linear systems. Comput. Methods Appl. Mech. Eng. **284**, 230–264 (2015)

55. M. Dörfel, B. Jüttler, B. Simeon, Adaptive isogeometric analysis by local h-refinement with T-splines. Comput. Methods Appl. Mech. Eng. **199**, 264–275 (2010)

56. J.A. Evans, T.J.R. Hughes, Isogeometric divergence-conforming B-splines for the Darcy-Stokes-Brinkman equations. Math. Models Methods Appl. Sci. **23**, 671–741 (2013)

57. J.A. Evans, T.J.R. Hughes, Isogeometric divergence-conforming B-splines for the Steady Navier-Stokes Equations. Math. Models Methods Appl. Sci. **23**, 1421–1478 (2013)

58. J.A. Evans, T.J.R. Hughes, Isogeometric divergence-conforming B-splines for the unsteady Navier-Stokes equations. J. Comput. Phys. **241**, 141–167 (2013)

59. J.A. Evans, Y. Bazilevs, I. Babuška, T.J.R. Hughes, n-widths, sup-infs, and optimality ratios for the k-version of the isogeometic finite element method. Comput. Methods Appl. Mech. Eng. **198**, 1726–1741 (2009)

60. K.P.S. Gahalaut, J.K. Kraus, S.K. Tomar, Multigrid methods for isogeometric discretization. Comput. Methods Appl. Mech. Eng. **253**, 413–425 (2013)

61. L. Gao, Kronecker products on preconditioning, Ph.D. thesis, King Abdullah University of Science and Technology, 2013

62. H. Gómez, V. Calo, Y. Bazilevs, T.J.R. Hughes, Isogeometric analysis of the Cahn–Hilliard phase-field model. Comput. Methods Appl. Mech. Eng. **197**, 4333–4352 (2008)

63. H. Gómez, T.J.R. Hughes, X. Nogueira, V.M. Calo, Isogeometric analysis of the isothermal Navier-Stokes-Korteweg equations. Comput. Methods Appl. Mech. Eng. **199**, 1828–1840 (2010)
64. P.L. Gould, *Introduction to Linear Elasticity* (Springer, Berlin, 1994)
65. M. Hillman, J. Chen, Y. Bazilevs, Variationally consistent domain integration for isogeometric analysis. Comput. Methods Appl. Mech. Eng. **284**, 521–540 (2015)
66. R. Hiptmair, Finite elements in computational electromagnetism. Acta Numer. **11**, 237–339 (2002)
67. C. Hofreither, S. Takacs, W. Zulehner, A robust multigrid method for isogeometric analysis using boundary correction. Tech. Rep., NFN (2015)
68. S. Hosseini, J.J. Remmers, C.V. Verhoosel, R. Borst, An isogeometric solid-like shell element for nonlinear analysis. Int. J. Numer. Methods Eng. **95**, 238–256 (2013)
69. T.J.R. Hughes, G. Sangalli, Mathematics of isogeometric analysis: a conspectus, in *Encyclopedia of Computational Mechanics* ed. by S. Erwin, R. de Borst, T.J.R. Hughes, 2nd edn. (Wiley, Hoboken, 2017)
70. T.J.R. Hughes, J.A. Cottrell, Y. Bazilevs, Isogeometric analysis: CAD, finite elements, NURBS, exact geometry and mesh refinement. Comput. Methods Appl. Mech. Eng. **194**, 4135–4195 (2005)
71. T.J.R. Hughes, A. Reali, G. Sangalli, Duality and unified analysis of discrete approximations in structural dynamics and wave propagation: comparison of p-method finite elements with k-method NURBS. Comput. Methods Appl. Mech. Eng. **197**, 4104–4124 (2008)
72. T.J.R. Hughes, A. Reali, G. Sangalli, Efficient quadrature for NURBS-based isogeometric analysis. Comput. Methods Appl. Mech. Eng. **199**, 301–313 (2010)
73. T.J.R. Hughes, J.A. Evans, A. Reali, Finite element and NURBS approximations of eigenvalue, boundary-value, and initial-value problems. Comput. Methods Appl. Mech. Eng. **272**, 290–320 (2014)
74. Z. Johan, T.J.R. Hughes, A globally convergent matrix-free algorithm for implicit time-marching schemes arising in finite element analysis in fluids. Comput. Methods Appl. Mech. Eng. **87**, 281–304 (1991)
75. K.A. Johannessen, Optimal quadrature for univariate and tensor product splines. Comput. Methods Appl. Mech. Eng. **316**, 84–99 (2017)
76. B. Jüttler, A. Mantzaflaris, R. Perl, M. Rumpf, On isogeometric subdivision methods for PDEs on surfaces. Technical Report (2015). arXiV:1503.03730
77. M. Kapl, V. Vitrih, B. Jüttler, K. Birner, Isogeometric analysis with geometrically continuous functions on two-patch geometries. Comput. Math. Appl. **70**, 1518–1538 (2015)
78. J. Kiendl, K.U. Bletzinger, J. Linhard, R. Wuchner, Isogeometric shell analysis with Kirchhoff–Love elements. Comput. Methods Appl. Mech. Eng. **198**, 3902–3914 (2009)
79. J. Kiendl, Y. Bazilevs, M.C. Hsu, R. Wüchner, K.U. Bletzinger, The bending strip method for isogeometric analysis of Kirchhoff–Love shell structures comprised of multiple patches. Comput. Methods Appl. Mech. Eng. **199**, 2403–2416 (2010)
80. R.C. Kirby, Fast simplicial finite element algorithms using Bernstein polynomials. Numer. Math. **117**, 631–652 (2011)
81. S.K. Kleiss, C. Pechstein, B. Jüttler, S. Tomar, IETI–isogeometric tearing and interconnecting. Comput. Methods Appl. Mech. Eng. **247–248**, 201–215 (2012)
82. D. Liu, J. Hoschek, $GC1$ continuity conditions between adjacent rectangular and triangular Bézier surface patches. Comput. Aided Des. **21**, 194–200 (1989)
83. R.E. Lynch, J.R. Rice, D.H. Thomas, Direct solution of partial difference equations by tensor product methods. Numer. Math. **6**, 185–199 (1964)
84. A. Mantzaflaris, B. Jüttler, B. Khoromskij, U. Langer, Matrix generation in isogeometric analysis by low rank tensor approximation, in *Curves and Surfaces: 8th International Conference*, Paris, 2014
85. B. Marussig, T.J.R. Hughes, A review of trimming in isogeometric analysis: challenges, data exchange and simulation aspects. Arch. Comput. Methods Eng. **24**, 1–69 (2017)

86. P. Monk, *Finite Element Methods for Maxwell's Equations* (Oxford University Press, Oxford, 2003)
87. B. Mourrain, R. Vidunas, N. Villamizar, Dimension and bases for geometrically continuous splines on surfaces of arbitrary topology. Comput. Aided Geom. Des. **45**, 108–133 (2016)
88. T. Nguyen, J. Peters, Refinable C^1 spline elements for irregular quad layout. Comput. Aided Geom. Des. **43**, 123–130 (2016)
89. T. Nguyen, K. Karčiauskas, J. Peters, A comparative study of several classical, discrete differential and isogeometric methods for solving Poisson's equation on the disk. Axioms **3**, 280–299 (2014)
90. J. Peters, Smooth interpolation of a mesh of curves. Constr. Approx. **7**, 221–246 (1991)
91. L. Piegl, W. Tiller, *The NURBS Book* (Springer, New York, 1997)
92. P. Rasetarinera, M.Y. Hussaini, An efficient implicit discontinuous spectral Galerkin method. J. Comput. Phys. **172**, 718–738 (2001)
93. D.F. Rogers, *An Introduction to NURBS: With Historical Perspective* (Morgan Kaufmann, San Francisco, 2001)
94. T. Rüberg, F. Cirak, Subdivision-stabilised immersed B-spline finite elements for moving boundary flows. Comput. Methods Appl. Mech. Eng. **209–212**, 266–283 (2012)
95. M. Ruess, D. Schillinger, Y. Bazilevs, V. Varduhn, E. Rank, Weakly enforced essential boundary conditions for NURBS-embedded and trimmed NURBS geometries on the basis of the finite cell method. Int. J. Numer. Methods Eng. **95**, 811–846 (2013)
96. G. Sangalli, M. Tani, Isogeometric preconditioners based on fast solvers for the Sylvester equation. SIAM J. Sci. Comput. **38**, A3644–A3671 (2016)
97. G. Sangalli, M. Tani, Matrix-free isogeometric analysis: the computationally efficient k-method (2017). ArXiv e-print 1712.08565
98. G. Sangalli, T. Takacs, R. Vázquez, Unstructured spline spaces for isogeometric analysis based on spline manifolds. Comput. Aided Geom. Des. **47**, 61–82 (2016)
99. D. Schillinger, L. Dedè, M.A. Scott, J.A. Evans, M.J. Borden, E. Rank, T.R. Hughes, An isogeometric design-through-analysis methodology based on adaptive hierarchical refinement of NURBS, immersed boundary methods, and T-spline CAD surfaces. Comput. Methods Appl. Mech. Eng. **249–252**, 116–150 (2012)
100. D. Schillinger, J.A. Evans, A. Reali, M.A. Scott, T.J.R. Hughes, Isogeometric collocation: cost comparison with Galerkin methods and extension to adaptive hierarchical NURBS discretizations. Comput. Methods Appl. Mech. Eng. **267**, 170–232 (2013)
101. D. Schillinger, S.J. Hossain, T.J.R. Hughes, Reduced Bézier element quadrature rules for quadratic and cubic splines in isogeometric analysis. Comput. Methods Appl. Mech. Eng. **277**, 1–45 (2014)
102. L.L. Schumaker, *Spline Functions: Basic Theory*, 3rd edn. (Cambridge University Press, Cambridge, 2007)
103. C. Schwab, *p- and hp-Finite Element Methods* (The Clarendon Press/Oxford University Press, New York, 1998)
104. M.A. Scott, X. Li, T.W. Sederberg, T.J.R. Hughes, Local refinement of analysis-suitable T-splines. Comput. Methods Appl. Mech. Eng. **213–216**, 206–222 (2012)
105. M.A. Scott, R.N. Simpson, J.A. Evans, S. Lipton, S.P.A. Bordas, T.J.R. Hughes, T.W. Sederberg, Isogeometric boundary element analysis using unstructured T-splines. Comput. Methods Appl. Mech. Eng. **254**, 197–221 (2013)
106. M.A. Scott, D.C. Thomas, E.J. Evans, Isogeometric spline forests. Comput. Methods Appl. Mech. Eng. **269**, 222–264 (2014)
107. T. Sederberg, J. Zheng, A. Bakenov, A. Nasri, T-splines and T-NURCCSs. ACM Trans. Graph. **22**, 477–484 (2003)
108. T. Sederberg, D. Cardon, G. Finnigan, N. North, J. Zheng, T. Lyche, T-spline simplification and local refinement. ACM Trans. Graph. **23**, 276–283 (2004)
109. T.W. Sederberg, G.T. Finnigan, X. Li, H. Lin, H. Ipson, Watertight trimmed NURBS. ACM Trans. Graph. **27**, 79 (2008)

110. V. Simoncini, Computational methods for linear matrix equations. SIAM Rev. **58**(3), 377–441 (2013)

111. G. Strang, G.J. Fix, *An Analysis of the Finite Element Method*, vol. 212 (Prentice-Hall, Englewood Cliffs, 1973)

112. T. Takacs, Construction of smooth isogeometric function spaces on singularly parameterized domains, in *Curves and Surfaces* (Springer, Berlin, 2014), pp. 433–451

113. T. Takacs, B. Jüttler, Existence of stiffness matrix integrals for singularly parameterized domains in isogeometric analysis. Comput. Methods Appl. Mech. Eng. **200**, 3568–3582 (2011)

114. T. Takacs, B. Jüttler, H^2 regularity properties of singular parameterizations in isogeometric analysis. Graph. Model. **74**, 361–372 (2012)

115. S. Takacs, T. Takacs, Approximation error estimates and inverse inequalities for B-splines of maximum smoothness. Math. Models Methods Appl. Sci. **26**, 1411–1445 (2016)

116. T. Takacs, B. Jüttler, O. Scherzer, Derivatives of isogeometric functions on n-dimensional rational patches in R^d. Comput. Aided Geom. Des. **31**, 567–581 (2014)

117. M. Tani, A preconditioning strategy for linear systems arising from nonsymmetric schemes in isogeometric analysis. Comput. Math. Appl. **74**, 1690–1702 (2017)

118. R.L. Taylor, Isogeometric analysis of nearly incompressible solids. Int. J. Numer. Methods Eng. **87**, 273–288 (2011)

119. D. Toshniwal, H. Speleers, R.R. Hiemstra, T.J.R. Hughes, Multi-degree smooth polar splines: a framework for geometric modeling and isogeometric analysis. Comput. Methods Appl. Mech. Eng. **316**, 1005–1061 (2017)

120. D. Toshniwal, H. Speleers, T.J.R. Hughes, Smooth cubic spline spaces on unstructured quadrilateral meshes with particular emphasis on extraordinary points: geometric design and isogeometric analysis considerations. Comput. Methods Appl. Mech. Eng. **327**, 411–458 (2017)

121. H.M. Tufo, P.F. Fischer, Terascale spectral element algorithms and implementations, in *Proceedings of the 1999 ACM/IEEE Conference on Supercomputing* (ACM, New York, 1999), p. 68

122. R. van Nieuwpoort, Solving Poisson's equation with dataflow computing. Master's thesis, Delft University of Technology, 2017

123. T.M. van Opstal, J. Yan, C. Coley, J.A. Evans, T. Kvamsdal, Y. Bazilevs, Isogeometric divergence-conforming variational multiscale formulation of incompressible turbulent flows. Comput. Methods Appl. Mech. Eng. **316**, 859–879 (2017)

124. R. Vázquez, A new design for the implementation of isogeometric analysis in Octave and Matlab: GeoPDEs 3.0. Tech. Rep., IMATI Report Series (2016)

125. R. Vázquez, A new design for the implementation of isogeometric analysis in Octave and Matlab: GeoPDEs 3.0. Comput. Math. Appl. **72**, 523–554 (2016)

126. C.V. Verhoosel, M.A. Scott, R. De Borst, T.J.R. Hughes, An isogeometric approach to cohesive zone modeling. Int. J. Numer. Methods Eng. **87**, 336–360 (2011)

127. C.V. Verhoosel, M.A. Scott, T.J.R. Hughes, R. de Borst, An isogeometric analysis approach to gradient damage models. Int. J. Numer. Methods Eng. **86**, 115–134 (2011)

128. A.V. Vuong, C. Giannelli, B. Jüttler, B. Simeon, A hierarchical approach to adaptive local refinement in isogeometric analysis. Comput. Methods Appl. Mech. Eng. **200**, 3554–3567 (2011)

LECTURE NOTES IN MATHEMATICS

Editors in Chief: J.-M. Morel, B. Teissier;

Editorial Policy

1. Lecture Notes aim to report new developments in all areas of mathematics and their applications – quickly, informally and at a high level. Mathematical texts analysing new developments in modelling and numerical simulation are welcome.

   Manuscripts should be reasonably self-contained and rounded off. Thus they may, and often will, present not only results of the author but also related work by other people. They may be based on specialised lecture courses. Furthermore, the manuscripts should provide sufficient motivation, examples and applications. This clearly distinguishes Lecture Notes from journal articles or technical reports which normally are very concise. Articles intended for a journal but too long to be accepted by most journals, usually do not have this "lecture notes" character. For similar reasons it is unusual for doctoral theses to be accepted for the Lecture Notes series, though habilitation theses may be appropriate.

2. Besides monographs, multi-author manuscripts resulting from SUMMER SCHOOLS or similar INTENSIVE COURSES are welcome, provided their objective was held to present an active mathematical topic to an audience at the beginning or intermediate graduate level (a list of participants should be provided).

   The resulting manuscript should not be just a collection of course notes, but should require advance planning and coordination among the main lecturers. The subject matter should dictate the structure of the book. This structure should be motivated and explained in a scientific introduction, and the notation, references, index and formulation of results should be, if possible, unified by the editors. Each contribution should have an abstract and an introduction referring to the other contributions. In other words, more preparatory work must go into a multi-authored volume than simply assembling a disparate collection of papers, communicated at the event.

3. Manuscripts should be submitted either online at www.editorialmanager.com/lnm to Springer's mathematics editorial in Heidelberg, or electronically to one of the series editors. Authors should be aware that incomplete or insufficiently close-to-final manuscripts almost always result in longer refereeing times and nevertheless unclear referees' recommendations, making further refereeing of a final draft necessary. The strict minimum amount of material that will be considered should include a detailed outline describing the planned contents of each chapter, a bibliography and several sample chapters. Parallel submission of a manuscript to another publisher while under consideration for LNM is not acceptable and can lead to rejection.

4. In general, **monographs** will be sent out to at least 2 external referees for evaluation.

   A final decision to publish can be made only on the basis of the complete manuscript, however a refereeing process leading to a preliminary decision can be based on a pre-final or incomplete manuscript.

   Volume Editors of **multi-author works** are expected to arrange for the refereeing, to the usual scientific standards, of the individual contributions. If the resulting reports can be forwarded to the LNM Editorial Board, this is very helpful. If no reports are forwarded or if other questions remain unclear in respect of homogeneity etc, the series editors may wish to consult external referees for an overall evaluation of the volume.

5. Manuscripts should in general be submitted in English. Final manuscripts should contain at least 100 pages of mathematical text and should always include

   – a table of contents;
   – an informative introduction, with adequate motivation and perhaps some historical remarks: it should be accessible to a reader not intimately familiar with the topic treated;
   – a subject index: as a rule this is genuinely helpful for the reader.
   – For evaluation purposes, manuscripts should be submitted as pdf files.

6. Careful preparation of the manuscripts will help keep production time short besides ensuring satisfactory appearance of the finished book in print and online. After acceptance of the manuscript authors will be asked to prepare the final LaTeX source files (see LaTeX templates online: https://www.springer.com/gb/authors-editors/book-authors-editors/manuscriptpreparation/5636) plus the corresponding pdf- or zipped ps-file. The LaTeX source files are essential for producing the full-text online version of the book, see http://link.springer.com/bookseries/304 for the existing online volumes of LNM). The technical production of a Lecture Notes volume takes approximately 12 weeks. Additional instructions, if necessary, are available on request from lnm@springer.com.

7. Authors receive a total of 30 free copies of their volume and free access to their book on SpringerLink, but no royalties. They are entitled to a discount of 33.3 % on the price of Springer books purchased for their personal use, if ordering directly from Springer.

8. Commitment to publish is made by a *Publishing Agreement*; contributing authors of multiauthor books are requested to sign a *Consent to Publish form*. Springer-Verlag registers the copyright for each volume. Authors are free to reuse material contained in their LNM volumes in later publications: a brief written (or e-mail) request for formal permission is sufficient.

Addresses:
Professor Jean-Michel Morel, CMLA, École Normale Supérieure de Cachan, France
E-mail: moreljeanmichel@gmail.com

Professor Bernard Teissier, Equipe Géométrie et Dynamique,
Institut de Mathématiques de Jussieu – Paris Rive Gauche, Paris, France
E-mail: bernard.teissier@imj-prg.fr

Springer: Ute McCrory, Mathematics, Heidelberg, Germany,
E-mail: lnm@springer.com

Printed in the United States
By Bookmasters